F.M.T. Brazier, Kees Nieuwenhuis, Gregor Pavlin, Martijn Warnier, and Costin Badica (Eds.)

Intelligent Distributed Computing V

Studies in Computational Intelligence, Volume 382

Editor-in-Chief
Prof. Janusz Kacprzyk
Systems Research Institute
Polish Academy of Sciences
ul. Newelska 6
01-447 Warsaw
Poland
E-mail: kacprzyk@ibspan.waw.pl

Further volumes of this series can be found on our homepage: springer.com

Vol. 360. Mikhail Moshkov and Beata Zielosko
Combinatorial Machine Learning, 2011
ISBN 978-3-642-20994-9

Vol. 361. Vincenzo Pallotta, Alessandro Soro, and Eloisa Vargiu (Eds.)
Advances in Distributed Agent-Based Retrieval Tools, 2011
ISBN 978-3-642-21383-0

Vol. 362. Pascal Bouvry, Horacio González-Vélez, and Joanna Kolodziej (Eds.)
Intelligent Decision Systems in Large-Scale Distributed Environments, 2011
ISBN 978-3-642-21270-3

Vol. 363. Kishan G. Mehrotra, Chilukuri Mohan, Jae C. Oh, Pramod K. Varshney, and Moonis Ali (Eds.)
Developing Concepts in Applied Intelligence, 2011
ISBN 978-3-642-21331-1

Vol. 364. Roger Lee (Ed.)
Computer and Information Science, 2011
ISBN 978-3-642-21377-9

Vol. 365. Roger Lee (Ed.)
Computers, Networks, Systems, and Industrial Engineering 2011, 2011
ISBN 978-3-642-21374-8

Vol. 366. Mario Köppen, Gerald Schaefer, and Ajith Abraham (Eds.)
Intelligent Computational Optimization in Engineering, 2011
ISBN 978-3-642-21704-3

Vol. 367. Gabriel Luque and Enrique Alba
Parallel Genetic Algorithms, 2011
ISBN 978-3-642-22083-8

Vol. 368. Roger Lee (Ed.)
Software Engineering, Artificial Intelligence, Networking and Parallel/Distributed Computing 2011, 2011
ISBN 978-3-642-22287-0

Vol. 369. Dominik Ryżko, Piotr Gawrysiak, Henryk Rybinski, and Marzena Kryszkiewicz (Eds.)
Emerging Intelligent Technologies in Industry, 2011
ISBN 978-3-642-22731-8

Vol. 370. Alexander Mehler, Kai-Uwe Kühnberger, Henning Lobin, Harald Lüngen, Angelika Storrer, and Andreas Witt (Eds.)
Modeling, Learning, and Processing of Text Technological Data Structures, 2011
ISBN 978-3-642-22612-0

Vol. 371. Leonid Perlovsky, Ross Deming, and Roman Ilin (Eds.)
Emotional Cognitive Neural Algorithms with Engineering Applications, 2011
ISBN 978-3-642-22829-2

Vol. 372. António E. Ruano and Annamária R. Várkonyi-Kóczy (Eds.)
New Advances in Intelligent Signal Processing, 2011
ISBN 978-3-642-11738-1

Vol. 373. Oleg Okun, Giorgio Valentini, and Matteo Re (Eds.)
Ensembles in Machine Learning Applications, 2011
ISBN 978-3-642-22909-1

Vol. 374. Dimitri Plemenos and Georgios Miaoulis (Eds.)
Intelligent Computer Graphics 2011, 2011
ISBN 978-3-642-22906-0

Vol. 375. Marenglen Biba and Fatos Xhafa (Eds.)
Learning Structure and Schemas from Documents, 2011
ISBN 978-3-642-22912-1

Vol. 376. Toyohide Watanabe and Lakhmi C. Jain (Eds.)
Innovations in Intelligent Machines – 2, 2011
ISBN 978-3-642-23189-6

Vol. 377. Roger Lee (Ed.)
Software Engineering Research, Management and Applications 2011, 2011
ISBN 978-3-642-23201-5

Vol. 378. János Fodor, Ryszard Klempous, and Carmen Paz Suárez Araujo (Eds.)
Recent Advances in Intelligent Engineering Systems, 2011
ISBN 978-3-642-23228-2

Vol. 379. Ferrante Neri, Carlos Cotta, and Pablo Moscato (Eds.)
Handbook of Memetic Algorithms, 2011
ISBN 978-3-642-23246-6

Vol. 380. Anthony Brabazon, Michael O'Neill, and Dietmar Maringer (Eds.)
Natural Computing in Computational Finance, 2011
ISBN 978-3-642-23335-7

Vol. 381. Radosław Katarzyniak, Tzu-Fu Chiu, Chao-Fu Hong, and Ngoc Thanh Nguyen (Eds.)
Semantic Methods for Knowledge Management and Communication, 2011
ISBN 978-3-642-23417-0

Vol. 382. F.M.T. Brazier, Kees Nieuwenhuis, Gregor Pavlin, Martijn Warnier, and Costin Badica (Eds.)
Intelligent Distributed Computing V, 2011
ISBN 978-3-642-24012-6

F.M.T. Brazier, Kees Nieuwenhuis, Gregor Pavlin,
Martijn Warnier, and Costin Badica (Eds.)

Intelligent Distributed Computing V

Proceedings of the 5th International Symposium on Intelligent Distributed Computing – IDC 2011, Delft, The Netherlands – October 2011

Springer

Editors

F.M.T. Brazier
Delft University of Technology
Faculty of Technology, Policy and Management
Intelligent Interactive Dynamic Systems
Section Systems Engineering
Jaffalaan 5
2628BX Delft
The Netherlands
E-mail: F.M.Brazier@tudelft.nl

Kees Nieuwenhuis
P.O. Box 90
2600 AB Delft
The Netherlands
E-mail: Kees.Nieuwenhuis@D-CIS.NL

Gregor Pavlin
P.O. Box 90
2600 AB Delft
The Netherlands
E-mail: Gregor.Pavlin@D-CIS.NL

Martijn Warnier
Delft University of Technology
Faculty of Technology, Policy and Management
Section Systems Engineering
Jaffalaan 5
2628BX Delft
The Netherlands
E-mail: M.E.Warnier@tudelft.nl

Costin Badica
University of Craiova
Faculty of Automatics,
Computers and Electronics
Software Engineering Department
Bvd. Decebal Nr. 107
200440 Craiova
Romania
E-mail: costin.badica@software.ucv.ro

ISBN 978-3-642-27033-8 ISBN 978-3-642-24013-3 (ebook)

DOI 10.1007/978-3-642-24013-3

Studies in Computational Intelligence ISSN 1860-949X

Typeset & Cover Design: Scientific Publishing Services Pvt. Ltd., Chennai, India.

Printed on acid-free paper

9 8 7 6 5 4 3 2 1

springer.com

Preface

The emergent field of Intelligent Distributed Computing focuses on the development of a new generation of intelligent distributed systems. It faces the challenges of adapting and combining research in the fields of Intelligent Computing and Distributed Computing. Intelligent Computing develops methods and technology ranging from classical artificial intelligence, computational intelligence and multi-agent systems to game theory. The field of Distributed Computing develops methods and technology to build systems that are composed of collaborating components. Theoretical foundations and practical applications of Intelligent Distributed Computing set the premises for the new generation of intelligent distributed information systems.

Intelligent Distributed Computing – IDC'2011 continues the tradition of the IDC Symposium Series that was started as an initiative of research groups from: (i) *Systems Research Institute, Polish Academy of Sciences in Warsaw, Poland* and (ii) *Software Engineering Department of the University of Craiova, Craiova, Romania.* IDC aims at bringing together researchers and practitioners involved in all aspects of Intelligent Distributed Computing. IDC is interested in works that are relevant for both Distributed Computing and Intelligent Computing, with scientific merit in at least one of these two areas. IDC'2011 was the fifth event in the series and was hosted by *D-CIS Lab / Thales Research & Technology* and *Delft University of Technology* in *Delft, The Netherlands* during October 5-7, 2011. IDC'2011 had a special interest in novel architectures and methods which facilitate intelligent distributed solutions to complex problems by combining human cognitive capabilities and automated processing.

IDC'2011 was collocated with: (i) AgentScape workshop; (ii) 3nd International Workshop on Multi-Agent Systems Technology and Semantics, MASTS'2011.

The material published in this book is divided into three main parts: (i) 3 invited contributions; (ii) 26 contributions of IDC'2011 participants; and (iii) 4 contributions of MASTS'2011 participants.

The response to IDC'2011 call for paper was generous. We received 42 submissions from 17 countries (we counted the country of each coauthor for each submitted paper). Each submission was carefully reviewed by at least 3 members of the IDC'2011 Program Committee. Acceptance and publication were judged based on the relevance to the symposium themes, clarity of presentation, originality and accuracy of results and proposed solutions. Finally 15 regular papers and 11 short papers were selected for presentation and were included in this volume, resulting in acceptance rates of 35.71 % for regular papers and 61.90 % for short papers. Additionally, MASTS'2011 workshop received 7 submissions. After the careful review (each submission was reviewed by at least 3 members of the MASTS'2011 Program Committee), 4 papers were selected for presentation and were included in this book.

The 33 contributions published in this book address many topics related to theory and applications of intelligent distributed computing and multi-agent systems, including: adaptive and autonomous distributed systems, agent programming, ambient assisted living systems, business process modeling and verification, cloud computing, coalition formation, decision support systems, distributed optimization and constraint satisfaction, gesture recognition, intelligent energy management in WSNs, intelligent logistics, machine learning, mobile agents, parallel and distributed computational intelligence, parallel evolutionary computing, trust metrics and security, scheduling in distributed heterogenous computing environments, semantic Web service composition, social simulation, and software agents for WSNs.

We would like to thank to Janusz Kacprzyk, editor of *Studies in Computational Intelligence* series and member of the Steering Committee for his continuous support and encouragement for the development of the IDC Symposium Series. We would like to thank to the IDC'2011 and MASTS'2011 Program Committee members for their work in promoting the event and refereeing submissions and also to all colleagues who submitted papers to IDC'2011 and MASTS'2011. We deeply appreciate the efforts of our invited speakers Katia Sycara, Amnon Meisels, and Emadeldeen Elsayed Ahmed Elakehal and thank them for their interesting lectures. Special thanks also go to organizers of MASTS'2011: Adina Magda Florea, John Jules Meyer, and Amal El Fallah Seghrouchni. Finally, we appreciate the efforts of local organizers on behalf of *D-CIS Lab / Thales Research & Technology* and *Delft University of Technology* in *Delft, The Netherlands* for hosting and organizing IDC'2011 and MASTS'2011.

Delft, Craiova
July 2011

Frances Brazier
Kees Nieuwenhuis
Gregor Pavlin
Martijn Warnier
Costin Bădică

Organization

Organizers

D-CIS Lab / Thales Research & Technology, The Netherlands
Systems Engineering Section, Faculty of Technology, Policy and Management, Delft University of Technology, The Netherlands
Software Engineering Department, Faculty of Automation, Computers and Electronics, University of Craiova, Romania

Conference Chairs

Frances Brazier	Delft University of Technology, The Netherlands
Kees Nieuwenhuis	D-CIS Lab / Thales Research & Technology, The Netherlands

Steering Committee

Janusz Kacprzyk	Polish Academy of Sciences, Poland
Frances Brazier	Delft University of Technology, The Netherlands
Costin Bădică	University of Craiova, Romania
Michele Malgeri	Università di Catania, Italia
Kees Nieuwenhuis	D-CIS Lab / Thales Research & Technology, The Netherlands
George A. Papadopoulos	University of Cyprus, Cyprus
Marcin Paprzycki	Polish Academy of Sciences, Poland

Invited Speakers

Katia Sycara	Carnegie Mellon University, USA
Amnon Meisels	Ben-Gurion University, Israel
Emadeldeen Elsayed Ahmed Elakehal	The Book Depository, UK

Program Committee Chairs

Gregor Pavlin	D-CIS Lab / Thales Research & Technology, The Netherlands
Costin Bădică	University of Craiova, Romania
Martijn Warnier	Delft University of Technology, The Netherlands

Program Committee

Salvador Abreu	Universidade de Évora and CENTRIA FCT/UNL, Portugal
Răzvan Andonie	Central Washigton University, USA and University of Transylvania, Romania
Amar Balla	LCMS-ESI Ecole nationale Supérieure d'Informatique, Algeria
Nick Bassiliades	Aristotle University of Thessaloniki, Greece
Doina Bein	ARL, Pennsylvania State University, USA
Nik Bessis	University of Derby, UK
Lars Braubach	University of Hamburg, Germany
Frances Brazier	Delft University of Technology, The Netherlands
Dumitru Dan Burdescu	University of Craiova, Romania
Giacomo Cabri	University of Modena and Reggio Emilia, Italy
David Camacho	Universidad Autonoma de Madrid, Spain
Vincenza Carchiolo	University of Catania, Italy
Jen-Yao Chung	IBM T.J. Watson Research Center, USA
Gabriel Ciobanu	"A.I. Cuza" University of Iaşi, Romania
Phan Cong-Vinh	Centre for Applied Formal Methods, London South Bank University, UK
Valentin Cristea	"Politehnica" University of Bucharest, Romania
Paul Davidsson	Blekinge Institute of Technology, Sweden
Beniamino Di Martino	Second University of Naples, Italy
Luminiţa Dumitriu	"Dunarea de Jos" University of Galaţi, Romania
George Eleftherakis	CITY International Faculty, University of Sheffield, Greece

Amal El Fallah Seghrouchni	LIP6 - University of Pierre and Marie Curie, France
Vadim A. Ermolayev	Zaporozhye National University, Ukraine
Mostafa Ezziyyani	Université Abdelmalek Essaâdi, Maroc
Adina Magda Florea	"Politehnica" University of Bucharest, Romania
Maria Ganzha	Elblag University of Humanities and Economics, Poland
Nathan Griffiths	University of Warwick, UK
Michael Hartung	University of Leipzig, Germany
Barna Laszlo Iantovics	Petru Maior University of Targu Mures, Romania
Mirjana Ivanović	University of Novi Sad, Serbia
Jason J. Jung	Yeungnam University, South Korea
Barbara Keplicz	Institute of Informatics, Warsaw University and Institute of Computer Science, Polish Academy of Sciences, Poland
Igor Kotenko	Russian Academy of Sciences, Russia
Dariusz Krol	Wroclaw University of Technology, Poland
Ioan Alfred Leţia	Technical University of Cluj-Napoca, Romania
Alessandro Longheu	University of Catania, Italy
Heitor Silverio Lopes	Federal University of Technology - Parana, Brazil
José Machado	University of Minho, Portugal
Giuseppe Mangioni	University of Catania, Italy
Yannis Manolopoulos	Aristotle University of Thessaloniki, Greece
Viviana Mascardi	University of Genova, Italy
Amnon Meisels	Ben-Gurion University, Israel
Ronaldo Menezes	Florida Institute of Technology, USA
Mihai Mocanu	University of Craiova, Romania
Grzegorz J. Nalepa	AGH University of Science and Technology, Kraków, Poland
Alexandros Nanopoulos	Institute of Computer Science, University of Hildesheim, Germany
Viorel Negru	Western University of Timişoara, Romania
Kees Nieuwenhuis	D-CIS/Thales Research & Technology, The Netherlands
Peter Noerr	MuseGlobal, Inc., USA
Andrea Omicini	University of Bologna, Italy
Mihaela Oprea	University Petroleum-Gas of Ploieşti, Romania
Gregor Pavlin	D-CIS Lab / Thales Research & Technology, The Netherlands
Dana Petcu	Western University of Timişoara, Romania

Laurent Perrussel	IRIT - Université de Toulouse, France
Radu-Emil Precup	"Politehnica" University of Timişoara, Romania
Shahram Rahimi	Southern Illinois University, USA
Ioan Salomie	Technical University of Cluj-Napoca, Romania
Murat Sensoy	University of Aberdeen, UK
Safeeullah Soomro	Yanbu University College, Saudi Arabia
Rainer Unland	University of Duisburg-Essen, Germany
Laurent Vercouter	Ecole des Mines de St-Etienne, France
Lucian Vinţan	Academy of Technical Sciences, Romania
Martijn Warnier	Delft University of Technology, The Netherlands
Niek Wijngaards	D-CIS Lab / Thales Research & Technology, The Netherlands
Franz Wotawa	Graz University of Technology, Austria
Filip Zavoral	Charles University, Czech Republic

Additional Reviewers

Prochazka Ales
Ionut Anghel
Pasquale Cantiello
Chidambaram Chidambaram
Claudia Chituc
Maxim Davidovsky
Sylvain Doctor
Adnan Hashmi
Cédric Herpson
Efstratios Kontopoulos
Peter Kopacek
Georgios Meditskos
Rosaldo Rossetti
Vasile Sinescu
Efthymia Tsamoura
Cioara Tudor

Organizing Committee

Elvira Popescu	University of Craiova, Romania
Mihnea Scafes	University of Craiova, Romania
Bernard van Veelen	D-CIS Lab / Thales Research & Technology, The Netherlands
Niek Wijngaards	D-CIS Lab / Thales Research & Technology, The Netherlands

MASTS'2011 Workshop Chairs

Adina Magda Florea	University "Politehnica" of Bucharest, Romania
John Jules Meyer	Universiteit Utrecht, The Netherlands
Amal El Fallah Seghrouchni	Université Pierre & Marie Curie, France

MASTS'2011 Workshop Program Committee

Costin Bădică	University of Craiova, Romania
Olivier Boissier	ENS des Mines Saint-Etienne, France
Mehdi Dastani	Utrecht University, The Netherlands
Amal El Fallah Seghrouchn	Université Pierre & Marie Curie, France
Adina Magda Florea	University Politehnica of Bucharest, Romania
Maria Gazha	Polish Academy of Sciences, Poland
Fabrice Guillet	Polytech'Nantes, France
Pascale Kuntz-Cosperec	Polytech'Nantes, France
John Jules Meyer	Utrecht University, The Netherlands
Andrei-Horia Mogos	University Politehnica of Bucharest, Romania
Viorel Negru	West University of Timisoara, Romania
Sascha Ossowski	University Rey Juan Carlos, Spain
Marcin Paprzycki	Polish Academy of Science, Poland
Tiberiu Stratulat	Université Montpellier 2, France
Adriana Tapus	ENSTA ParisTech, France
Stefan Trausan-Matu	University Politehnica of Bucharest, Romania
Laurent Vercouter	ENS des Mines de Saint Etienne, France
Gerard Vreeswijk	Utrecht University, The Netherlands
Antoine Zimmermann	LIRIS Lyon, France

Contents

Part III: Parallel and Distributed Computational Intelligence

Part IV: Autonomous and Adaptive Distributed Systems

Part V: Pervasive and Cloud Computing

Part IX: MASTS'2011

Part I
Invited Papers

Dynamics of Information Propagation in Large Heterogeneous Networks

Katia Sycara

Abstract. Large scale networked systems that include heterogeneous entities, eg humans and computational entities are becoming increasingly prevalent. Prominent applications include the Internet, large scale disaster relief and network centric warfare. In such systems, large heterogeneous coordinating entities exchange uncertain information to obtain and increase situation awareness. Uncertain and possibly conflicting sensor data is shared across a peer-to-peer network. Not every team member will have direct access to sensors and team members will be influenced mostly by their neighbors in the network with whom they communicate directly. In this talk I will present our work on the dynamics and emergent behaviors of belief propagation in such large networks. Unlike past work, the nodes in the networks we study are autonomous and actively fuse information they receive. Nodes can change their beliefs as they receive additional information over time.

A key property of the system is that it exhibits qualitatively different dynamics and system performance over different ranges of system parameters. In one particular range, the system exhibits behavior known as scale-invariant dynamics which we empirically find to correspond to dramatically improved system performance. I will present results on the emergent belief propagation dynamics in those systems, mathematical characterization of the systems behavior and distributed algorithms for adapting the network behaviors to steer the whole system to areas of optimized performance.

Katia Sycara
Robotics Institute, Carnegie Mellon University, Pittsburg, USA
e-mail: katia@cs.cmu.edu

F.M.T. Brazier et al. (Eds.): Intelligent Distributed Computing V, SCI 382, p. 3.
springerlink.com

Distributed Search by Constrained Agents

Amnon Meisels

The investigation of Distributed Constraints Satisfaction Problems (*DCSPs*) has started a little more than a decade ago. It focuses on constraints satisfaction problems (*CSPs*) that are distributed among multiple agents. Imagine a large University that includes many departments. The weekly schedule of classes is generated by each department, scheduling its classes and teachers for the whole semester. A weekly schedule is a typical constraint satisfaction problem. Class meetings are variables and the time-slots of the week are the domain of values that have to be assigned to classes in order to generate a schedule. The fundamental constraints of timetabling require that two classes taught by the same teacher have to be assigned different time-slots. Another common constraint is to require that two meetings of the same class will be assigned to different days of the week. In the University as a whole, the departments can be thought of as agents that generate their departmental weekly schedules. The weekly schedules of different departments are constrained by the fact that there are students that select classes from these departments. This generates constraints between departments and the generic scenario of a *distributed CSP*. Agents own parts of the global problem (e.g. departmental schedules) and cooperate in search for a global solution in which the constraints between departments are satisfied. In order to solve such a distributed problem, all agents must cooperate in a global search process. Search algorithms for a distributed problem operate by agents performing assignments to their variables and exchanging messages in order to check their assignments against those of constraining agents.

When a distributed set of Agents run the search for a globally consistent solution (or a global optimum, in *distributed optimization* problems in they need to coordinate their operations by distributed means. Even when the global search algorithm

Amnon Meisels
Ben Gurion University of the Negev, P.O.B 653, Be'er Sheva 84105, Israel
e-mail: am@cs.bgu.ac.il

F.M.T. Brazier et al. (Eds.): Intelligent Distributed Computing V, SCI 382, pp. 5–9.
springerlink.com

resembles simple backtracking, problems of concurrency pop up. Is the distributed algorithm complete (e.g., does it return a solution if one exists) ? Is it free of deadlocks ? Does it necessarily terminate. All of these are typically complex problems that relate to the distributed nature of distributed search algorithms.

This talk is dedicated to the presentation of distributed search algorithms and to a comparative study of their performance. It presents both distributed constraints satifaction problems and distributed constraints optimization problems (*DCOPs*). Problems of constraints optimization (*COPs*) arise in cases where constraints have costs (values or weights) and the goal is to find a minimal-cost solution. An important class of *DCOP*s is that of unsolvable *DCSP*s, where the goal is to find a complete assignment that has a minimal number of constraint violations (conflicts). This family of problems has been termed *DisMaxCSP* and has been studied extensively as an example family of *COP*s in the centralized context.

Distributed constraints optimization problems - *DCOP*s - will be introduced and the needed methods of search for their solution. In general, complete search methods for *DCOP*s are based on Branch and Bound algorithms. The asynchronous branch and bound algorithm for *DCOP*s was published in 2005 - the Asynchronous Distributed OPTimization algorithm (*ADOPT*) [2]. ADOPT performs assignments and distributes information among agents in a completely asynchronous manner. It turns out that similarly to the case for *DCSP*s, it is interesting to investigate the potential of asynchronous processing. In other words, to explore algorithmic options that use asynchronous distributed processing combined with sequential assignments. This ensures the processing of consistent partial solutions and avoids redundant (asynchronous) computations relating to intrinsically inconsistent partial solutions [1].

Let us try to get some feeling of the distributed nature of search. Consider a 4-Queens problem in which the 4 queens are independent agents. The 4-Queens problem is probably well known. It requires to put 4 queens on a 4x4 chess board, so that none of the queens threatens any other. In searching for such a solution (4 non-threatening positioned queens) one needs to go over all possible positions of all queens over the 4x4 board. Let us try to imagine a distributed scenario, one of many possible. Each agent/queen can position itself at any square. Queens check for the compatibility of their positions with the positions of other queens by exchanging messages with them. For simplicity let us assume that each agent only selects positions in one specific row. This will simplify the search and will enable a partial search space, where each agent has only 4 possible positions, which still contains all solutions. Let us assume that each agent positions itself initially as in figure 1. Next, each agent starts to exchange messages with the other agents in order to check its compatibility with all agents' positions. When conflicting positions are encountered, agents change their positions and attempt to achieve compatibility.

This example is extremely naive in that it does not consider an *algorithm* for achieving a solution (e.g., a distributed search algorithm). Instead, it tries to play around with a distributed scenario in order to ger some feeling for the behavior of such scenarios. Assuming no specific algorithm, we'll just continue for a few steps. Let us take the case that each agent sends messages informing all relevant agents about its position. For the 4-Queens problem every agent is constrained by every

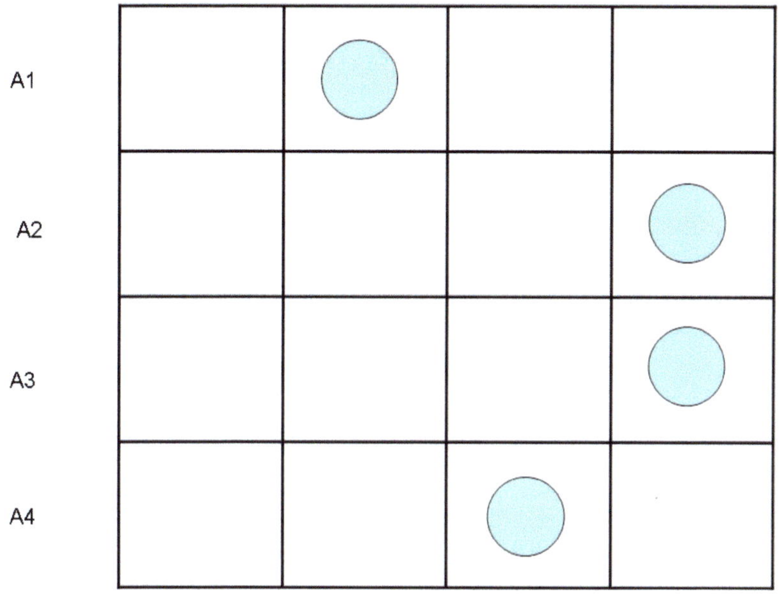

Fig. 1 Starting positions for the Distributed 4-Queens problem

other agent. So, each agent sends a message to the other three informing them of its position. After all messages have been sent and received, all agents know the positions of all other agents.

Observe agent A_1. It finds out that it is involved in a single conflict, with agent A_3. Agent A_2 is involved in a single conflict with agent A_3 and this is also the number of conflicts discovered by agent A_4. Agent A_3 is the only agent that finds out that it is involved in 2 conflicts - with agents A_2 and A_4. Each agent has to decide what to do in this situation. Remember that this is only an ilustrative example of a *distributed search problem* and not of a *search algorithm*. Therefore, we are just exploring possible actions of agents. Let us assume that agents that are involved in a conflict decide on the party that changes its position by giving priority to the agent that has more conflicts. Among each pair of conflicting agents it is agent A_3 that has more conflicts than its conflicting partner. So, agent A_3 decides to change its position and it finds a position that has no conflicts for itself. It moves its position to that conflict-free square and the result is the state in figure 2. This is actually a solution and none of the agents needs to change anything.

Please note that there are very many different ways to construct the protocol of the agents during search. Agents could exchange messages about their *intentions* to select a value, to give one extreme example. They could wait for all others to select values and then choose their own. All of these would form different distributed search algorithms. My talk presents several approaches to design such search

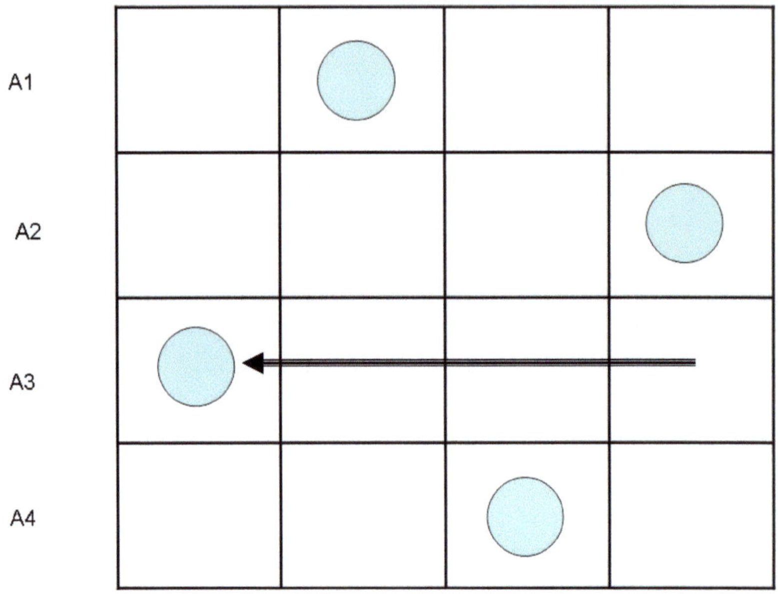

Fig. 2 The resulting move of agent A_3

algorithms and demonstrates that such algorithms use protocols that have important features. A distributed search algorithm must terminate. It must return a solution if such a solution exists and otherwise return a no-solution message. These features and additional ones are essential to distributed search algorithms and will be discussed in detail for all presented algorithms.

In the above very simple example there were an exponential many ways to *describe* the run of the agents. We could have thought about agent A_3 making its decision after it received the message from A_2, after a received message from A_4, before all messages arrived, and many more possibilities. All of these options form the many ways in which a distributed computation can be performed. Analyzing the *behavior* of distributed search algorithms is therefore complex. Since any design of search algorithms needs tools for measuring their performance and for comparing their behavior, we will dwell on methods for measuring distributed search algorithms in. The resulting measures will be used for comparisons of the algorithms that will be described in the talk.

All DCSP search problems are NP-Complete and consequently the worst case run-time of distributed search algorithms is exponential in the number of variables in the worst case. The acceptable way of comparing different distributed search algorithms and heurstics for these algorithms is to evaluate their performance empirically.

References

1. Gershman, A., Meisels, A., Zivan, R.: Asynchronous forward bounding. J. of Artificial Intelligence Research 34, 25–46 (2009)
2. Modi, P.J., Shen, W., Tambe, M., Yokoo, M.: ADOPT: Asynchronous distributed constraints optimization with quality guarantees. Artificial Intelligence 161(1-2), 149–180 (2005)

A Practical Method for Developing Multi Agent Systems: APMDMAS

Emad Eldeen Elakehal and Julian Padget

Abstract. While Multi Agent Systems (MAS) attracted a great deal of attention in the field of software engineering, with its promises of capturing complex systems, they remain far away from commercial popularity mainly due to the lack of a MAS methodology that is accessible for commercial developers. In this paper we present a practical method for developing MAS that we believe will enable not only software developers but also business people beyond the academic community to design and develop MAS easily.

1 Introduction and Problem Statement

The main aim in Multi Agent Systems is to provide principles for the building of complex distributed systems that involve one or multiple agents and to take advantage of the mechanisms for cooperation and coordination of these agents' behaviours. However, building multi-agent applications for complex and distributed systems is not an easy task [5], add to that the development of industrial-strength applications requires the availability of software engineering methodologies. Although, there are some good MAS development methodologies such as those in section 2 these are all not enough as none of them stands out as a comprehensive methodology. Also MAS exhibit all traditional problems of distributed and concurrent systems, and the additional difficulties that arise from flexibility requirements and sophisticated interactions [12], all of which results in having a real difficulty to define MAS development methodology. According to Luck et al [7] "One of the most fundamental obstacles to the take-up of agent technology is the lack of mature software development methodologies for agent-based systems."

Emad Eldeen Elakehal · Julian Padget
University of Bath, Bath, UK
e-mail: emad@bookdepository.co.uk, jap@cs.bath.ac.uk

F.M.T. Brazier et al. (Eds.): Intelligent Distributed Computing V, SCI 382, pp. 11–20.
springerlink.com

A close look at the existing MAS development methodologies reveals that none of them has become the main stream method to use, for a wide range of reasons. Here we list those we regard as the most crucial:

1. None of the current methodologies support inexperienced developers; they all require good knowledge of agent concepts so the developers need to specify all semantic components of their agents. This could be the main reason why commercial applications are rarely found to be developed using the MAS paradigm.
2. The absence of an holistic view of system logic and its cognitive aspects, that leads to some confusion and ambiguity in both analysis and design phases.
3. None of them is a comprehensive methodology that supports all development life cycle phases. Some of them offer only design and analysis tools but none for deployment, while others offer theory without supporting tools.
4. There is an obvious gap between the design models and the existing implementation languages [10], which leads to great difficulty for the programmers try to map the complex designs into executable code.
5. Most of the current methodologies do not include an implementation phase and the ones that do, such as Tropos [2], its implementation language does not explain how to implement beliefs, goals and plans, nor reasoning about agent communication.
6. Finally, the lack of a full formal representation of MAS concepts, even accounting for the work by Wooldridge [13], and Luck [8], neither of which can be considered complete. Even though a partial approach may be effective, the question remains, which concepts to formalize? And what is the best way to specify and describe them?

Our proposed methodology is meant to solve most if not all of these issues with the aim to become more accessible to a wider range of academics and software engineers.

In the following sections we present an overview of the proposed methodology and give some details of its three phases with the inclusion of a sample diagram of each model. Then we draw some conclusions and highlight possible future work.

2 Related Work

1. **GAIA Methodology:** The GAIA [14] is a general methodology that supports both micro (agent structure) and macro (organisational structure) development of agent systems. It was proposed by Wooldridge et al in 2000 and subsequently extended by Zambonelli et al. to support open multi-agent systems [15]. It has two phases that cover the analysis and design only. GAIA is a very lengthy and complex and it lacks an implementation phase.

2. **MaSE Methodology:** Multiagent Systems Engineering [3] covers the full development life cycle from an initial system specification to system implementation. It has two phases that contain seven steps in total and offers a tool that supports all phases. MaSE does not enforce any particular implementation platform, but it is hard to follow for inexperienced users.
3. **Prometheus Methodology:** Prometheus [9] aims to be suitable for non-expert users to design and develop MAS. The methodology has three phases: System Specification, Architectural Design and Detailed Design. Prometheus has a tool that supports the design process, as well as consistency checking and documentation generation. Although Prometheus is more practical than other approaches it does not connect the system model to any execution platform.
4. **TROPOS:** Tropos [2] distinguishes itself from other methodologies by giving great attention to the requirements analysis where all stake-holders requirements and intentions are identified then analysed. The modelling process consists of five phases and uses JACK for the implementation, the developers would need to map the concepts in their design into JCK five constructs. Tropos offer some guidelines to help in this process, but it seems very lengthy and complex.

3 APMDMAS Methodology Overview

APMDMAS consists of three phases that cover the full life cycle of multi agent software development. The first phase focusses on **System Requirements** gathering: it allows the system designer to describe many possible use case scenarios as well as to define a high level system goals' specification. It has two diagram types; the **System Goals Diagram** and **Use Cases Diagrams**.

The second phase focusses on **Detailed Analysis and Design**; during this phase the system requirements can be transformed into a fully modelled system. Each diagram contributes to the building of the system **Meta Model**, that is the basis for generating a full MAS code for one or more target execution languages/platforms. The system designer can start this phase either from the business process (BP) view or the system participant view. BP requires the completion of **Specific Business Process Models** and **Basic Business Process Models** diagrams. Experienced users with a good knowledge of the multi-agent paradigm can alternatively start from the **System Participants Models** and the definition of their entities **(Agent–Actor–Service–Environment)** as well as the definition communication components **(Protocol–Message)** alongside the usage or definition of **(Goals–Plans–Norms–Beliefs)**.

Finally, the third phase is the implementation and execution phase where the user can verify the system design and export the **Meta Model** file as RDF or choose to generate code in one of the supported execution languages/middlewares such as Jason, AgentScape etc.

In the following sections we describe each of these phases and their modelling diagrams/components briefly and give some examples.

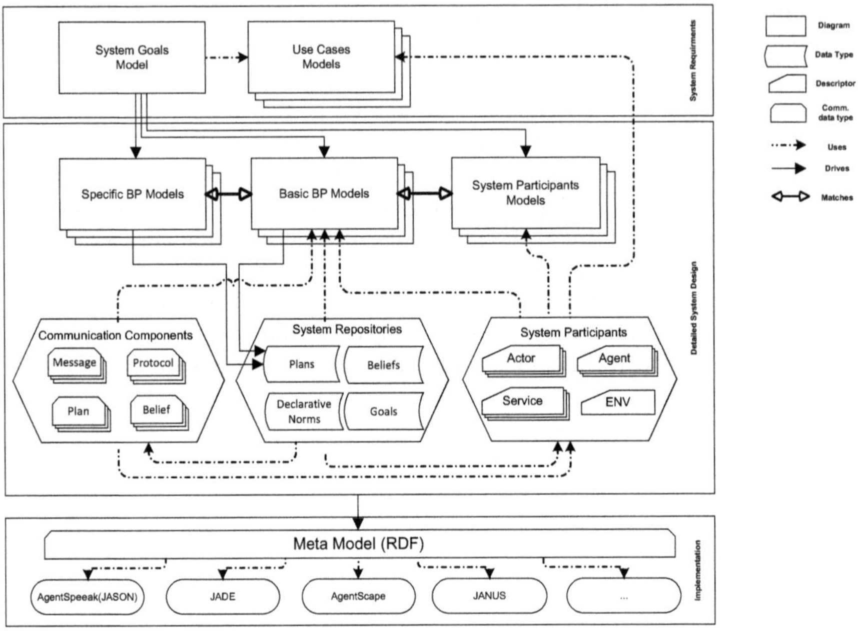

Fig. 1 APMDMAS Overview

3.1 System Requirements Phase

The main aim of this phase is to describe the system functions in terms of use cases after identifying the main system goals. There are only two models to be created during this phase: **System Goals Model** and **Use Cases Models**.

3.1.1 System Goals Model

Every system should have a set of goals; these are simply the motives behind building such a system: the system designer does not need at this stage to specify system goals in great detail, instead the goals hierarchy should be built till it reaches the level where every goal can be fulfilled by *only* one basic business process. System goals are the drivers of all diagrams of the next phase.

The system goal is basically the system status it is wished to achieve. The system goals definition is not to be confused with the common agent goals: in our model the system goals are procedural, in other words the goal name is similar to a method in a traditional programming language. This is very useful to divide—if we take a top to bottom approach—the system from one unit to a group of functions. At the same time to see in a simple way how particular group of actions would lead to fulfilling one big system function. Figure 2 left shows a sample System Goals Model.

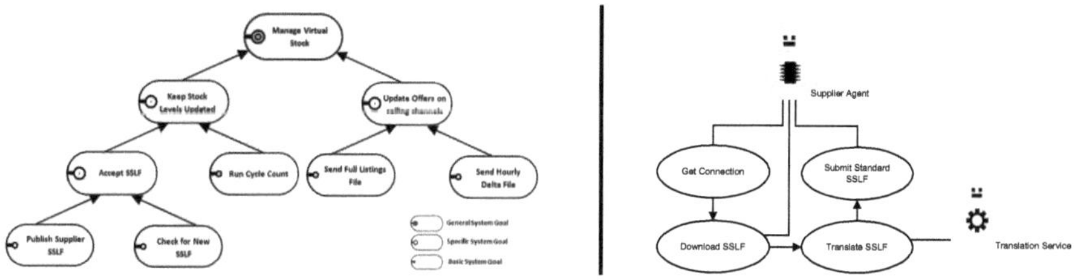

Fig. 2 System Goals Diagram [left] and Use Case Diagram (Publish Supplier Stock Levels File (SSLF) [right]

The system goals model contains three types of goals (i) **General System Goal:** Any system should have *only* one General System Goal, this is the widest reason for building such a system. (ii) **Specific System Goals:** These are more functional goals that can be achieved by one or more business processes; each Specific System Goal can have a number of sub-goals. (iii) **Basic System Goals:** These are the end leaves in the goals tree, they cannot have sub-goals.

3.1.2 Use Cases Models

Use cases are simply a clarification of some or all the system functionalities, in this step the system designer can create some models of the most important functions for future reference. The use cases are used in our methodology to help the system designer to think through the different functions of the system and possible issues to be considered. The use case diagram normally shows how different system participants interact, or which steps they take to carry out a system function.

Figure 2 shows a sample use case diagram, where there are two system participants: a software agent (Supplier) and software service (Translator), and four functions. The arrows show the sequence of execution and the connectors between the agent and the function define the responsibility.

3.2 Detailed System Design Phase

The aim of the Detailed System Design phase is to define all the system components, their detailed structure and the ways they can interact with each other. There are three different diagram types in this phase; **Specific Business Process Models**, **Basic Business Process Models**, and **System Participants Models**. To complete these diagrams the system designer needs to define/use different types of supporting entities which are held in the form of repositories, or standard descriptors.

The system designer starts this phase either by (i) modelling the system participants; this requires some experience and familiarity with MAS concepts, or by (ii) modelling the Business Processes; this is the more common approach for business users who may not be able to define system agents and their plans etc.

3.2.1 Business Process Models (BPM)

Generally, business process modelling is a way of representing organizational processes so that these processes can be understood, analysed, improved and enacted. The drawback to most BPM techniques is their procedural nature, which can lead to over-specification of the process, and the need to introduce decision points into execution that are hard to know in advance and unsuited to MAS modelling. We use declarative style modelling to describe our BPs using the Declarative Service Flow Language (DecSerFlow) [1]. More details of this are given in section 3.2.2.

Business Process Models are derived directly from the system goals and they are used to describe and identify the steps needed to achieve one or more of the system goals, these steps forming the system plans. For each **Specific System Goal** there is at least one BPM. Each Sub-Specific Goal is represented as an **Activity** inside its Super Goal BPM. Business Process Models are either **Specific Business Process**—that is, derived from a Specific System Goal—or **Basic Business Process**, that describes a Basic System Goal.

3.2.2 Modelling BPs and Specifying System Norms Using DecSerFlow

According to Jennings [6] Commitments and Conventions Hypothesis: all coordination mechanisms can ultimately be reduced to (join) commitments and their associated (social) conventions. Introducing conventions to the system participants' interactions can be achieved through one of three approaches (i) reducing the set of possible options by restricting and hard coding all these conventions in all agents, (ii) enforcing these conventions at the protocol level that all system participants follow so there is no way for the agent to violate the conventions even if it tries to, or (iii) Using the norms to only influence the systems participants behaviour as suggested by Dignum et al [4].

We adopt a declarative style for modelling our BPs, namely DecSerFlow as proposed by Aalst and Pesic [1], which offers an effective way to describe loosely-structured processes. So instead of describing the process as a directed graph where the process is a sequence of activities and the focus of the design is on "***HOW***", the system designer specifies "***WHAT***" by adding constraints in the activities' model as well as rules to be followed during execution time. For constraints specification, DecSerFlow uses LTL (Linear Temporal Logic) as underlying formal language and these constraints are given as templates, i.e. as relationships between two (or more) whatsoever activities. Each constraint template is expressed as an LTL formula. We use DecSerFlow notation and its underlying LTL formal representation.

The system designer can add the convention norms in one of the following ways: (i) at the business process level, the designer may include any number of activities alongside the business process activities and enforce any relation he might see necessary among the activities, or (ii) at the activity level, where the designer may choose to add the convention norms as preconditions of any number of activities; in this way the system participant would not be able to execute such activities in the absence of the satisfaction of that precondition.

3.2.3 Specific Business Process

Each system goal is realised through one specific BP, which is a collection of sub-processes or activities that normally lead to the achievement of that specific goal.

Figure 3 shows a sample diagram of "Accept SSLF" Specific Business Process, which has two activities (A) "Check for New SSLF" that is a sub-process to achieve the "Check for New SSLF" Specific System Goal and (B) "Publish Supplier SSLF" Basic Business Process to achieve "Publish Supplier SSLF" Basic System Goal. Both activities can run an arbitrary number of times, however the **Succession relationship** requires that every execution of activity A should be followed by the execution of activity B and each activity B should be preceded by activity A. That relationship is formally expressed in LTL as: $\square(A \Rightarrow \Diamond(B)) \wedge \Diamond(B) \Rightarrow ((\neg B) \sqcup A)$.

3.2.4 Basic Business Process

A Basic Business Processes is the most detailed BP model; it can contain any number of plans to achieve ONLY one Basic System Goal. The Basic Business Process diagram comprises a set of activities. Figure 3 shows a diagram for the "Publish SSLF" Basic Business Process, which has five possible activities; each activity is done by one or more system participants. Each activity has its pre-conditions and post-conditions, there is no need to specify the execution sequence, because the activity whose pre-conditions are met should start automatically. "Get connection" has no pre-conditions which means it should start as soon as this "Publish SSLF" Business Process starts. There are two activities named "Download SSLF", each of which has the same post-conditions but different pre-condition. During execution, based on the available resources, the supplier agent can download the new SSLF from either a FTP or an Email account. To avoid duplication of execution of this activity there is the **not co-existence** relationship that means ONLY one of the two tasks "A" or "B" can be executed, but not both. The not co-existence relation is expressed in LTL as: $\Diamond(A) \Longleftrightarrow \neg(\Diamond(B))$.

3.2.5 System Participants Models

System Participants Models are equivalent in context to Detailed Business Process Models. They offer a different view of the process by describing the detailed activities from the participants' perspective. They define also how activity owners communicate with other participants. System Participants Diagram includes one box for each system participant (**Agent–Actor–Service**) and one for the **Environment**, that allows for the definition of any external event caused by other system participants.

Figure 4 shows a sample system participants diagram of Publish SSLF Basic BP. There are two Software Agents (Supplier Agent and Central Virtual Stock Agent) and one Software Service (Translation Service) and the Environment. The **Software Agent** is a piece of automated software that has its own set of goals expressed as states that it tries to achieve continuously. It holds its knowledge as a belief set and

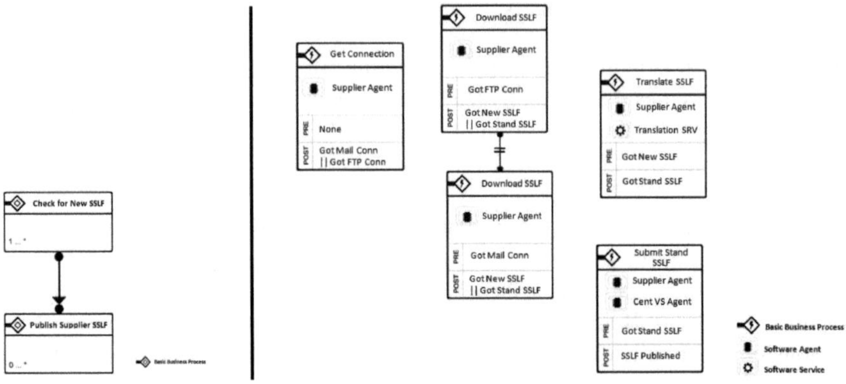

Fig. 3 Specific Business Process Diagram (Accept SSLF) [left] and Basic Business Process Diagram (Publish SSLF Business Process) [right]

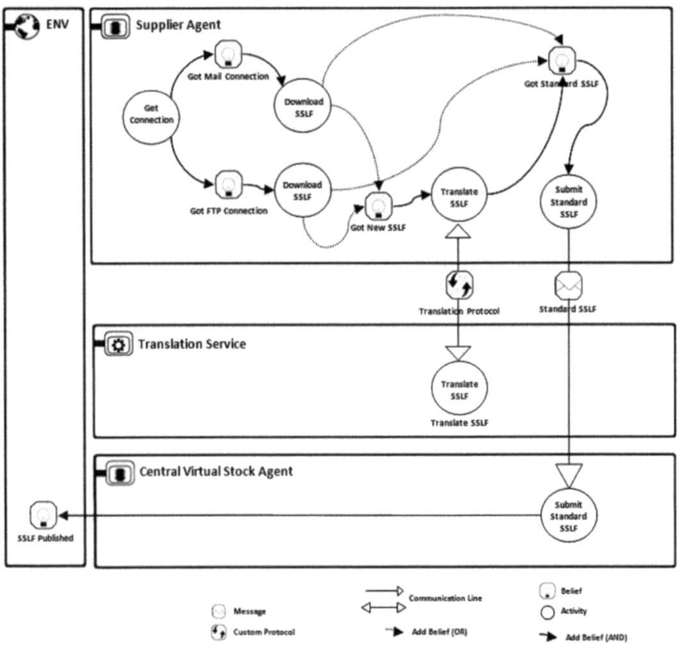

Fig. 4 System Participant Diagram (Publish SSLF)

it is able to define dynamically new goals and update its belief set as well as define the needed steps (plans) to achieve its goals. The software agent is situated within an **Environment** that allows the agent to carry out its dynamic actions (plans), the environment also facilitates the ways in which the agent might need to communicate with other software entities sharing the same environment. We adopt also the concept of a human system participant (**Actor**), as proposed by [11] to allow for modelling a participatory team of software agents and human actors. This view is found to be more practical to support real life scenarios where some decisions are necessarily assigned to humans to make. Finally, the system participant can be a **Software Service** which is a piece of software that has a set of related functionalities together with policies to control its usage and is able to respond to any relevant requests from other software entities in a reactive manner.

System participants communicate using a **Communication Protocol**, which is a set of rules determining the format and transmission of a sequence of data in the form of messages between two or more system participants. APMDMAS offers a number of pre-defined (**Native Protocols**), as well as allowing the user to define (**Custom Protocols**). The communication protocol can have any number of messages of either of two types: (***Inform Message*** and ***Request Message***).

During the detailed system design phase the user can define each entity from scratch or link it to a definition file. All entity definitions are stored in the system repositories that hold ***System and Agent Plans***, ***Environment and Agents' beliefs***, ***System and Agents' goals***, as well as all ***system processes, communications protocols, system declarative norms***.

3.3 *Implementation Phase*

The third and last phase of APMDMAS is focused on the verification and consistency check across all system models. The Verified system model can be exported into one **Meta Model (RDF)** . That meta model is used to generate code for one of the supported execution Languages, Platforms or Middlewares. We are currently developing tools to support the full cycle of APMDMAS methodology including the generation of the executable code.

4 Conclusion and Future Work

We have described briefly the key features of APMDMAS methodology. A methodology that aims at overcoming the issues we have found with current MAS methodologies and aimed at attracting a wider range of users to adapt MAS concepts in commercial settings. The clear and well defined steps should help the users describe any small scale MAS with ease and make MAS concepts accessible and easy to comprehend by business users as well as academics. The methodology covers most common MAS concepts and allows to describe the system formally for verification and implementation purposes.

A set of tools is currently being developed to support all phases of APMDMAS, and future work includes establishing the most appropriate means for specifying the system norms, describing system and agent plans to support dynamic planning, and deployment methods for distributed MAS systems.

References

1. van der Aalst, W.M.P., Pesic, M.: DecSerFlow: Towards a truly declarative service flow language. In: Proc. 3rd Int. Workshop on Web Services and Formal Methods. Springer, Heidelberg (2006)
2. Bresciani, P., Giorgini, P., Giunchiglia, F., Mylopoulos, J., Perini, A.: Tropos: an agent oriented software development methodology. J. Autonomous and Multi-Agents (2003)
3. DeLoach, S.A., Wood, M.F., Sparkman, C.H.: Multiagent systems engineering. Int. Journal of Software Engineering and Knowledge Engineering 11(3), 231–258 (2001)
4. Degnum, F., Morley, D., Sonenberg, E.A.: Towards socially spphisticated BDI agents. In: DEXA Workshop, pp. 1134–1140 (2000)
5. Edmunson, S.A., Botterbusch, R.D., Bigelow, T.A.: Application of System Modelling to the Development of Complex Systems. In: 11th IEEE/AIAA Proceedings of the Digital Avionics Systems Conference, vol. (5-8), pp. 138–142 (1992)
6. Jennings, N.R.: Commitments and conventions: The foundation of coordination in multi-agent systems. The Knowledge Engineering Review 8(3), 223–250 (1993)
7. Luck, M., McBurney, P., Preist, C.: Agent Technology: Enabling Next Generation Computing (A Roadmap for Agent Based Computing). AgentLink (2003)
8. Luck, M., Griffiths, N., d'Inverno, M.: From agent theory to agent construction: A case study. In: Jennings, N.R., Wooldridge, M.J., Müller, J.P. (eds.) ECAI-WS 1996 and ATAL 1996. LNCS, vol. 1193, pp. 49–64. Springer, Heidelberg (1997)
9. Padgham, L., Winikoff, M.: Prometheus: A methodology for developing intelligent agents. In: Giunchiglia, F., Odell, J.J., Weiss, G. (eds.) AOSE 2002. LNCS, vol. 2585, pp. 174–185. Springer, Heidelberg (2003)
10. Sudeikat, J., Braubach, L., Pokahr, A., Lamersdorf, W.: Evaluation of Agent–Oriented Software Methodologies – Examination of the Gap Between Modeling and Platform. In: Odell, J.J., Giorgini, P., Müller, J.P. (eds.) AOSE 2004. LNCS, vol. 3382, pp. 126–141. Springer, Heidelberg (2005)
11. Wijngaards, N.J.E., Kempen, M., Smit, A., Nieuwenhuis, K.: Towards Sustained Team Effectiveness. In: Boissier, O., Padget, J., Dignum, V., Lindemann, G., Matson, E., Ossowski, S., Sichman, J.S., Vázquez-Salceda, J. (eds.) ANIREM 2005 and OOOP 2005. LNCS (LNAI), vol. 3913, pp. 35–47. Springer, Heidelberg (2006)
12. Wood, M.F., DeLoach, S.A.: An Overview of the Multiagent Systems Engineering Methodology. In: Ciancarini, P., Wooldridge, M.J. (eds.) AOSE 2000. LNCS (LNAI), vol. 1957, pp. 207–221. Springer, Heidelberg (2001)
13. Wooldridge, M.: The Logical Modelling of Computational Multi-Agent Systems, PhD thesis, Department of Computation, UMIST, Manchester, UK (1992)
14. Wooldridge, M.J., Jennings, N.R., Kinny, D.: The Gaia methodology for agent-oriented analysis and design. Journal of Autonomous Agents and Multi-Agent Systems 3(3), 285–312 (2000)
15. Zambonelli, F., Jennings, N.R., Wooldridge, M.: Developing multi-agent systems: The Gaia methodology. ACM Transactions on Software Engineering and Methodology 12(3), 317–370 (2003)

Part II
Constraints and Optimization

Delegate MAS for Large Scale and Dynamic PDP: A Case Study

Shaza Hanif, Rinde R.S. van Lon, Ning Gui, and Tom Holvoet

Abstract. Pickup and Delivery Problems (PDPs) have received significant research interest in the past decades. Their industrial relevance has stimulated the study of various types of solutions. Both centralized solutions, using discrete optimization techniques, as well as distributed, multi-agent system (MAS) solutions, have proven their merits. However, real PDP problems today are more and more characterized by (1) dynamism - in terms of tasks, service time, vehicle availability, infrastructure availability, and (2) their large scale - in terms of the geographical field of operation, the number of pickup and delivery tasks and vehicles. A combination of both characteristics brings unsolved challenges.

Delegate MAS is a coordination mechanism that could prove to be valuable for constructing a decentralized solution for dynamic and large scale PDP problems. In this paper, we illustrate a solution based on delegate MAS for solving PDP. Our solution enables different agents to dynamically collect and disseminate local information and make decisions in a fully decentralized way. We applied our approach to a concrete case study. Experimental results indicate the suitability of the approach for dynamic and large scale PDP problems.

1 Introduction

Pickup and Delivery Problems (PDPs) [4] constitute a well-known family of optimization problems in logistics, faced by package transportation companies. In a PDP, loads have to be transported form origins to destinations by a set of vehicles. Many approaches have been proposed for addressing numerous variants of the problem for more than half a century [4, 5].

Shaza Hanif · Rinde R.S. van Lon · Ning Gui · Tom Holvoet
Katholiek Universiteit Leuven , Dept. Computer Science,
Celestijnenlaan 200A, B-3001 Heverlee
e-mail: `{Shaza.Hanif,Rinde.vanLon,Ning.Gui,Tom.Holvoet}@cs.kuleuven.be`

F.M.T. Brazier et al. (Eds.): Intelligent Distributed Computing V, SCI 382, pp. 23–33.
springerlink.com

Traditional approaches tackle the problem by applying discrete optimization approaches (DOA) [20].These approaches require formulating the problem in a mathematical model, in which constraints and objective functions are mathematically expressed. These yield static solutions; when the problem changes, the mathematical model needs updating, and potentially the selection of the DOA to solve the problem needs to be reconsidered.

In practice, a PDP is typically both a dynamic and large scale problem. New transportation requests might arrive, accidents or traffic jams may require vehicles to change their schedules, service time may vary. Such dynamism cannot be handled effectively by a solution based on a DOA. Additionally, the scalability requirement creates extra challenges - managing a PDP for large numbers of vehicles, packages in a large geographical environment add considerable stress to any solution.

In order to handle such a dynamic environment in a scalable way, improvements of DOA, often known as on-line algorithms [6, 7], have been proposed. These algorithms use smart heuristics and exploit the problem structure for fast computation times to yield a feasible solution. Powel et al. highlights that as these improvements are based on a static solution, when the problem is highly dynamic and large in scale, such solution approaches become infeasible [18].

We argue that a solution to dynamic problems should embrace the dynamism in the environment and consider dynamism as an integral part of the original problem. This is contrary to the DOA, which integrate the dynamism by modifying a static solution. Decentralized, Multi-agent System (MAS) [22] based solutions incorporate dynamism explicitly in more systematic ways. Jennings indicates that *'Agent-based computing has the potential to significantly improve our ability to model, design and build complex, distributed software systems'* [19]. These solutions provide modeling abstractions that are decentralized in nature and are more suitable for presenting scalable solutions for real world systems [21].

Delegate MAS [12] is a coordination mechanism that is well suited for coordination and control applications. It takes dynamism as well as large scale as a primary concern. The mechanism is being studied with good results in various domains of coordination and control applications, including traffic [13, 14] and manufacturing control [12]. In this paper, we present a 'delegate MAS' based solution for solving PDP. In our solution the system entities, vehicles and packages, are modeled as agents. The agents coordinate using delegate MAS to explore the environment around them, collecting locally available knowledge. Moreover, the agents disseminate information to those parts of the system for which their information is relevant.

Two distinguished features make the solution suitable for large scale and dynamic PDP: (i) The vehicles and packages are modeled as individual agents that coordinate using locally available information only. This enables a scalable solution since it is not required to maintain accurate global information for coordination. (ii) The delegate MAS allows the solution to cope with dynamic changes through indirect communication via the environment.

The remainder of the paper is structured as follows: Section 2 discusses the related work from two different aspects, DOA and MAS based approaches. Section

3 and 4 describe our delegate MAS model for solving PDP and the preparatory experiments respectively. Finally, section 5 concludes the paper.

2 Related Work

In the last decade, various taxonomical papers have appeared [4, 5, 6, 16], which classify the diverse approaches proposed for solving multiple variants of PDP. Here, we differentiate between DOA, and MAS based approaches for solving PDP.

2.1 Discrete Optimization Approaches for Solving PDP

A wide variety of solutions are based on DOA [4, 5]. So-called 'exact DOAs', which guarantee optimal solutions, yield high computational complexity. Therefore, other advanced solutions like heuristics, meta-heuristics and hyper-heuristics are being used [20]. All these approaches are very useful for solving static problems but can not effectively handle the continual change in a dynamic problem.

Online algorithms use a combination of DOA approaches for dealing with dynamism [6]. Several approaches solve the problem by manipulating data available at a certain moment in time using a mathematical model. As new data becomes available, instead of recomputing the mathematical model, they re-optimize the solution using (meta-)heuristics [15, 16, 17]. Another meta-heuristics solution for dynamic PDP is the Ant Colony Optimization (ACO) based approach [20]. For instance, Rizoli et al. [8] presents an online ACO algorithm in which dynamic change is incorporated by evaporating old information and reinstating the new change in the schedule.

While DOAs for dynamic PDP practically handle the dynamism in the environment, they have a clearly identifiable role of a *dispatcher*. They assume that all information about the problem is known, and calculate optimal routes for serving the set of requests available at a certain amount of time. As a centralized approach, DOAs face scalability challenges in terms of number of vehicles, number of transportation requests or geographical expansion.

2.2 MultiAgent Systems Based Approaches for Solving PDP

MASs consist of a set of autonomous software entities that collaborate to attain overall system objectives. As such, MASs provide the modeling abstractions for building decentralized solutions [19], without requiring a mathematical model.

Competitive results were obtained for optimization of logistics in industry [1, 2, 3, 11].

For solving dynamic PDP, Fisher et al. a MAS based solution [10] in which tasks are allocated to trucks and company agents using an extension of Contract Net Protocol (CNP) [22]. They compute a basic solution using CNP which is significantly improved by using auction protocols [22].

Kouki et al. also present an online solution [9] that performs re-optimization of routes (calculated by a mathematical model) for individual vehicle agents. It uses CNP to intelligently deal with dynamic environments. Both solutions are based on re-optimizing the whole solution for adaptability and may not be feasible for highly dynamic environments. In another MAS based solution [3], a centralized dispatcher allocates incoming requests to regional dispatchers. It adapts to the dynamic environment but its centralized nature limits its scalability.

A common disadvantage of above mentioned solutions is their reliance on centrally maintained global information. This means that the role of *dispatcher* is still evident, despite that MASs do not require to maintain centralized global information [19]. This centralistic approach faces problems in dealing with large scale PDP.

3 Delegate MAS for PDP

The term 'delegate MAS' was coined in [12] after the mechanism was studied in context of manufacturing control. Subsequent research [23, 24] provided more insight in the mechanism. Applications of delegate MAS in different domains show that it is an effective, multipurpose agent coordination mechanism [12, 13, 14].

The purpose of this paper is to evaluate the feasibility of a delegate MAS-based solution for providing a scalable, decentralized solution for dynamic PDPs.

3.1 Problem Description

In this case study we consider the 'dynamic pickup and delivery problem with time windows' (DPDPTW) [4]. The DPDPTW can be described as follows. New arriving pick-up-and-delivery tasks need to be allocated to a truck that can serve the request. A task is comprised of a pickup and delivery location and a delivery time window. In our problem dynamism is caused by continually arriving requests. Based on Marrow's [21] concept of scalability, we refer to a solution as scalable if the efficiency of the solution remains proportional to the problem scale.

The problem requires to consider the following concerns. First, all customer requests need to be satisfied by delivering the packages without violating time windows. Second, the total distance traveled by the trucks needs to be minimized.

3.2 A Delegate MAS Solution

In this section, we describe a decentralized solution for solving DPDPTW. This solution is designed to be adaptive and scalable.

Our solution consists of two types of primary agents: truck agents and package agents. The agents are *situated* in the road infrastructure (the *environment*) [25], which offers localized information sharing for agents. Agents can perceive the environment and act based on the locally (in time and space) available information.

Truck agents can travel over the road infrastructure. A truck agent is responsible for finding nearby packages, for acquiring information from package agents about possible paths and for selecting a path. A package arrives dynamically and its status may change over time: when its deadline approaches, it becomes a 'critical package', having a higher priority for pick up. In order to maintain updated information, packages are responsible to periodically inform other nearby package agents about package related information, including package destination and time windows. Packages also manage information about feasible delivery paths and make decisions when multiple truck agents want to pick up a certain package.

In order to achieve coordination between different agents, our approach uses several types of delegate MASs to coordinate agents' interactions. A delegate MAS is a light-weight agent, which performs delegated tasks for their primary agent [24]. This design allows the implementation of auxiliary tasks separated from primary agent logic. To distinguish them from the primary agents, situated light-weight agents in a delegate MAS are called ants. In order to enable such ant based coordination between agents, we assume that there are smart devices installed at pickup and delivery location which have computation and communication capabilities. Three different types of delegate MASs are used in our solution.

1. Feasibility Ants. In order to spread local information to other agents, package agents delegate this functionality to a *feasibility delegate MAS*. Figure 1 shows how they travel to the neighboring destinations of packages within a certain range and leave information that points back to their sender package. By using feasibility ants, package agents can share their information in a decentralized way and construct local package maps (paths) which will be used for other delegate MAS.

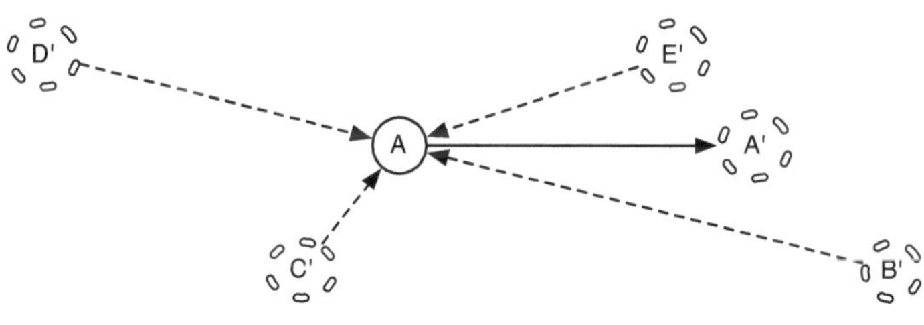

Fig. 1 An abstract representation of behavior of feasibility ants: the package agent A (with destination A') has sent feasibility ants to package destinations B', C', D' and E'. Each of these destinations has information that points back to A.

2. Path Exploration Ants. Path exploration ants are dispatched by truck agents, their goal is to find an optimal delivery path. An exploration ant travels over the network based on the information shared by the feasibility ants. In Figure 2 an example of the behavior of path exploration ants is illustrated.

Typically, path exploration ants are sent out several hops ahead as this reduces the probability of getting stuck in local optima (e.g. going after the package which is closest by). When it reaches a pre-specified number of hops, this ant will return to the truck agent with accumulated information about its path. A truck agent sends

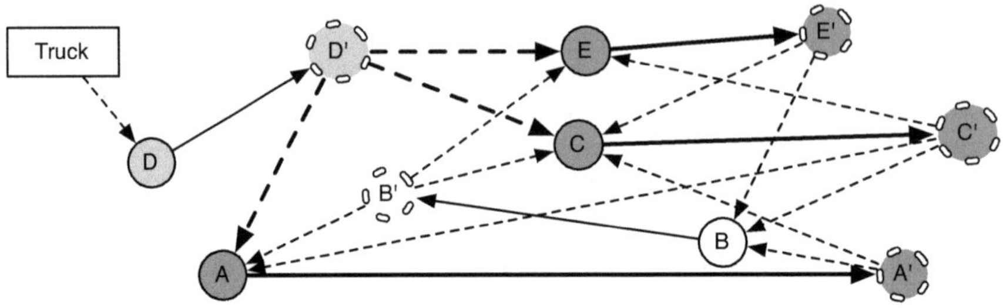

Fig. 2 An abstract representation of behavior of exploration ants on our solution for DPDPTW: a truck has indicated that it will pickup package D and deliver it at D'. Exploration ants are dispatched from D to D' to explore possible future paths from there. At D', information (spread by feasibility ants) about packages A, C and E is known. Thus the exploration ants follow the arrows towards these packages to explore these paths. The explored paths are drawn in bold. Note that in this example the ants consider only one package ahead (one hop).

multiple exploration ants, which provide a number of possible paths to the truck agent. The truck agents chooses its path by using a heuristic with the following objectives, in descending order of importance:

1. Delivering all packages without violating time windows.
2. If it is not possible to comply to a time window, minimization of lateness.
3. Minimization of the total distance traveled by the trucks.

When multiple trucks are planning the delivery of the same package, a conflict arises. This conflict is resolved by using a similar heuristic for comparing paths, packages can decide to be planned in a path of higher quality.

3. Intention Ants. Truck agents delegate the task of negotiation and reservation with the package agent to intention ants. Once a truck decides to go to a package, intention ants are regularly sent to the package agents to make and continually reinstate a reservation.

Agent decision logic. As described earlier, truck agents make decisions about path selection by using heuristics. This decision determines the order at which packages should be serviced. In Figure 3 a heuristic and three helper functions are introduced. The implementation conforms to the heuristic objectives and is used in the experiments described in section 4. The pseudocode first considers time window violations, then lateness followed by distance traveled.

In order to deal with dynamic environments, ants are sent out periodically. Package information and reservations are expired after a certain period of time by examining their time stamps. The *adaptation rate* controls these aspects of our solution.

Our solution embraces the dynamism in the problem by periodically disseminating local information, expiring outdated information, and by collecting the recent information through ants. We are able to tune *adaptation rate* which determines

$criticality(path)$ = // average number of 'critical' packages in path. A package is critical when its deadline is approaching.
$lateness(path)$ = // average lateness for all packages in path.
$unpaidDistance(path)$ = // average unpaid distance for all packages in path. Unpaid distance is the distance that has to be traveled from one delivery location to a pickup location of another package.

```
compare(path1, path2) {
    if criticality(path1) ≠ criticality(path2)
        return argmax(criticality(path1), criticality(path2))
    if lateness(path1) ≠ lateness(path2)
        return argmin(lateness(path1), lateness(path2)
    return argmin(unpaidDistance(path1), unpaidDistance(path2))
}
```

Fig. 3 Agent decision heuristic.

the solution's responsiveness to changes in the environment. Since agents in our solution can make decisions only based on local information, the solution is decentralized. This avoids the requirement of maintaining centralized global information which allows the solution to be scalable.

4 Experiments

In this section we present the setup and results of two simulation experiments. The goals of our experiments are to evaluate the adaptability and scalability of our solution in a simulation.

Adaptability experiment. In this experiment, we investigate the performance of our solution in environments with a different degree of dynamism. We compare two instances of our solution on an identical problem scenario: 'fast adaptation' and 'slow adaptation' with a different value for their *adaptation rate*. As indicated by the name, 'fast adaptation' adapts faster to dynamism in the problem compared to 'slow adaptation' which corresponds to a more static solution.

Scalability experiment. This experiment is designed to investigate whether our solution 'scales', i.e. a scaled problem should result in a proportionally decrease in performance of our solution. Two scenarios of different scale are used, one 'small' scenario of the Leuven city map and one 'big' scenario using a map of Brussels.

Experiment setup. In our experiments the road structure is based on OpenStreetMap[1] data. We use the road structure of two cities in Belgium: Brussels and Leuven. The adaptability experiment uses only the Leuven map, while the scalability experiment uses both maps. Each experiment is done for a range of trucks from 7 until 14. This range of trucks is chosen since for less than 7 trucks, the solution cannot handle the packages properly which results in inferior performance. Above this range, the performance of the solution does not improve significantly anymore. Each experiment is

[1] `www.openstreetmap.org`

repeated 20 times with uniform randomly distributed package arrival time and pickup and delivery location. A package deadline is set to a multiple of the travel time to deliver the package ($package\ deadline = 15 \cdot deliverTime$). All packages are made available within 12 hours after the start time. A simulation is finished when all packages are delivered. This means that each simulation will take at least 12 hours of simulation time, a maximum duration depends solely on the performance of the solution. The experiments were run using a simulator developed specifically for this project. Computing a simulation generally takes about 30 seconds to 5 minutes, depending on the size of the map and the other experimental settings.

In the following table the Leuven and Brussels scenarios are characterized.

	Leuven	**Brussels**	**Factor**
Number of packages	200	800	4
Number of trucks (n=7..14)	n	$4 \cdot n$	4
Accumulated road length (two way)	968.09 km	12946.78 km	13.37
Mean distance between package origin and destination	8.94 km	28.75 km	3.22

In order to compare the results between these two maps, we increased the package delivery deadline for the Brussels scenario by 3.22, based on the difference in distance between package origin and destination as shown in the table above.

Experiment results. For both experiments the same two dependent variables are measured:

lateness The average lateness of the delivery of all packages (in minutes).
distance The average distance traveled by all trucks (in kilometer).

Figure 4(a) shows lateness of both the fast adapting and the slow adapting solution. On average the fast adapting solution performs better as compared to the slow adapting solution. The variance of the fast adapting solution is smaller, indicating that its performance is more predictable. This is to be expected because the slow adapting (more static) solution generally suffers from any dynamism in the problem, which causes a greater variance in outcome. This variance can be explained by the global behavior of MASs which depends on the interaction of many agents. However, the better performance of the fast adapting solution has a cost of communication overhead since it sends ants more frequently. Part of our future work is to investigate how to make a balance between these two factors. Figure 4(b) shows that the fast adapting solution consistently performs better with respect to distance traveled compared to the slow adapting solution. The outcome of this experiment can also be used for determining the preferred number of trucks in a scenario given an objective.

Figure 5(a) shows the average package lateness for the scalability experiment. It can be seen that for the Brussels scenario the lateness is descending with the increase in number of trucks. It is shown that starting at about 40 trucks the lateness can be reduced close to 0 for the Brussels scenario. This shows that to multiply the number of trucks with 4 is not enough, which can be explained by the differences in topology

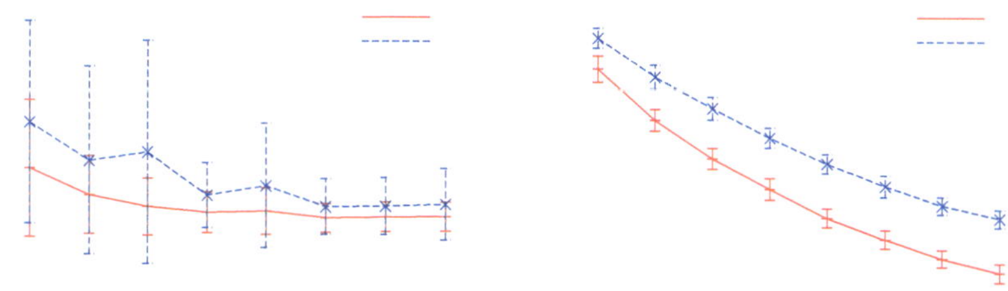

(a) Average package lateness. (b) Average distance traveled per truck.

Fig. 4 Adaptability experiment: comparison of results for the solution with fast adaptation and slow adaptation.

between the two maps. Note that the results for the Leuven map are actually the same as the fast adapting solution which was plotted in Figure 4(a). Figure 5(b) shows the average distance traveled per truck for the scalability experiment. The difference in traveled distance between the Leuven and Brussels map is about a factor of 2. In both Figure 5(a) and Figure 5(b) a consistent increase in performance is shown. This illustrates that when the scale of the problem is increased, our solution performs well with respect to lateness and distance traveled.

Analyzing the scalability of a solution is complicated since it is nearly impossible to properly scale a problem instance using a single factor. In our experiments, we scaled the problem by taking a similar but bigger city and choosing an arbitrary scale factor. The experiment results indicate that our solution has consistent performance on both scales.

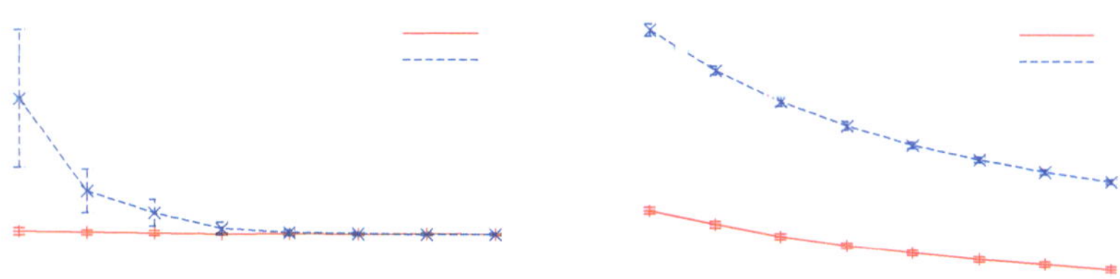

(a) Average package lateness. (b) Average distance traveled per truck.

Fig. 5 Scalability experiment: comparison of results for the Leuven and Brussels maps.

5 Conclusions

In this paper, we presented a delegate MAS solution for solving dynamic PDP. The agents interact through lightweight ants that explore a local environment and perform actions on behalf of their originators. There are two main features of the approach. First, the control is decentralized. Agents make decisions based on their localized view. This offers scalability, as it is not required to maintain accurate global information like in previous approaches. Second, agents coordinate with each other through the environment, which enables the solution to dynamically adapt to the changing environment.

We applied our approach to a DPDPTW and performed a set of experiments. Simulation results show that our approach is capable of adapting to a dynamic environment. The solution shows consistent performance in terms of lateness and distance traveled, supporting the claim on scalability. If the necessary infrastructure is available (i.e. smart devices with communication capabilities, at pickup and target locations, and at vehicles), the solution can be physically deployed and operated as a distributed software system.

In our future experiments, we will incorporate additional problem constraints covering additional details i.e. traffic jams or failures encountered by trucks, different patterns of request arrivals, dynamically changing request time windows etc.. We already performed experiments for truck failures, which showed encouraging results. We also plan to extend this solution for other variants of PDPs.

References

1. Caridi, M., Cavalieri, S.: Multi-agent Systems in Production Planning and Control: an Overview. Production Planning and Control 15(2), 106–118 (2004)
2. Glaschenko, A., Ivaschenko, A., Rzevski, G., Skobelev, P.: Multi-Agent Real Time Scheduling System for Taxi Companies. In: Autonomous Agents and Multi-Agent Systems, AAMAS (2009)
3. Dorer, K., Calisti, M.: An Adaptive Solution to Dynamic Transport Optimization. In: Proc. of 4th Int. Conf. on Autonomous Agents and Multi-Agent Systems, AAMAS (2005)
4. Parragh, S.N., Doerner, K.F., Harti, R.F.: A survey on PDP Part II: Transportation between Pickup and Delivery Locations. Journal fr Betriebswirtschaft 58(2), 81–117 (2008)
5. Eksioglu, B., Vural, A.V., Reisman, A.: The Vehicle Routing Problem: A Taxonomic Review. Computers and Industrial Engineering 57(4), 1472–1483 (2009)
6. Berbeglia, G., Cordeau, J.F., Laporte, G.: Dynamic Pickup and Delivery Problems. European Journal of Operational Research 202, 8–15 (2010)
7. Hentenryck, P.V., Bent, R.W.: Online Stochastic Combinatorial Optimization. Operations Research 52, 977–987 (2004)
8. Rizoli, A.E., Montemanni, R., Lucibello, E., Gambardella, L.M.: Ant Colony Optimization for Real World Vehicle Routing Problems. Swarm Intelligence 1(2), 135–151 (2007)
9. Kouki, Z., Chaar, B.F., Ksouri, M.: Extended CNP Framework for the dynamic Pickup and Delivery Problem Solving. In: Artificial Intelligence and Innovations III 296 (2009)

10. Fischer, K., Muller, J., Pischel, M.: Cooperative Transportation Scheduling: An Application domain for DAI. Journal of Applied Artificial Intelligence (1995)
11. Kozlak, J., Creput, J.C., Hilaire, V., Koukam, A.: Multi-agent Approach to Dynamic PDP with Uncertain Knowledge about Future Transport Demands. Fundamenta Informaticae 71 (2006)
12. Holvoet, T., Valckenaers, P.: Exploiting the Environment for Coordinating Agent Intentions. In: Weyns, D., Van Dyke Parunak, H., Michel, F. (eds.) E4MAS 2006. LNCS (LNAI), vol. 4389, pp. 51–66. Springer, Heidelberg (2007)
13. Claes, R., Holvoet, T.: Maintaining a Distributed Symbiotic Relationship using Delegate Multi agent Systems. In: Proc. of Winter Simulation Conference (2010)
14. Weyns, D., Holvoet, T., Helleboogh, A.: Anticipatory Vehicle Routing Using Delegate Multi-Agent Systems. In: Proc. of IEEE intelligent Transportation Systems Conference (2007)
15. Gendreau, M., Guertin, F., Potvin, J.Y., Seguin, R.: Neighborhood Search Heuristics for a Dynamic Vehicle Dispatching Problem with Pick-ups and Deliveries. Transportation Research Part C 14, 157–174 (2005)
16. Psaraftis, H.N.: Dynamic Vehicle Routing: Status and Prospects. Annals of Operations Research 61, 143–164 (1995)
17. Gutenschwager, K., Niklaus, C., Vob, S.: Dispatching of an Electric Monorail System: Applying Metaheuristics to an Online PDP. Transportation Science 38, 434–446 (2004)
18. Powell, W.B., Towns, M.T., Marar, A.: On the Value of Optimal Myopic Solutions for Dynamic Routing and Scheduling Problems in the Presence of user Noncompliance. Transportation Science 34(1), 67–85 (2000)
19. Jennings, N.R.: On Agent-based Software Engineering. Artificial Intelligence 117, 277–296 (2000)
20. Burke, E.K., Kendall, G. (eds.): Search Methodologies: Introductory Tutorials in Optimization and Decision Support Techniques. Springer, Heidelberg (2005)
21. Marrow, P.: Scalability in Multi-Agent Systems: The DIET Project. In: Workshop on Infrastructure for Agents. MAS and Scalable Multi-Agent Systems at Autonomous Agents (2001)
22. Wooldridge, M. (ed.): An Introduction to MultiAgent Systems. Wiley, Chichester (2009)
23. Van Dyke Parunak, H., Brueckner, S.A., Weyns, D., Holvoet, T., Verstraete, P., Valckenaers, P.: E Pluribus Unum: Polyagent and Delegate MAS Architectures. In: Antunes, L., Paolucci, M., Norling, E. (eds.) MABS 2007. LNCS (LNAI), vol. 5003, pp. 36–51. Springer, Heidelberg (2008)
24. Holvoet, T., Weyns, D., Valckenaers, P.: Patterns of Delegate MAS. In: Third IEEE International Conference on Self-Adaptive and Self-Organizing Systems (2009)
25. Weyns, D., Holvoet, T.: From Reactive Robotics to Situated Multiagent Systems: A Historical Perspective on the Role of the Environment in Multiagent Systems. In: Workshop on Engineering Societies in the Agents World, pp. 31–56 (2005)

SOCIAL DCOP - Social Choice in Distributed Constraints Optimization

Arnon Netzer and Amnon Meisels

Abstract. Distributed Social Constraints Optimization Problems (DSCOPs) are DCOPs in which the global objective function for optimization incorporates a social welfare function (SWF). DSCOPs have individual, non-binary and asymmetric constraints and thus form a unique DCOP class. DSCOPs provide a natural framework for agents that compute their costs individually and are therefore self-interested. The concept of social welfare is discussed and SWFs are presented. An important aspect of DSCOPs and of social objective functions is their ability to support distributed hill climbing algorithms. The DSCOP hill climbing algorithm is presented and tested experimentally on multi agent pickup and delivery problems. It turns out to improve the distribution of utility gains among agents, while loosing very little in global gain.

1 Introduction

Distributed Constraint Optimization Problems (DCOPs) form a powerful framework for distributed problem solving that has a wide range of application in Multi Agent Systems. Typical examples include mobile sensing networks[10],graphical games [Maheswaran et al. (2004b)Maheswaran, Tambe, Bowring, Pearce, and Varakantham] and synchronization of traffic lights[8]. In a standard DCOP agents need to assign values in a way that would minimize (or maximize) the sum of the resulting constraints costs. A DCOP algorithm searches the combinatorial assignments space, in order to find the combination of assignments that will produce this minimum [Meisels(2007), 6, 15].

Arnon Netzer · Amnon Meisels
Ben Gurion University of the Negev, P.O.B 653, Be'er Sheva 84105, Israel
e-mail: {netzerar,am}@cs.bgu.ac.il

F.M.T. Brazier et al. (Eds.): Intelligent Distributed Computing V, SCI 382, pp. 35–47.
springerlink.com

Consider a multi agent system in which each agent needs to solve its own Pickup and Deliver Problem (PDP). Packages need to be picked up and delivered under a set of conditions, each package adds a fixed utility, and every distance traveled adds some given cost (for extensive surveys on PDPs see [1]).

In some cases an agent will be better off if another agent will pick one of the packages assigned to it. Figure 1 shows a two agent system. The packages of one agent are marked by small triangles and of the other agent by small circles. It is easy to see that all of the triangle packages are located at the left side of the area and all of the circled packages except one are on the right hand side. This may lead to a situation where the added trip to reach the leftmost package, for the circled agent, may cost more than its gain.

Since each package must be picked up, an agent can only discard a package if another agent picks it up. The problem is that no cost (or utility) can be assigned directly to the package. The amount of utility change that arises from the transfer of a package from one agent to another depends on the other packages that need to be picked up by that agent. In fact, calculating this utility may be a non trivial Planning problem, that needs to be solved by the receiving agent.

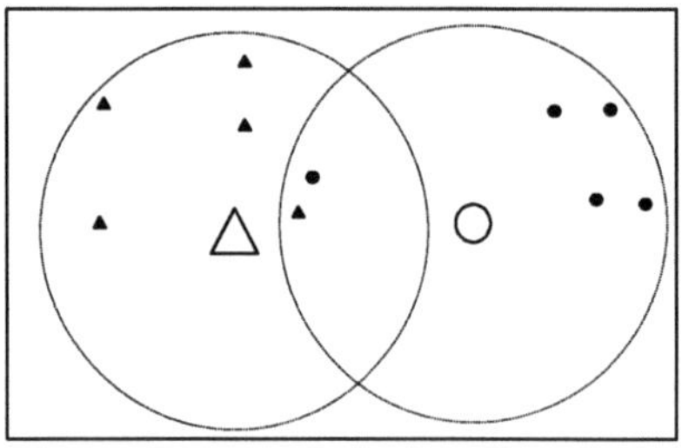

Fig. 1 A two agents Pickup and Deliver problem

Further more, agents that transfer a package from one to the other cannot be only motivated by the global objective of minimizing or maximizing the sum of all utilities. This aspect of the standard DCOP model seems to be unrealistic. A more realistic approach would be to use an objective function that incorporates some individual gains of the participating agents.

The present paper proposes a DCOP model that attempts to answer the above two problems. It generalizes the standard DCOP model and addresses self-interested agents. The proposed model discards two common assumptions of the standard DCOP model. First, it assigns costs or utilities to states and not to constraints. Second, it replaces the standard objective function that minimizes the sum of all costs by a Social Welfare function that incorporate fairness and equality measures.

In Distributed Social Constraints Optimization Problems (DSCOPs) every agent autonomously computes its utility, based on its assignment and on the assignments of agents it is constrained with. The individual costs and gains of agents represent self-interested agents that may not find it natural to search cooperatively for a solution that maximizes the sum of all utilities as a target function.

Recently [3] proposed the use of Leximin as a target function for DCOPs, attempting to increase fairness. Their discussion was limited to DCOPs with hard constraints and unary costs only. Further more, the Leximin SWF may not not be the best SWF for such purpose [16].

The proposed model of DSCOPs uses target functions that incorporate both efficiency and equality. The present study proposes objective functions that arise from known social welfare functions (SWFs) that have been studied extensively in the field of Economics (cf. [2, 21]). These ideas are discussed in Section 3 and some of the key Social Welfare Functions (SWF) are reformulated for multi agent systems.

A social local search algorithm for DSCOPs is presented in sectio 4. The behavior of the proposed framework is tested on an experimental framework of truck delivery problems (Section 5).

2 Distributed Social Constraint Optimization Problems

DSCOPs are based on the following assumptions:

- Each agent can compute its utility based on its assignment and the assignments of its constraining agents.
- The method of computation of utility is private to each agent.
- Cooperation is expressed by a faithful search of agents for the optimum of the objective function.

More formally, a *DSCOP* is a tuple $< \mathscr{A}, \mathscr{X}, \mathscr{D}, \mathscr{N} >$. $\mathscr{A}$ is a finite set of agents $A_1, A_2, ..., A_n$. $\mathscr{X}$ is a finite set of variables $X_1, X_2, ..., X_m$. Each variable is held by a single agent (an agent may hold more than one variable). $\mathscr{D}$ is a set of domains D_1, $D_2, ..., D_m$. Each domain D_i contains a finite set of values which can be assigned to variable X_i. $\mathscr{N}$ is a list of neighbors N_1, $N_2, ..., N_m$. Each list N_i contains a list of neighbors $n_1, n_2, ... n_{k_i}$ which includes all the agents that influence the utility evaluation of agent A_i.

An *assignment* is a pair, a variable and a value from that variable's domain. A complete assignment is an assignment for every variable in X. A *utility profile* is the collection of all the utilities calculated by the agents for a given complete assignment. An optimal solution is a complete assignment whose utility profile optimizes the objective function.

The DSCOP model can be viewed as a generalization of Asymmetric DCOPs (ADCOPs) [7]. ADCOPs incorporate a different cost for each agent that is involved in a constraint. Unlike ADCOPs, the proposed model of DSCOPs does not assume the costs of constraints of any given agent to be additive. The underlying framework of distributed social COPs (DSCOPs) does not impose any restriction on the form of computation of the utility of each agent. All costs of an assignment of a DSCOP are represented as individual costs (utilities) of individual agents in a given compound assignment.

Consider the distributed delivery example of section 1 (Figure 1). One can represent it by a K-ary constraint associated with every given package that enforces the pickup of the package by some agent. The utility change for a given agent, due to the addition or the removal of a package depends on the other packages that need to be picked up by that agent. The utility also depends on parameters of the PDP of the specific agent (number of trucks, sizes of trucks, etc.).

Different assignments of packages result in different PDPs for each agent. A solution of the problem is a redistribution of the packages between the agents, that will optimize a given target function on the utilities calculated by solving the PDPs of individual agents.

Interestingly, this type of problem was addressed in [17] in the form of a DCOP. The DCOP of [17] defines the domain of each agent to be all possible combinations of packages it can pick up and includes constraints to ensure the delivery of all packages. Such a representation has domains of size that is exponential in the number of packages that can be picked up by an agent. It also does not take into account the self interest of the agents (e.g., it maximizes the sum of all personal utilities).

An agent who travels a lot and picks up many packages is more likely to gain from picking up additional packages. In contrast, an agent that picks up a small number of packages may have a smaller probability of traveling near the new package. Agents that are likely to lose have little or no incentive to cooperate in search for the optimal assignment that maximizes the sum of all gains.

The DSCOP model enables the use of alternative goal (or objective) functions, that increase fairness and equality among the utilities of different agents. The view of the present paper is that fairness, or the equality of gains of individual agents, is the key for securing the cooperation of agents in search for an optimal solution.

3 Social Welfare Functions

The Social Welfare Function (SWF) was introduced by Abram Bergson [2] and is used as a way to rank the possible states of a society. For a DCOP, a SWF can serve as the objective function that the agents are optimizing.

Definition 1. Given a finite set of agents $\mathscr{A} = (a_1,...,a_n)$, $u_i = utility(a_i)$, and a utility profile $\mathscr{U} = (u_1,...,u_n)$, a Social Welfare Function maps $\mathscr{U}$ to the real numbers: $SFW(\mathscr{U}) \rightarrow \Re$.

The objective function used in standard DCOPs [15] is the max-sum (or min-sum) which is the Utilitarian SWF. The Utilitarian SWF is an extreme example in which only the state of the whole group of agents is considered, with no consideration for the welfare of any specific agent. This may lead to situations in which the sum of all the costs of all the agents is minimal but specific agents carry an un-proportionate part of the cost. For example, according to the Utilitarian SWF a cost distribution between two agents of $(u_1 = 100, u_2 = 0)$ has the same score as a cost distribution of $(u_1 = 50, u_2 = 50)$.

The Egalitarian SWF (max min) [9] forms another extreme, in which only the welfare of the worst well-off (the poorest) agent is considered. The Egalitarian SWF makes sure that the worst-off agent will be in as good a state as it can be, but, it does not reflect the situation of all the other agents. For example, according to the

Egalitarian SWF a cost distribution between two agents of $(u_1 = 100, u_2 = 0)$ has the same score as a cost distribution of $(u_1 = 50, u_2 = 0)$.

3.1 *Pareto Optimality*

A fundamental concept in social welfare is Pareto optimality. Given two utility profiles, one is said to be Pareto better than the other if at least one agent prefers the former, and no agent prefer the latter [18]

Definition 2. Given a finite set of agents $\mathscr{A} = (a_1, ..., a_n)$, $u_i = utility(a_i)$, a utility profile $\mathscr{U} = (u_1, ..., u_n)$ is Pareto better than $\mathscr{U}' = (u_1', ..., u_n')$ iff $\forall i : u_i \geq u_i'$ and $\exists j : u_j > u_j'$.

A utility profile is Pareto optimal if no other utility profile is Pareto better than it. The collection of all the Pareto optimal utility profiles is the Pareto frontier. Note that Pareto optimality of a utility profile is a characteristic of the profile itself, and is independent of any specific SWF. It is desirable that a SWF would prefer a Pareto better utility profile over a Pareto inferior one.

It is easy to see that the Utilitarian SWF is Pareto optimal, since an increase in an agent utility, ceteris paribus, will increase the sum of all agents utilities. The Egalitarian SWF acts differently than the utilitarian SWF. An increase in the utility of an agent that is not the worst off, will not change the SWF evaluation of the utility profile. An adaptation of the Egalitarian SWF called Leximin [5] solves this problem.

3.2 *Fairness and Equality in SWF*

The basic property for introducing fairness into a SWF is the *Pigou-Dalton transfer principle* [19, 4]. According to this principle, a transfer of utility between one agent and another is desirable as long as it decreases the difference between the utilities of the two agents, making them more equal.

Definition 3. A SWF satisfies the *Pigou-Dalton transfer principle* if given a finite set of agents $\mathscr{A} = (a_1, ..., a_n)$, $u_i = utility(a_i)$, a utility profile $\mathscr{U} = (u_1, ..., u_n)$, and $\mathscr{U}' = (u_1', ..., u_n')$, $\forall k \neq i, j : u_k = u_k'$, $u_i + u_j = u_i' + u_j'$ and $|u_i - u_j| > |u_i' - u_j'|$ then $\mathscr{U}'$ is preferred over $\mathscr{U}$

It is easy to see that neither the Utilitarian SWF nor the Egalitarian SWF (or Leximin) satisfy the *Pigou-Dalton transfer principle*. Perhaps the most straight forward SWF that does satisfy the *Pigou-Dalton transfer principle* is the Nash SWF [13]. Which is the multiplication of all agents utilities ($SWF_{Nash} = \prod_{i=1}^{n} u_i$).

Though the Nash SWF is Pareto optimal and satisfies the *Pigou-Dalton transfer principle*, it is considered to be very biased toward making the poorest better off, at the expense of the total well being of the society.

Sen [21] defined a family of Social Welfare Functions that incorporate both the *inequality* and the *efficiency* measures. In particular the Social Welfare Function based on the Gini index [20] $W_{Gini} = \mu(1-G)$ where μ is the average utility and G is the Gini index. Another example is the Theil based SWF: $W_{Theil} = \mu * e^{-T}$ where T is the Theil Index $T = \frac{1}{N}\sum_{i=1}^{N}(\frac{x_i}{\bar{x}} \cdot ln\frac{x_i}{\bar{x}})$. The present paper proposes the use of such a SWF for DSCOPs, in order to attain the right balance between the self interest of agents and the global interest (e.g. efficiency).

4 Hill Climbing for DSCOP

Any hill climbing local search algorithm checks whether an assignment change will increase the overall evaluation of the objective function. In standard DCOP it is enough to ensure that when an agent replaces its assignment, none of its neighbors will replace their assignments at the same time [22, 7].

Consider the example in Figure 2. In this example every circle is an agent and every line denotes a constraint between two agents. Assume that after each agent computed its best possible assignment change (possibly using additional information from its neighbors) agents A and C discover that they can make the best assignment change out of all their neighbors. Due to the additivity of constraint costs in standard DCOPs, there is no problem for agents A and C to replace their assignment at the same time.

This is no longer true for DSCOPs, due to the fact that utilities in DSCOP are calculated by each agent and the utility is not necessarily additive. Consider agent B in Figure 2. It is possible that both an assignment change by A and an assignment change by C will decrease B's utility.

Fig. 2 DSCOP Neighborhood example

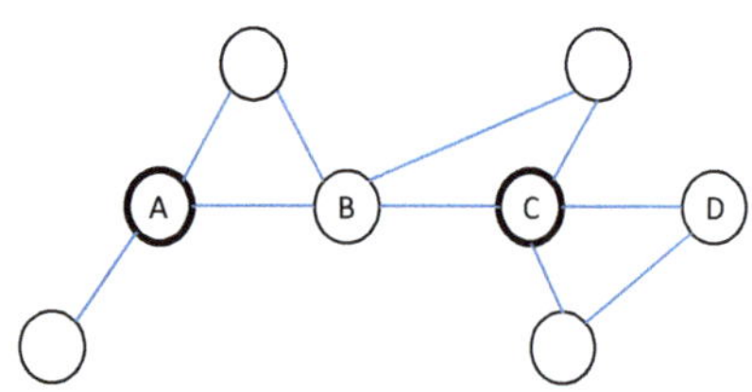

For DSCOPs it is not enough to ensure that none of an agents neighbors are changing assignment if the agent does. One must make sure that none of the neighbor-neighbors change their assignment as well. In the example of Figure 2 that means that agents A and D can move simultaneously, but agents A and C cannot.

Figure 3 presents the pseudo code for a cycle-based hill climbing local search algorithm for solving DSCOPs. The proposed algorithm is independent of the specific objective function. This algorithm will be slightly adapted for the specific DSCOP of the experimental evaluation in section 5.

The proposed algorithm in Figure 3 is a greedy hill climber. Each step taken by an agent not only improves the global target function evaluation, it is also guarantied to be the "best" possible step in its neighborhood, as in the Max Gain Message (MGM) algorithm [11].

The algorithm includes a preprocess stage in which all agents construct a list of all their second order neighbors. This list includes all of their neighbors and all of their neighbors' neighbors (lines 1–4 in pre-process). The list is termed *b_neighbors*.

At the beginning of the main procedure (line 1), each agent selects an assignment. Since the assignments of neighbors are unknown at this stage, the assignment is selected by considering only the unary part of the utility. This assignment is then sent to all neighbors, and all neighbors assignments are received (lines 2,3).

The termination condition (line 4) may be a given number of cycles, or some threshold on progress. Until the termination condition is met all the following steps are repeated.

Each agent calculates its utility according to its assignment and its neighbors assignments (line 5). Each agent selects its best assignment (line 6) taking into account the assignments of all its neighbors. A new utility is calculated assuming no assignment changes in the neighbors (line 7). This assignment is sent to all neighbors (line 8) and the proposed assignments are collected (line 9). For every neighbor that wants to replace its assignment, the agent computes its new utility assuming only this neighbor will change its assignment (line 11). This utility is sent back to the specific neighbor (line 12).

The utility reports of all neighbors are calculated based on the agent's proposed assignment change (assuming it is the only one changing its assignment in the neighborhood - line 13). If the new assignment improves the local evaluation of the target function (line 14), the agent sends the amount of improvement (line 15) to all its *b_neighbors* list.

The agent with the largest improvement makes the move by updating its assignment and utility (lines 19,20).

5 Experimental Evaluation

Experiments where performed on a multi agent system in which every agent has a simple PDP. Each agent has one truck with limitless capacity that needs to get out of the agent offices, travel through all packages and return to the agent offices (see section 2).

This problem needs a slightly modified algorithm from the generic one described in 4. This is because an agent cannot decide unilaterally on an assignment change. In the present case an assignment change means that an agent is interested in transferring a package to another agent and it must find the other agent that will take this package.

pre-process:
1. send Neighbors list to all neighbors
2. collect Neighbors list from all neighbors
3. *b_neighbors* ← all neighbors
4. *b_neighbors* ← all neighbors neighbors

Main Procedure:
1. *current_assignment* ← ChoosewBestAssignment()
2. send *current_assignment* to neighbors
3. collect neighbors assignments
4. **while** (no termination condition is met)
5. *utility* ← CalcUtility(current_assignment)
6. *assignment* ← ChoosewBestAssignment()
7. *utility_new* ← CalcUtility(assignment)
8. send assignment to neighbors
9. collect neighbors assignments
10. **foreach** (neighbor *n*)
11. u_n ← utility assuming new *n* assignment
12. send u_n to neighbor *n*
13. collect neighbors utilities based on *assignment*
14. **if** *assignment* improves Target Function
15. *utility_diff* ← *utility* − *utility_new*
16. send *utility_diff* to all b_neighbors
17. collect *utility_diff* from all b_neighbors
18. **if** *utility_diff* > *utility_diffs* off all *b_neighbors*
19. *current_assignment* ← *assignment*
20. *utility* ← *utility_new*

Fig. 3 Greedy Hill Climbing for DSCOPs

5.1 DSCOP for PDPs

Figure 4 presents the hill climbing local search algorithm for the multi agent PDP system. The preprocess stage of the algorithm is identical to the general one (in Figure 3), and results in the *b_neighbor* list of each agent.

The agents then check a termination condition (line 4) which may be a given number of cycles, or the detection whether progress is below a given threshold. Until the termination condition is met all following steps are repeated.

Utility is calculated by solving the PDP with the packages owned by the agent (line 3). The utility is sent to all neighbors (line 4), and all neighbors utilities are collected (line 5). The agent finds if and what package it wants to hand over to someone else by solving the PDPs for all options of removing a single package from the packages it owns (line 6). This package is offered to all relevant neighbors (line 7).

For each package that was offered to a given agent, that agent calculates the resulting potential gain (line 10). The computed gain is sent back to the offering agent (line 11).

Each agent collects all the responses (line 12) and decides which (if any) of the neighbors will get the package (line 13). The agent then calculates the gain (according to the target function) from transferring the package to the selected neighbor (line 14), and sends this gain to all *b_neighbors* (line 15).

Agents collect the potential gains from all *b_neighbors* (line 16). The offered package is removed from the agent's package list (line 18) and sent to the selected neighbor whose added gain is largest (line 19). The selected agent gets a message and adds the package to its package list (line 22)

Main Procedure:

```
1.  current_assignment ← all packages belonging to this agent
2.  while (no termination condition is met)
3.    utility ← SolvePDP(current_assignment)
4.    send utility to neighbors
5.    collect neighbors utilities
6.    suggested_package ← FindWhatPackageToGive()
7.    send suggested_package to all neighbors that can pick it
8.    collect neighbors suggested packages
9.    foreach (neighbor n)
10.     u_n ← SolvePDP(current_assignment + Package_n)
11.     send u_n to neighbor n
12.   collect neighbors responds to suggested_package
13.   choose best neighbor to give the package to.
14.   utility_diff ← improvement due to the package transfer.
15.   send utility_diff to all b_neighbors.
16.   collect utility_diff from all b_neighbors
17.   if utility_diff > utility_diffs off all b_neighbors
18.     remove suggested_package from current_assignment
19.     send suggested_package to selected neighbor
20.   else
21.     if received Package from neighbor
22.       add Package to current_assignment
```

Fig. 4 DSCOP for the PDP problem

5.2 *Experimental Setup and Results*

The setup for the experiment includes an area represented by a grid of $100X100$ in which 150 agents are randomly placed. Each agent has a single truck and is given an area defined by a circle centered at the agent location with a Radius of 10. Each agent picks up packages within its circle, and assigned five packages that are randomly placed inside its area. The Pickup and Delivery Problems were solved using a straight forward $A*$ algorithm. The results where accumulated and averaged over 10 randomly generated problems.

Each problem was solved with two different objective functions. MaxSum (utilitarian) and the Theil based SWF. Table 1 presents the comparison of the MaxSum and Theil SWF with respect to three different attributes: The averaged utility, the Theil Index and the Egalitarian utility.

Table 1 Theil SWF vs. MaxSum (utilitarian)

	Theil SWF	MaxSum
Average Utility	32.4467	33.4996
Theil Index	0.03455	0.1199
Egalitarian utility	13.406	1.8407

It is easy to see that for a small sacrifice in the overall average utility (less then 3.5%), one can get an improvement by a factor of 3.5 in the Theil index (a Theil index closer to 0 represents a more equal solution). This means that the utility is spread more evenly among the agents. The Egalitarian utility is also improved by a factor of 7.2. That means that the "worst off" agent is in a much better state.

The above results can be further illustrated by inspecting the histogram of the utilities of the individual agents (Figure 5). It is clear that when using MaxSum the histogram has a much wider spread. Setting the Theil SWF as the target function results in a much narrower histogram, illustrating the higher level of equality in the distribution of utility among agents.

Fig. 5 Histogram of the utility of Agents

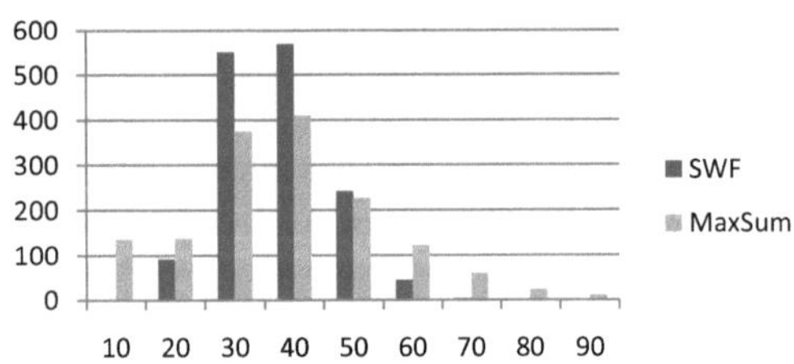

Figure 6 is a histogram of the number of packages that are assigned to each agent at the end of the algorithm execution. Again we see the same pattern of having a much more centered histogram when using Theil SWF as the target function.

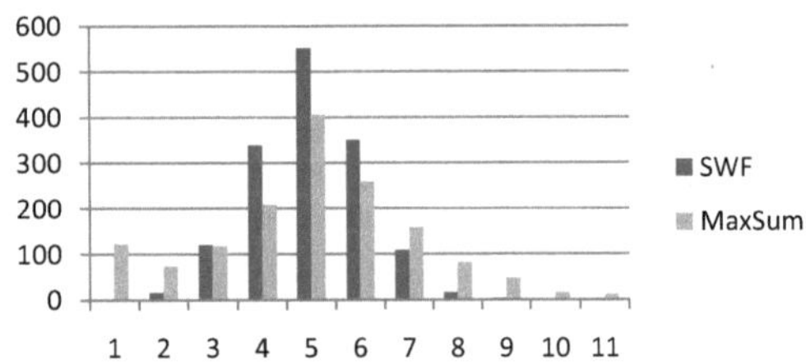

Fig. 6 Histogram of the number of Packages per Agent

The use of MaxSum as the objective function results in some agents picking up a large number of packages. Since the solution complexity of the PDP in the general case is exponential in the number of packages, solving the problem with MaxSum as the target function results in a much higher run time than using the Theil SWF as a target function. In fact, this type of problem took over a hundred times more CPU time for MaxSum than for the Theil SWF.

Figure 7 shows a histogram of the utility gain of agents. The use of MaxSum produces many low gains. In contrast, Theil SWF produces a relatively small number of agents that lose utility.

Fig. 7 Histogram of the utility gain per Agent

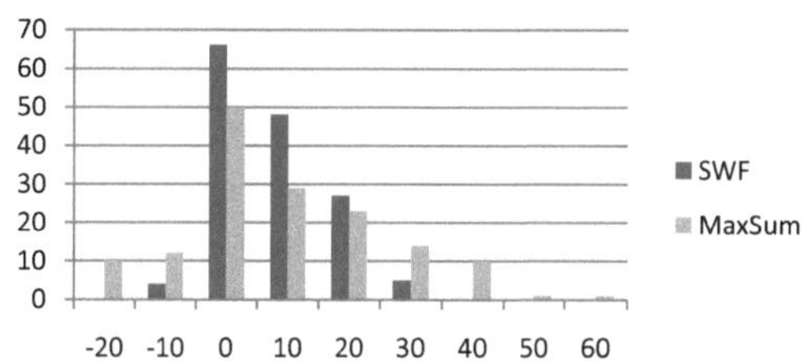

In order to see the effect of the different target functions when the initial state of the agents is unbalanced, a test was performed with a nonuniform number of packages per agent. 120 agent were assigned 4 packages each(*small* agents), and 30 were assigned 9 (*large* agents). The results are in Table 2.

Table 2 Agents gains for non uniform initial state

	Theil SWF	MaxSum
Average Gain of *large* agents	-2.054	12.932
Average Gain of *small* agents	6.021	3.312

Using MaxSum as target function results in each of the *large* agents gaining, on the average, almost four times the amount of utility gained by each the small agents. A *large* agent that travels more, has a higher probability of gaining from adding a package, due to higher probability of traveling near its location. When using the Theil based SWF, the effect is reversed. The *large* agents lose utility, while the *small* ones are gaining almost twice the amount they gained in the MaxSum scenario. This demonstrates the equality enhancing characteristic of the Theil based SWF.

6 Conclusions

A new framework - Distributed Social Constraints Optimization Problems (DSCOP) - is introduced for multi agent cooperation in a constrained problem. In DSCOPs each agent calculates its utility based on its assignment and the assignments of all of its constraining agents. Individual utilities, based on states and computed by each agent naturally represent self-interested agents. The cooperation of such agents in finding an optimal solution for the DSCOP needs to be based on an attractive social choice.

A reasonable social choice can be based on the optimization of objective functions that incorporate equality index based Social Welfare Functions. This idea is investigated and a simple objective function is proposed - the Theil index.

The present study designs a hill climbing algorithm for DSCOPs that is based on the Theil index as objective function. In order to investigate the behavior of DSCOPs and the proposed local search method, experiments were performed on a multi agent Pickup and Delivery Problem (PDP). In these problems each agent has its own PDP to solve. It turns out that cooperation between agents (handing over unwanted packages to other agents) can improve the utility of agents.

The experiments demonstrated that for a small trade-off in the overall utility, the use of the Theil based SWF objective function achieves a much more attractive solution. The distribution of utility among agents becomes more uniform and the agent with the lowest utility (Egalitarian) has a much higher utility than in the solution of the same problem with the MaxSum objective function.

Experiments with a non uniform initial distribution of agents' wealth demonstrated the tendency to become less equal when optimizing for MaxSum, making the "rich" richer. Using the Theil SWF on the other hand, had the opposite effect. It lowered the utility of the "rich" agents and increased the utility of the smaller ("poorer") ones, arriving at efficiency of 95 percent.

References

1. Berbeglia, G., Cordeau, J., Gribkovskaia, I., Laporte, G.: Static pickup and delivery problems: a classification scheme and survey. TOP: An Official Journal of the Spanish Society of Statistics and Operations Research 15(1), 1–31 (2007)
2. Bergson, A.: A reformulation of certain aspects of welfare economics. The Quarterly Journal of Economics 52(2), 310–334 (1938)
3. Bouveret, S., Lemaître, M.: Computing leximin-optimal solutions in constraint networks. Artif. Intell. 173(2), 343–364 (2009)
4. Dalton, H.: The measurement of the inequality of income. The Economic Journal 30, 348–361 (1920)
5. D'Aspremont, C., Gevers, L.: Equity and the informational basis of collective choice. The Review of Economic Studies 44(2), 199–209 (1977)
6. Gershman, A., Meisels, A., Zivan, R.: Asynchronous forward bounding. J. of Artificial Intelligence Research 34, 25–46 (2009)
7. Grubshtein, A., Zivan, R., Grinshpoun, T., Meisels, A.: Local search for distributed asymmetric optimization. In: Proc. AAMAS 2010, pp. 1015–1022 (2010)
8. Junges, R., Bazzan, A.L.C.: Evaluating the performance of dcop algorithms in a real world, dynamic problem. In: Proc. AAMAS 2008, pp. 599–606 (2008)
9. Kolm, S.C.: Justice et equite. Centre National de la Recherche, Paris (1972)
10. Lisý, V., Zivan, R., Sycara, K.P., Pechoucek, M.: Deception in networks of mobile sensing agents. In: AAMAS 2010, pp. 1031–1038 (2010)

11. Maheswaran, R.T., Pearce, J.P., Tambe, M.: Distributed algorithms for DCOP: A graphical-game-based approach. In: Proc. Parallel and Distributed Computing Systems (PDCS), pp. 432–439 (September 2004)
12. Maheswaran, R.T., Tambe, M., Bowring, E., Pearce, J.P., Varakantham, P.: Taking DCOP to the real world: Efficient complete solutions for distributed multi-event scheduling. In: Proc. AAMAS 2004, New York, pp. 310–317 (2004)
13. Mamoru, K., Kenjiro, N.: The nash social welfare function. Econometrica 47(2), 423–435 (1979)
14. Meisels, A.: Distributed Search by Constrained Agents: Algorithms, Performance, Communication. Springer, Heidelberg (2007)
15. Modi, P.J., Shen, W., Tambe, M., Yokoo, M.: ADOPT: asynchronous distributed constraints optimization with quality guarantees. Artificial Intelligence 161(1-2), 149–180 (2005)
16. Ng, Y.-K.: Bentham or bergson? finite sensibility, utility functions and social welfare functions. The Review of Economic Studies 42(4) (1975)
17. Ottens, B., Faltings, B.: Coordination agent plans trough distributed constraint optimization. In: Proceedings of the Multi Agent Planning Workshop MASPLAN 2008, Sydney, Australia (September 2008)
18. Pareto, V.: Cours d'economie politique. Librairie Droz, Geneva (1964)
19. Pigou, A.C.: Wealth and welfare. Macmillan, London (1912)
20. Sen, A.: Real national income. The Review of Economic Studies 43, 19–39 (1976)
21. Sen, A.: The welfare basis of real income comparisons: A survey. Journal of Economic Literature 17(1), 1–45 (1979)
22. Zhang, W., Xing, Z., Wang, G., Wittenburg, L.: Distributed stochastic search and distributed breakout: properties, comparishon and applications to constraints optimization problems in sensor networks. Artificial Intelligence 161(1-2), 55–88 (2005)

A Distributed Cooperative Approach for Optimizing a Family of Network Games

Alon Grubshtein and Amnon Meisels

Abstract. The present study considers a distributed cooperative approach for network problems where agents have personal preferences over outcomes. Such problems can be described by Asymmetric Constraints where the joint action of agents yields different gains to each participant Grubshtein et al. (2010). The proposed method constructs and solves an Asymmetric Distributed Constraints Optimization Problem whose solutions guarantee a minimal gain for each agent, which is at least as high as the agents' Bayesian equilibrium gain. The paper focuses on a special class of Network Games and proves that the proposed method produces optimal results in terms of the number of agents whose gain improves over their equilibrium gain and that the resulting solutions are Pareto Efficient. Extensive empirical evaluation of the studied network problem shows that the number of improving agents is not negligible and that under some configurations up to 70% of the agents improve their gain while none of the agents receive a payoff lower than their equilibrium gain.

1 Introduction

A key attribute of any Multi Agent System (MAS) is the level of collaboration agents are expected to adopt. When agents share a common goal it is natural for them to follow a fully cooperative protocol. A common goal can be the election of a leader, finding shortest routing paths, or searching for a globally optimal solution to a combinatorial problem (cf. Meisels (2007)). When the agents have different and conflicting goals, competition and self interest are expected.

Distributed Constraints Satisfaction or Optimization Problems (DCSPs or DCOPs) are important examples of the cooperative model, where a set of agents attempt to

Alon Grubshtein · Amnon Meisels
Computer Science Dept., Ben Gurion University of the Negev
e-mail: {alongrub, am}@cs.bgu.ac.il

F.M.T. Brazier et al. (Eds.): Intelligent Distributed Computing V, SCI 382, pp. 49–62.
springerlink.com

satisfy or optimize a common *global* objective Maheswaran et al. (2004a); Baba et al. (2010); Grubshtein et al. (2010). The Asymmetric Distributed Constrains Optimization Problems (ADCOPs) model provides a generalization of DCOPs where constrained agents gain differently (asymmetric gains) Grubshtein et al. (2010). Such asymmetric constraint networks can be thought of as "**cooperative** network games". In the context of a DCOP, it is natural to expect cooperative behavior when agents are interested in a global objective (e.g. the *value* of the objective) and to expect selfish or competitive behavior when agents are interested only in their personal gains.

An example of a competitive (or selfish) scenario are Network Games Jackson (2008); Galeotti et al. (2010) that assume that participants are only willing to take actions which improve (or do not worsen) their gains. In these situations it is customary to find an equilibrium state from which no agents would care to deviate Nisan et al. (2007). These equilibria states are not necessarily efficient with respect to the global objective Roughgarden (2005); Nisan et al. (2007) or even with respect to the personal gains of the participants (e.g. "the prisoners' dilemma") Monderer and Tennenholtz (2009).

The present study focuses on a *cooperative search* approach for a class of network games Galeotti et al. (2010). The proposed approach is based on the construction of an ADCOP whose cooperative solution guarantees a minimal gain to each agent. To accommodate the natural selfish behavior of the participating agents, the proposed cooperative solution guarantees a *personal gain* for every participant the is not lower than the agent's game theoretic equilibrium gain. That is, the proposed approach uses the standard assumptions on cooperation of ADCOP (and DCOP) agents, but is capable of bounding the personal gain of its solution according to a game theoretic result. Thus, this approach extends the common approach taken when cooperatively solving MAS problems Maheswaran et al. (2004b); Baba et al. (2010); Modi and Veloso (2004); Wallace and Freuder (2005). It applies the same set of assumptions needed to cooperatively solve a multi agent system but provides a different, user oriented, solution concept. It is important to note that the proposed ADCOP solution benefits all agents simultaneously - gains are at least as high and often higher than the equilibrium gains.

This study, is to our best knowledge, a first attempt to use a *distributed search method* that guarantees an improvement in the personal gain of agents over their equilibrium (game theoretic) gain. It uses the definition of the *Cost of Cooperation* Grubshtein and Meisels (2010), which defines a measure of individual loss from cooperation. An ADCOP representation of the given network game is presented along with a proof that in its optimal solutions this cost is non positive (i.e. no agent stands to lose). We further prove that no other solution improves the gain of a greater number of agents. Empirical evaluation on a set of randomly generated network problems demonstrates that the fraction of improving agents can be quite large (up to 70%).

2 Download / Free Ride Network Interaction

Consider a set of agents wishing to download large amounts of information from a remote location. Agents may receive the information from other agents in their local neighborhood (if these neighbors have it) or directly download it from a central hub. The process of downloading data from the hub requires significant bandwidth and degrades performance (resource consumption, battery, etc). In contrast, sharing information with peers does not result in any significant impact on the agents. Information exchange between agents is limited to directly connected peers only (no transfer of information to second degree neighbors).

The entire interaction can be captured by a graph $G = (V, E)$ where each edge e_{ij} specifies a neighborhood relationship and a vertex $i \in V$ represents an agent. x_i denotes the action (e.g., assignment) made by agent i and $x_N(i)$ is the joint action taken by all of i's neighbors. [1] The degree of node i will be denoted by d_i.

The network itself (the graph $G = (V, E)$) is not known to all agents. The total number of participants $n = |V|$ is known and so is the (fixed) probability p for an edge to exist between any two vertices. Such graphs $G = (n, p)$ are known as Poisson Random Graphs or an Erdös - Rényi network Jackson (2008).

The gain from receiving information can vary and depends on the willingness of neighbors to share information and on some predefined maximal limit value, k. The cost of downloading it is c ($c < 1$). Agents can either download the relevant information from the hub (assign action $x_i = \mathscr{D}$, also denoted $x_i = 1$) or wait for one of their neighbors to download it (assign action $x_i = \mathscr{F}$, or $x_i = 0$). The payoffs are a function of the agent's action and of its peers:

$$u_i(x_i, x_N(i)) = \min\Big(\sum_{j \in \{i\} \cup N(i)} x_j,\ k\Big) - x_i \cdot c$$

That is, if the agent exerts effort and downloads the information its gain is counted as the number of downloading agents in its vicinity (including the agent itself) up to a limit of k, minus c. If, on the other hand, it does not download its gain depends on the number of agents in its vicinity that do download.

This interaction models the local provision of information (or a local public good) where the agents' actions are termed "strategic substitutes". A simpler variant is given as an example by Galeoti et al. Galeotti et al. (2010) who recently studied the effect of the network on behavior and payoff in network games.

2.1 *Solving the D/F Interaction*

Given the set of agents in the D/F interaction, what should the joint assignment be? While there are numerous possible responses which are naturally affected by the agents' cooperation level, we first consider the two most important ones representing two opposing extremes.

[1] We specifically refrain from the common x_{-i} notation to represent a joint action by all players but i, to emphasize that player i is only affected by the set of its neighbors.

Fully Cooperative Agents - In the fully cooperative setting, individual gains of participants are often used to compute the value of some social, or global welfare function. Two extreme examples of Social Welfare Functions (SWFs) are the "Utilitarian" and "Egalitarian" SWF Moulin (1991). Mathematically speaking, the utilitarian approach seeks to maximize the sum of individual utilities, while the egalitarian approach seeks to equalize individual utilities.

The fully cooperative approach was successfully applied to numerous multi agent problems. By casting the problem to a distributed constraint problem – defining agents, variables, domains and constraints – one can invoke any of the DCR algorithms and find the optimal (or near optimal) solution to the problem at hand.

In the D/F game, the optimal cooperative solution, maximizing the sum of all participants' gains (or the maximal average gain of each one) when $k = 1$ is the Minimal Dominating Set (MDS). A dominating set of a graph $G = (V,E)$ is a set of nodes $S \subseteq V$ such that every node is either in S, or else a neighbor of at least one element of S. Finding such a set in a distributed fashion can be readily achieved by formulating the problem as a Distributed Constraint Optimization Problem (DCOP) in which constraints incur an extreme cost to joint assignments where an agent does not belong to the set S or is a neighbor to one that is Meisels (2007).

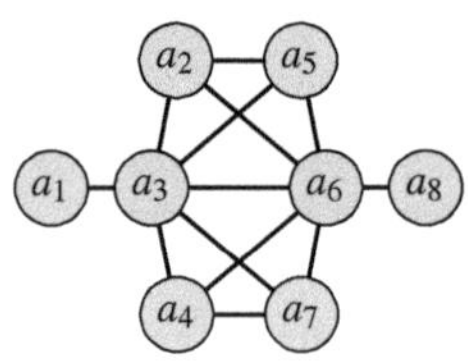

Fig. 1 An example graph with $n = 8$ and $p = 0.45$

The MDS of the graph $G = (8,0.45)$ illustrated in Figure 1 includes only two vertices. In one such optimal solution a_3 and a_6 are members of the dominating set and when $c = 0.7$, their gain is 0.3, the gain of all other participants is 1 and the aggregated social gain is 6.6. However, despite the high average gain of 0.825, one may argue that this solution is inadequate for agents with personal utility functions and that the lower gains of a_3 and a_6 are in contrast to their relative importance. In such cases one may consider a different solution concept.

Fully competitive agents - The common approach taken when self interested entities are present, is to invoke *Game Theory* Nisan et al. (2007). By accepting the fundamental axioms of game theory one is able to predict possible outcomes of an interaction. These outcomes are standardly assumed to be equilibrium points (in terms of agents' actions) from which no agent is willing to deviate. The best known solution concept is the Nash Equilibrium (NE) point, but there are many other equilibrium points which fit different aspects of the interaction (cf. Nisan et al. (2007)).

In a recent work by Galeotti et. al Galeotti et al. (2010) the D/F interaction is analyzed and a unique Bayesian Nash Equilibrium (BNE) is computed. Based on the global probability p for an edge to exist between two neighbors, each agent calculates the probability for a randomly selected neighbor to be of degree d. This probability value is denoted with $Q(d,p)$. Using this information Galeotti et. al define a parameter t which, in the case of $k=1$ is the smallest integer such that:

$$1-\left[1-\sum_{d=1}^{t} Q(d;p)\right]^{t} \geq 1-c$$

In the unique BNE for the D/F network game, any participant of degree $d \leq t$ selects strategy $\mathscr{D}$ and any participant of degree $d > t$ selects $\mathscr{F}$.

When applied to the interaction instance illustrated in Figure 1 (cost value of $c = 0.7$) one first calculates the threshold value $t = 2$ from the above equations. This value defines the assignment for each agent in the BNE solution: a_1 and a_8 will assign action $\mathscr{D}$ while all others will assign $\mathscr{F}$. Agents in this setting maximize expected gain which may result in low actual gain. In the BNE solution of the example problem agents a_2, a_4, a_5 and a_7's gain is actually 0, while both a_3 and a_6 actual gains are higher than their expected gains. The aggregated social gain in the game theoretic approach is 2.6 with an average gain of 0.325.

Although it is easy to see that one can cooperatively improve this BNE solution, reverting to the MDS solution will increase the overall sum of gains (average gain of agents) at the expense of a_3 and a_6. That is, it will penalize a_3 and a_6 and reduce their gains when acting with no cooperating protocol.

3 A Cooperative Framework for Agents with Personal Utility Functions

The cooperative and competitive approaches described in the previous section suffer from several problems. The cooperative approach blindly optimizes a pre-defined objective function with little or no care for the end cost of agents. This leads to a situation where results may be biased towards or against some agents who are assigned extreme costs. On the other hand, game theoretic solutions are often inefficient as in our example. There are numerous authors who claim that game theoretic solutions fail to faithfully represent reality or predict strategies of real world players Axelrod (1984).

The present paper focuses on the latter point. More specifically, we follow the ideas presented in Grubshtein and Meisels (2010) which define the Cost of Cooperation for agents:

Definition 1 (Cost of Cooperation (CoC)). Given a global objective function $f(x)$, the Cost of Cooperation of an agent is defined as the difference between the lowest gain that the agent can receive in any equilibrium of the underlying game (if any, otherwise zero) and the lowest gain this agent receives from a cooperative protocol's solution x that maximizes $f(x)$.

The CoC provides a measure of the agents' gain from cooperation. In the present paper, this measure is used to direct a cooperative search process by examining the expected actions and gains of self interested rational users and disallowing solutions which result in lower gains than those of the worst stable game – **for any of the agents**. One can view cases that may encourage cooperation as problems that have solutions with non positive CoC.

The present study proposes a method that constructs and solves an ADCOP for each D/F network game guaranteeing non positive CoC for all agents on the network. An ADCOP is an extension to the DCOP paradigm which addresses individual agent's gains and captures game-like interactions between agents Grubshtein et al. (2010).

Before proceeding to formally introducing the proposed ADCOP we define two useful terms. The first provides a quantitative assessment on the potential cooperative gain of an agent if the maximal gain k is very high:

$$\nu_i = d_i + 1 - \| \{a_j : a_j \text{ has at least k neighbors of degree} \leq \text{t and } d_j > t\} \|$$

The ν_i value is a count of agents in a_i's local surroundings (including itself) which will increase their gain by assigning $\mathcal{D}$. These agents can be of degrees greater than t, but still stand to gain by assigning $\mathcal{D}$ since they have yet to maximize their personal gains. If k is sufficiently large, $\nu_i - c$ is the maximal cooperative gain an agent can have (ν_i is the number of agents assigning $\mathcal{D}$ and not a utility value).

The second term which we define is μ_i, the maximal cooperative gain a_i can attain:

$$\mu_i = \min(\nu_i, k)$$

The μ_i value is used to represent the maximal cooperative gain a_i can attain based on its degree, its neighbors degree and the maximal gain k. One immediate observation is that when $\mu_i = \nu_i$, attaining this maximal gain, without the cooperation of the k gaining neighbors, is only possible when a_i assigns $x_i = \mathcal{D}$.

Having defined μ_i and ν_i, we can now proceed to present our main contribution, a general form ADCOP formulation that when cooperatively solved, provides a non positive CoC guarantee to all agents:

- A set of Agents $A = \{a_1, a_2, ..., a_n\}$ - each holds a single variable and corresponds to a single user.
- A set of Domains $D = \{D_1, D_2, ..., D_n\}$ for the variables held by the agents. Each domain consists of only two possible values: $\{\mathcal{D}, \mathcal{F}\}$.

- A set of asymmetric constraints C. Each constraint c_i is identified with a specific agent (e.g. agent i) and defined as a constraint of arity (d_i+1) over all of agent i's neighbors. The set of agents involved in constraint c_i is $A_{c_i} = a_i \cup x_N(i)$. That is, the ADCOP has n constraints and each agent contributes to d_i+1 different constraints. The costs associated with each constraint that add up to the total cost of c_i for agent i are summarized in the table in Figure 2.

Costs of c_i	
n^4	if $\|\mathscr{D}\| < \mu_i$ (where $\|\mathscr{D}\|$ denotes the number of agents that assign $\mathscr{D}$)
d_j	**for each** $a_j \in A_{c_i}$ with $d_j > t$ and less than μ_j neighbors with degree $\leq$ t that assign $\mathscr{D}$
$(n-1)n$	**for each** $a_j \in A_{c_i}$ with $d_j > t$ and at least μ_j neighbors with degree $\leq$ t that assign $\mathscr{D}$
1	**for each** $a_j \in A_{c_i}$ with $d_j \leq t$ which assigns $x_j = \mathscr{D}$

Fig. 2 Constraint's costs for each combination of assignments

Note: costs of the ADCOP's solutions (full or partial) are completely unrelated to the agents' gains.

Let us now proceed to prove that finding a minimal assignment (an optimal solution) for the above ADCOP results in gains which are at least as high as the *actual* gains resulting from applying the unique BNE strategies for all agents.

Let the optimal solution to the ADCOP be denoted by x. If, in a given solution x', the total number of agents connected to a_i that assign $\mathscr{D}$ (including a_i) is Δ, we will say that in this solution a_i has a Δ *neighborhood*. For example, the term *null neighborhood* defines a situation in which an agent and all of its connected peers assign $\mathscr{F}$ as their action.

Lemma 1. *In the optimal solution to the ADCOP there are no Δ neighborhoods with $\Delta < \mu_i$. That is, there are at least μ_i agents assigning $\mathscr{D}$ in each agent's local surrounding.*

Proof. Let m_i denote the number of agents assigning $\mathscr{D}$ in agent's i local neighborhood (including itself). Assume by negation that for some agent a_i, $m_i < \mu_i$ in x (the optimal ADCOP solution). This means that the cost of c_i (the constraint originating from a_i) is n^4. Let x' be a complete assignment which differs from x in the assignment of $m = \mu_i - m_i$ agents in A_{c_i}. That is, $x'_j = \mathscr{D}$ for each of the m' agents in i's local neighborhood. As a result, a_i has a μ_i neighborhood in x'.

One can compute a bound on the cost of x' by assuming that this change is imposed on high degree agents with and low-degree neighbors (i.e., incur a maximal cost on x'). That is, for each a_j in the set of m agents changing their assignment, assume that $d_j = n-1$ (where $n-1 > t$) and further assume sufficiently many neighbors $a_k \in x_N(j)$ of degree $d_k \leq t$.The resulting bound on the cost of x':

$$COST(x') = COST(x) - n^4 + m(n-1)n^2 = COST(x) + n^2\left[m(n-1) - n^2\right]$$

The cost of x' will be lower than that of x if:

$$\left[m(n-1) - n^2\right] < 0$$

$$m < n + 1 + \frac{1}{n-1}$$

Since $m < \mu_i \leq n$, we have $COST(x') < COST(x)$ which contradicts the assumption that x is the optimal ADCOP solution. □

Lemma 2. *The gain of agents a_i with degree $d_i \leq t$ in x is at least as high as their actual gain in the BNE.*

Proof. The maximal possible BNE gain of agent a_i with degree $d_i \leq t$ is $\mu_i - c$ (a_i assigns $\mathscr{D}$ in the BNE because $d_i \leq t$). Following Lemma 1 the ADCOP's optimal solution disallows for Δ neighborhoods with $\Delta < \mu_i$. Thus, a_i has at least a μ_i neighborhood in x, which implies that a_i's ADCOP gain is at least $\mu_i - c$. □

Lemma 3. *The gain of agents a_i with degree $d_i > t$ in x is at least as high as its actual gain in the BNE.*

Proof. The minimal ADCOP gain of an agent a_i can be bound by following Lemma 1. The number of agents which assign $\mathscr{D}$ in its local neighborhood is μ_i and hence its minimal gain is at least $\mu_i - c$.

Now, consider a_i's BNE gain. If k is sufficiently large, $k > \nu_i$ and $\mu_i = \min(\nu_i, k) = \nu_i$. However, since its degree is greater than t, a_i assigns $\mathscr{F}$ in the BNE which results in a slightly lower gain, never exceeding $\nu_i - 1 = \mu_i - 1 < \mu_i - c$.

If k's value is lower, $\mu_i = k$ and a_i BNE gain can become k. Since this is the only outcome which results in a higher gain than the ADCOP's gain, we need only prove that there is no setting which degrades a_i's gain from $\mu_i = k$ to $\mu_i - c = k - c$.

Let us assume that there exists an agent a_i with a BNE gain of k. We note that a_i's minimal ADCOP gain is $\mu_i - c = k - c$, and in this case its assignment in x must be $\mathscr{D}$. This assignment incurs a positive contribution on the ADCOP only when it prevents the existence of an m'_j neighborhood for at least one agent $a_j \in X_n(i)$ with $m'_j < \mu_j$ (since a_i's BNE gain is k there are at least k neighbors with degree lower than t and hence it is easy to see that a_i will never assign $\mathscr{D}$ to prevent an m'_i neighborhood for itself).

Consider the agent $a_j \in X_n(i)$ with $\|\mathscr{D}\| = m'_j < \mu_j$. There must exist at least one other agent a_h which can be assigned $\mathscr{D}$ instead of a_i (the agent a_i is not counted in ν_j on the one hand, and $\|\mathscr{D}\| < \mu_j$ on the other). This neighbor can be of degree $d_h > t$ but will have less than μ_h neighbors of degree t or lower – by the definition of ν_j and μ_j. The incurred cost of assigning $\mathscr{D}$ to this agent in the ADCOP is d_h.

Assume that a_i's assignment in x is $x_i = \mathscr{D}$. Further assume that it incurs a positive contribution on the ADCOP for all of a_i neighbors. We define the new solution x' which differs from x in the assignment of all a_h agents and bound its cost:

$$COST(x') = COST(X) - (d_i + 1)(n-1)n + (n-1)\cdot(n-1)\cdot d_i$$

(where we assume that each of the neighbors incur a maximum cost – its degree is $n-1$, the maximal possible degree.) This clearly implies that $COST(x') < COST(X)$ in contradiction to x's optimality and is therefore impossible. □

Theorem 1 (Gain guarantee). *In the optimal solution to the ADCOP described above for the D/F interaction, all agent's CoC is non positive with respect to the actual gains in the BNE.*

Proof. Since the BNE is unique, it also provides the lowest gain that each agent can receive in an equilibrium. Thus, by directly following Lemmas 1-3 our claim is proved. □

Figure 3 presents a randomly generated example of a D/F network of 50 participants with an average of 5 neighbors per participant $G = (50, 0.1)$. The cost of exerting effort (assign action $\mathscr{D}$) is $c = 0.7$ and the maximal gain an agent cat attain is $k = 1$. The example in Figure 3 demonstrates the differences in gains with respect to the BNE in two alternative solutions. One alternative applies and solves the proposed ADCOP. The other alternative is the socially optimal MDS solution. The Gray nodes represent agents which received higher gains than their BNE gain, Black nodes represent agents which did not improve their gains and finally White nodes represent lower gains than those attained in the BNE. Two important features are illustrated by this example:

1. There is a significant number of agents (nearly 40%) improving their gains in the ADCOP solution (Figure 3(a)) while none attain lower gains.
2. The socially (cooperative) optimal solution (Figure 3(b)) increases the gains of even more participants, however, it also reduces the gains of nearly 20% of the agents.

Having proved that the gains of agents can only improve, the following two theorems address the question of efficiency of the proposed scheme:

Theorem 2 (Pareto efficiency). *The results of the proposed method are Pareto Efficient. That is, there is no solution which further increases the gain of some agents without reducing the gain of at least a single agent.*

Proof. To prove our claim, consider the different gains an agent can attain. Following Lemma 1 there are no agents with gains lower than $\mu_i - c$ in the ADCOP solution. A gain of k (when $\mu_i = k \leq v_i$) can not be further improved. When $\mu_i = v_i$ an agents gain is $v_i - c = \mu_i - c$. This can be improved by requiring some of its neighbors an assignment of $\mathscr{D}$. However, by the definition of v_i, these agents are expected to gain k in both the ADCOP and BNE. Such an assignment results in lower

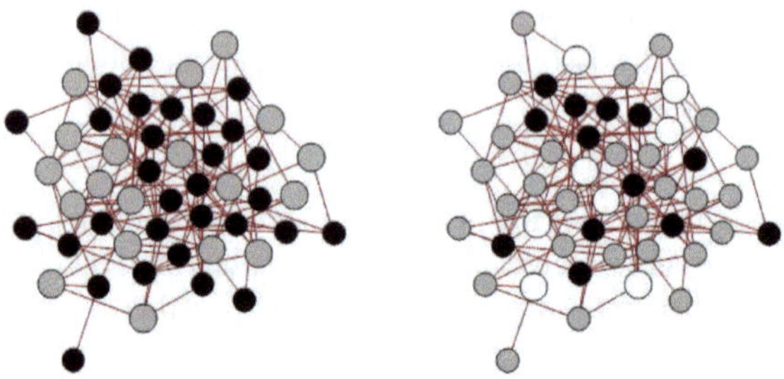

(a) ADCOP (b) MDS

Fig. 3 Differences in agents gains in a G=(50,0.1) interaction where the maximal gain is 1 ($k = 1$). Gray nodes represent improved gains, Black nodes represent unchanged gains and White nodes represent reduced gains. 3(a) The ADCOP solution; 3(b) the cooperative optimal solution (MDS).

gains for these agents and is therefore avoided. Finally, it is worth noting that when an agent's gain is $v_i - c$ assigning $\mathscr{F}$ instead of $\mathscr{D}$ reduces the agents gain from $v_i - c$ to $v_i - 1$. This stems directly from the definition of v_i and μ_i.

Thus we have shown that improving the ADCOP gain of an agent requires lowering the gain of another. That is, the ADCOP solution is Pareto Efficient. □

Theorem 3 (Efficiency). *The proposed solution is optimal in term of the number of agents that improve their gain. That is, there is no other solution in which a larger number of agents have negative CoC.*

Proof. Again we note that agents with a BNE gain of k have a maximal gain and that all agents with a BNE gain which is lower than $\mu_i - c$ improve their gain (Lemma 1). We also note that retaining BNE $m < \mu_i$ neighborhoods will not improve the gains of any agent (assigning $\mathscr{D}$ in any such Δ neighborhood can only increase gains).

Since every $\mathscr{D}$ assignment incurs a negative cost on the ADCOP solution, we are guaranteed that the ADCOP optimal solution will be the one with the minimal number of such assignments (and that the number of agents changing their assignment from $\mathscr{F}$ to $\mathscr{D}$ is minimal). Thus, we need only show that the maximal number of agents which assign $\mathscr{D}$ in the BNE change their assignment to $\mathscr{F}$.

Assume by negation that there exists a solution x' in which the number of agents changing their assignment to $\mathscr{F}$ is greater than that of the ADCOP's optimal solution x. If there was such a solution it would imply that the cost of x' is lower than that of x, and hence would contradict the optimality of x. The existence of such a solution is therefore impossible. □

4 Experimental Evaluation

The proofs presented in the previous section demonstrate that the proposed approach is capable of finding a joint action improving the gain of all agents who can do so without introducing a negative impact on others. However, this guarantee does not answer the qualitative question: what is the actual number of agents improving their gain?

To empirically evaluate the effectiveness of the proposed approach, one can use a large set of randomly generated interactions. These interactions are characterized by the interaction graph $G = (n, p)$, the maximal gain k, and the cost c associated with assigning the action $\mathscr{D}$.

Random graphs with 30 agents ($n = 30$) were generated for a variety of density values (p). For each density value the cost value c was varied while the maximum gain remained fixed at $k = 1$. The joint BNE assignment was calculated and used as the base line gain of agents. Results of each configuration were averaged over 50 instances and in each run the gain of the agents was compared to the gain reached by the ADCOP solution.

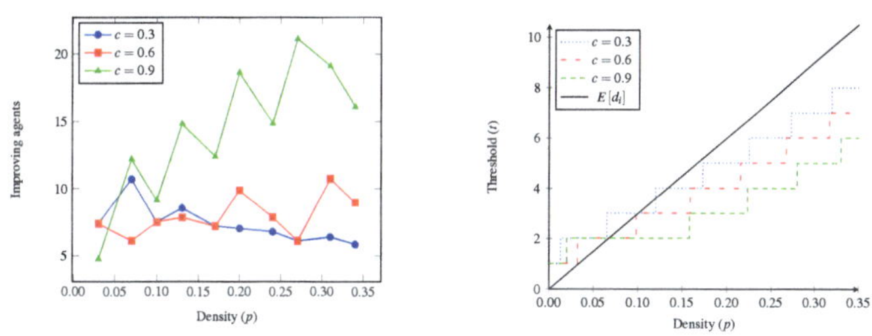

(a) Agents with improved gains (b) Threshold and mean deg. values

Fig. 4 (a) The average number of agents improving their gain under different density and cost values ($n = 30, k = 1$). (b) The threshold value t with respect to the mean degree value.

Figure 4(a) presents the average number of agents with negative CoC (these represent all agents who could potentially improve their gain without incurring costs on others). The results indicate that the proposed method indeed improves the gain of a significant number of agents and that this occurs for all values of density that were tried. As cost value increases the number of agents with increased gain dramatically increases. This is due to the rather moderate increase in the threshold values t occurring when costs are higher (Figure 4(b)). These values become significantly lower than the mean degree value ($E[d_i] = p \cdot n$) which in turn drives more agents to the assignment $\mathscr{F}$ in the BNE. The resulting joint BNE action has therefore higher chances of including null neighborhoods which are not present in the ADCOP solution. Indeed our run logs show a significant number of agents improving their gains from 0 to 1 implying that large number of null neighborhoods exist.

The qualitative assessment of the change in the gain of agents can also be examined for the cooperative optimal solution. In this case, one must also account for the fact that some agents may lose from this solution.

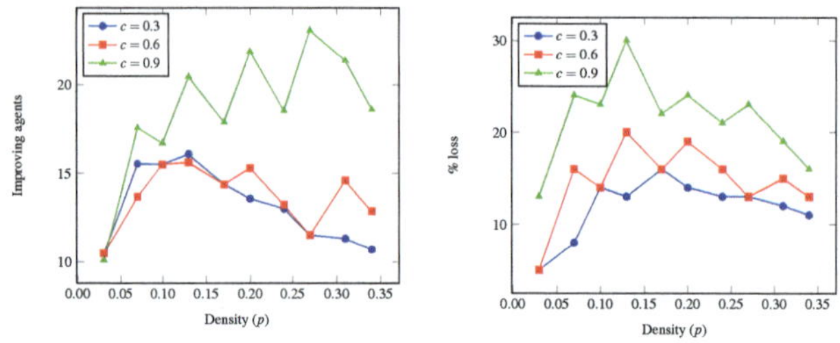

(a) MDS gains (number of agent) (b) MDS loss (percentage)

Fig. 5 The average gains and losses of agents in the cooperatively optimal solution – the Minimal DS ($k = 1$).

Figure 5 presents the average impact on the gain of agents for the same set of interactions when the joint action taken is the MDS solution. The trend on the number of agents with increased gains (Figure 5(a)) is somewhat similar to that of the ADCOP's, although this number is consistently larger for the MDS. However, when one examines the average loss (calculated as the ratio between the number of agents with gains lower than that of the BNE to the number of agents with non zero gains in the BNE) it is evident that the MDS incurs a high cost on agents: it will almost always result in a gain reduction of 10% - 30% of the agents with non zero BNE gains.

5 Discussion

A new approach to distributed cooperative optimization is proposed where agents interact in a family of network games and have personal utility functions. The proposed approach guarantees a minimal gain equivalent to the equilibrium gain of each participant (non positive Cost of Cooperation) by forming and cooperatively solving an Asymmetric DCOP.

The above guarantee is realized on a class of Network Games where agents gain by either free riding other agents' efforts or download information for themselves. Although any equilibrium point defines a non positive CoC solution, our proposed ADCOP is proved to yield Pareto Efficient solutions which also maximize the number of agents with improved gains over their game theoretic (e.g. equilibrium) gains.

The D/F interaction described in this paper can be used to approximate P2P networks and their interactions. although a re-calculation of the t value is required for each maximal gain value, the proposed ADCOP achieves non positive CoC solutions for any value of k.

Experimental evaluation of the D/F interaction under different configurations demonstrates the effectiveness of the proposed approach. A large fraction (up to 70%) of the agents benefit from it while none gain less than their game theoretic gain. In contrast, the socially maximizing Minimal Dominating Set solutions benefit an even larger fraction of the agents but do so at the expense of up to 30% of the agents whose gains become lower.

Finally, we note that a greedy hill climbing search starting from the game theoretic equilibrium state may also result in a Pareto Efficient solution, however, it is not guaranteed to find the solution benefiting the maximal number of agents[2].

Acknowledgements. The research was supported by the Lynn and William Frankel Center for Computer Sciences and by the Paul Ivanier Center for Robotics Research.

References

Axelrod, R.: The evolution of cooperation. Basic Books, New York (1984); ISBN 0465021212

Baba, S., Iwasaki, A., Yokoo, M., Silaghi, M., Hirayama, K., Matsui, T.: Cooperative problem solving against adversary: Quantified distributed constraint satisfaction problem. In: Proceedings of the 9th International Conference on Autonomous Agents and Multi-Agent Systems (AAMAS 2010), Toronto, Canada, pp. 781–788 (May 2010)

Galeotti, A., Goyal, S., Jackson, M.O., Vega-Redondo, F., Yariv, L.: Network games. Review of Economic Studies 77(1), 218–244 (2010)

Grubshtein, A., Meisels, A.: Cost of cooperation for scheduling meetings. Computer Science and Information Systems (ComSIS) 7(3), 551–567 (2010); ISSN 1820-0214

Grubshtein, A., Zivan, R., Meisels, A., Grinshpoun, T.: Local search for distributed asymmetric optimization. In: Proceedings of the 9th International Conference on Autonomous Agents and Multi-Agent Systems (AAMAS 2010), Toronto, Canada, pp. 1015–1022 (May 2010)

Jackson, M.O.: Social and Economic Networks. Princeton University Press, Princeton (2008)

Maheswaran, R.T., Pearce, J.P., Tambe, M.: Distributed algorithms for DCOP: A graphical-game-based approach. In: Proceedings of Parallel and Distributed Computing Systems PDCS, pp. 432–439 (September 2004)

Maheswaran, R.T., Tambe, M., Bowring, E., Pearce, J.P., Varakantham, P.: Taking DCOP to the real world: Efficient complete solutions for distributed multi-event scheduling. In: Kudenko, D., Kazakov, D., Alonso, E. (eds.) AAMAS 2004. LNCS (LNAI), vol. 3394, pp. 310–317. Springer, Heidelberg (2005)

Meisels, A.: Distributed Search by Constrained Agents: Algorithms, Performance, Communication. Springer, Heidelberg (1848); ISBN 1848000391

Modi, J., Veloso, M.: Multiagent meeting scheduling with rescheduling. In: Proc. 5th Workshop on Distributed Constraints Reasoning DCR 2004, Toronto (September 2004)

Monderer, D., Tennenholtz, M.: Strong mediated equilibrium. Artificial Intelligence 173(1), 180–195 (2009); ISSN 0004-3702

[2] Currently, there is no distributed hill climbing search algorithm for ADCOPs. Existing algorithms such as MGM or k-Opt lose their monotonicity property when asymmetric gains are involved Grubshtein et al. (2010).

Moulin, H.: Axioms of Cooperative Decision Making. Cambridge University Press, Cambridge (1991)

Nisan, N., Roughgarden, T., Tardos, É., Vazirani, V.V.: Algorithmic Game Theory. Cambridge University Press, Cambridge (2007); ISBN 0521872820

Roughgarden, T.: Selfish Routing and the Price of Anarchy. MIT Press, Cambridge (2005); ISBN 0262182432

Wallace, R.J., Freuder, E.: Constraint-based reasoning and privacy/efficiency tradeoffs in multi-agent problem solving. Artificial Intelligence 161(1-2), 209–228 (2005)

Cloud Storage and Online Bin Packing

Doina Bein, Wolfgang Bein, and Swathi Venigella

Abstract. We study the problem of allocating memory of servers in a data center based on online requests for storage. Given an online sequence of storage requests and a cost associated with serving the request by allocating space on a certain server one seeks to select the minimum number of servers as to minimize total cost. We use two different algorithms and propose a third algorithm. We show that our proposed algorithm performs better for large number of random requests in terms of the variance in the average number of servers.

1 Introduction

Cloud storage is the service provided by numerous entities to store and backup computer files. We study the cost of storing vast amounts of data on the servers in a data center and we propose a cost measurement together with an algorithm that minimizes such cost. The bin packing optimization problem seeks to minimizes the number of bins a given sequence of items needs to be packed. This problem can apply to storing data on hard-drives in a data centers as follows: In a data center, files need to be stored on hard-drives. Without loss of generality we assume that all the hard-drives have the same size (of 1 unit), since buying them in bulk, of the same size is much cheaper (and easier to be replaced) than buying them individually, with various sizes. In this case, the hard drive represents a bin that needs to receive items to be stored. The goal is to minimize the number of bins (or disks) used for storing the files. We note that many important applications in areas such as multi-processor scheduling, resource allocation, packet routing, paged computer

Doina Bein
The Pennsylvania State University, Applied Research Laboratory
e-mail: siona@psu.edu

Wolfgang Bein · Swathi Venigella
The University of Nevada, Las Vegas, School of Computer Science
e-mail: beinw@unlv.nevada.edu, swathiv99@gmail.com

F.M.T. Brazier et al. (Eds.): Intelligent Distributed Computing V, SCI 382, pp. 63–68.
springerlink.com

memory systems, storage, multiprocessor scheduling, stock cutting, loading trucks with certain weight capacities, creating file backup in removable media and technology mapping in FPGA implementation of custom hardware can be modeled as a bin packing optimization problem [1, 2, 3, 4].

We study various bin packing schemes which could be relevant for cloud computing as they can be used for minimizing the energy consolidation [5, 6] in data centers. In Section 2 we present basic notions related to the bin packing problem, give a number of approximation algorithms which solve this problem, and we propose a novel algorithm "k-Binary". Simulation results are presented in Section 3. Concluding remarks are in Section 4.

2 Online Bin Packing

The bin packing problem requires packing a set of objects into a finite number of bins of capacity V in a way that minimizes the number of bins used: Given a sequence of n items $L = \{1,2,,n\}$ with sizes $s_1, s_2, \ldots, s_n \in (0,1]$, one needs to find a partition of the items into sets of size 1 (called *bins*) so that the number of sets in the partition is minimized [7] and the sum of the sizes of the pieces assigned to any bin may not exceed its capacity. We say that an item that belongs to a given bin (set) is *packed* into this bin. A bin is *empty* if no item is packed into it, otherwise it is used. Since the goal is to minimize the number of bins used, we would like to find an algorithm which finds a number of bins that is a constant factor of the minimum possible number of bins, no matter what the input is. Bin packing is NP-hard, thus finding an exact solution for any given input can be done currently only in exponential time.

For a given input sequence L of n items, let $A(L)$ (or simply A) be the number of bins used by algorithm A on the input sequence L. Let $OPT(L)$ (or simply OPT) be the number of bins used by an optimal offline algorithm which knows the sequence L in advance. The asymptotic performance ratio for an algorithm A is

$$lim\ sup_{n\to\infty}\ sup_L\{\frac{A(L)}{OPT(L)}|OPT(L) = n\}$$

For the general bin packing algorithm we assume that the entire list and its item sizes are known before the packing begins. A common situation is where the items arrive in some order and must be assigned to some bin as soon as they arrive, without knowledge of the remaining items. A bin packing algorithm that can construct its packing under this online regime is called an *online bin packing algorithm*, for which items in the list must be processed in exactly the same order as they are given, one at a time. The online processing is difficult owing to the fact that unpredictable item sizes may appear. In general, the performance of an online bin packing algorithm is substantially affected by the permutation of items in a given list.

We present next the three most common and simplest online bin packing algorithms: Next Fit, First Fit, and Best Fit. The asymptotic performance ratio of these algorithms is $r_{NextFit} = 2$, while the $r_{FirstFit}$ and $r_{FirstFit}$ are both 1.7 [8].

Next Fit is a bounded space online algorithm in which the only partially filled bin that is open is the most recent one to be started. It uses one active bin into which it packs the input. Once the free space in this bin becomes too small to accommodate the next item, a new active bin is opened and the previous active bin is never used again. This process continues until there are no more elements. *First Fit* keeps all non-empty bins active and tries to pack every item in these bins before opening a new one. If no bin is found, it opens a new bin and puts the item in the new bin. *Best Fit* picks the bin with the least amount of free space in which it can still hold the current element. All unfilled bins are kept open until the end. Let $SA(n)$ denote the maximum number of storage locations (for active bins) needed by some algorithm A during the processing of the list L whose size is n, and refer to algorithm A as an $SA(n)$-space algorithm. Next Fit is an $O(1)$-space algorithm, since it involves only one active bin at all times; both First Fit and Best Fit are an $O(n)$-space algorithms, since they keep all non-empty bins active until the end of the processing [9].

There are other online algorithms, based on a non-uniform partitioning of interval space (0,1] into M sub-intervals, assuming that the size of all items is no larger than 1. We are given a list of items $L = \{a_1, a_2, \dots, a_n\}$, each with item size $s(a_i)$, $0 < s(a_i) \leq 1$. The interval (0, 1] that represents the content of a bin is partitioned into harmonic sub-intervals $I^M = \{(0, \frac{1}{M}], (\frac{1}{M}, \frac{1}{M-1}], \dots (1/2, 1]\}$ where M is a positive integer. Then we classify each item according to the size, i.e., if the item size is in the interval $I_k = (\frac{1}{k-1}, \frac{1}{k}]$, $k > 1$, then the item is called an I_k-item; if the item size is in the interval $I_M = (0, \frac{1}{M}]$ then the item is called I_M-item. We note than in one bin, no more than k I_k-items can be fit, $k < M$, and each I_k filled bin packs exactly k items, irrespective of the actual sizes of these items, $k < M$. Into an I_M bin at least M items can fit. We present two most known such algorithms: Harmonic [7] and Cardinality Constrained Harmonic [10]. Later in this section we present our proposed algorithm – also based on a different type of partitioning.

Algorithm Harmonic keeps active an unfilled bin of each type: one bin of I_1 type items, one bin of I_2 type items, etc., thus a total of M bins are active at each moment. When an item arrives, it is packed in the corresponding bin; if the bin is filled, then it is closed and another bin is open, that will store items of that type. Once all the items are packed, we add the number of bins closed and the number of bins unfilled. Harmonic is completely independent of the arriving order of items [7]. A disadvantage of Harmonic is that items of size larger than 1/2 are packed one per bin, possibly wasting a lot of space in each single bin. Harmonic runs in $O(n \log M)$ time and uses M storage spaces for the active bins. Lee and Lee [7] had shown that the worst-case performance ratio of the algorithm is not related to the partition number M. Therefore, M can be regarded as a constant, and hence Harmonic is $O(1)$-space and $O(n)$-time algorithm.

Algorithm Cardinality Constrained Harmonic imposes a limit of M on the number of items that can fit into I_M bins, irrespective of whether more items can still fit into that bin. Whenever M items are stored in an I_M-bin, we close the bin and open a new one. We note again that the exact size does not affect the packing process, since each bin will contain only items of a single type, irrespective of the exact size.

Algorithm 1. k-Binary$(M,N,L=\{a_1,a_2,\ldots,a_n\})$

```
Inputs: M,N,L = {a_1,a_2,...,a_n}).
Output: T.
Partition (0,1] into M sub-intervals I_k = (1/2^k, 1/2^(k-1)], k = 1,..M.
T = 0.
Initialize all B_k to 0, 1 <= k <= M.
for i = 1 to N do
    if 0 < a_i < 1/2^M then {a_i is an I_M-piece}
        if B_M + a_i > 1 then {the item a_i does not fit into the I_M-bin}
            Get a new I_M-bin: T = T + 1, B_M = 0.
        end if
        Place a_i into the open I_M-bin: B_M = B_M + a_i.
    else if ∃k, 1/2^(k-1) < a_i <= 1/2^k then {a_i is an I_k-piece, 1 <= k < M}
        if B_k + a_i > 1 then {the item a_i does not fit into the I_k-bin}
            Get a new I_k-bin, namely T = T + 1, B_k = 0.
        end if
        Place a_i into the open I_k-bin: B_k = B_k + a_i.
    end if
end for
Add to T the number of unfilled bins.
for i = 1 to k do
    if B_k > 0 then
        T = T + 1.
    end if
end for
Output T.
```

Algorithm k-Binary (see algorithm box 1) partitions the interval (0,1] into subintervals in which the endpoints are a negative power of 2: $(0,1]=\cup_{k=1}^{M} I_k$ where $I_k=(\frac{1}{2^k},\frac{1}{2^{k-1}}]$ and k, the number of partitions, is a positive integer. For example, if $k=3$ then the interval $(0,1]$ is partitioned into three intervals $(0,1/4]$, $(1/4,1/2]$, and $(1/2,1]$.

3 Simulation Results

Different sequences of requests produce different results when M, the number of partitions, varies. We compare the performance of the k-binary algorithm with the performance of Harmonic and Cardinality Constrained Harmonic. We measure the efficiency of the implemented algorithms using the average number of bins of any type used. For a given input, let N be the total number of items and M be the total number of partitions. Let us consider TB^1 to be the total number of bins used by Harmonic, TB^2 to be the total number of bins used by Cardinality Constrained Harmonic and TB^3 be the total number of bins used by k-Binary. For each algorithm we will compute how well the algorithm balances the bins of any type. For an algorithm

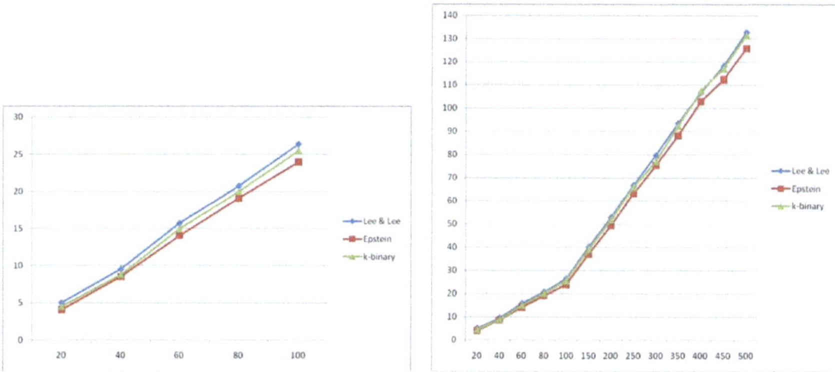

(a) Smaller number of items (b) Larger number of items

Fig. 1 Trend of the variance

j, $1 \le j \le 3$, let NB_k^j be the number of bins of I_k-type. If the algorithm j, $1 \le j \le 3$, balances the items well, then all NB_k^js should be roughly $\frac{TB^j}{M}$.

To compare how well an algorithm balances the types of bins used, we compute the variance VAR:

$$VAR = \sum_{k=1} M(NB_k^j - TB^j/M)^2 \qquad (1)$$

We generate random sequence of requests as follows. We first generate a sub-sequence of N random values in the interval $(0.01, 0.99)$, $L_1 = \{a_1, a_2, \ldots, a_N\}$, $0.01 \le a_i \le 0.99$ and we call this the *random sub-sequence*. We append to L_1 another sub-sequence of N values, $L2 = \{b_1, b_2, \ldots, b_N\}$, where $b_i = 1 - a_i$, and we call this as the *computed sub-sequence*. The random sequence contains the random sub-sequence followed by the computed sub-sequence.

We run extensive number of simulations by generating such random sequences of length $2N$ for various values of N and M. We observed that in some cases our proposed Algorithm k-Binary requires less number of bins than Harmonic, and there are other cases where k-Binary requires more bins than Harmonic. In almost all the cases, Cardinality Constrained Harmonic required at least the same number of bins as Harmonic, if not more. No trend has been observed as for values of N and M which algorithm, k-Binary or Harmonic, has the best performance, namely it gives the least number of bins.

We also compared the three algorithms based on the variance in the number of bins used, namely the value of VAR (Equation 1), for small or large sequences of random requests. When the number of items is relatively small, $N < 100$, the values for variance are plotted in Figure 1(a). When the number of items is relatively large, $10 < N < 1000$, the values for variance are plotted in Figure 1(b). In both figures, the x axis shows the number of bins and the y axis show the variance value obtained by each algorithm.

We note that for smaller number of items, there is not much variation between the three algorithms. The proposed Algorithm k-Binary is as efficient as Harmonic and Cardinality Constrained Harmonic. For larger number of items, we note that k-Binary has a lower variance than Harmonic but slightly higher variance than Cardinality Constrained. We can then conclude that the proposed k-binary algorithms behaves in average better than both previously proposed algorithms in terms of the variance on the type of bins used.

4 Conclusion and Future Work

The algorithm proposed by Lee and Lee has a competitive ratio which is very close to the optimum, but has a high variation on the types of bins used. The algorithm proposed by Epstein has a worse competitive ratio, i.e. uses more bins for packing the items than Lee and Lee, but it has a much smaller variation than Lee and Lee.

We propose an algorithm, k-Binary, that uses fewer bins than the algorithm proposed by Epstein, and a slightly more bins than the algorithm proposed by Lee and Lee. At the same time, k-Binary works better than both previously proposed algorithms. Modern computers have multiple processors, thus it might be of interest to provide algorithms that allow serving two or more requests at a time.

References

1. Ullman, J.D.: The performance of a memory allocation algorithm. Tech. Report 100, Princeton University, Princeton, NJ (1971)
2. Coffman Jr., E.G., Garey, M.R., Johnson, D.S.: Approximation algorithms for bin packing: An updated survey, pp. 46–93. PWS Publishing Co., USA (1997)
3. Coffman Jr., E.G., Csirik, J.: Performance guarantees for one-dimensional bin packing. ch. 32. Chapman & Hall, CRC (2007)
4. Csirik, J., Woeginger, G.J.: In: Fiat, A., Woeginger, G.J. (eds.). On-line packing and covering problems. ch. 7, pp. 147–177. Springer, Heidelberg (1998)
5. Li, B., Li, J., Huai, J., Wo, T., Li, Q., Zhong, L.: Enacloud: An energy-saving application live placement approach for cloud computing environments. In: IEEE International Conference on Cloud Computing (CLOUD), pp. 17–24 (2009)
6. Srikantaiah, S., Kansal, A., Zhao, F.: Energy aware consolidation for cloud computing. In: Conference on Power Aware Computing and Systems (HotPower), vol. 10 (2008)
7. Lee, C., Lee, D.: A simple on-line bin-packing algorithm. Journal of ACM 32, 562–572 (1985)
8. Seiden, S.S.: An optimal online algorithm for bounded space variable-sized bin packing. SIAM Journal on Discrete Mathematics 14, 2001 (2000)
9. Drake, M.: Bin packing (2006)
10. Seiden, S.S., Van Stee, R., Epstein, L.: New bounds for variable-sized online bin-packing. SIAM Journal on Computing, 455–469 (2003)

Part III
Parallel and Distributed Computational Intelligence

Multilevel Parallelization of Unsupervised Learning Algorithms in Pattern Recognition on a Roadrunner Architecture

Stefan-Gheorghe Pentiuc and Ioan Ungurean

Abstract. The aim of the paper is to present a solution to the NP hard problem of determining a partition of equivalence classes for a finite set of patterns. The system must learn the classification of the weighted patterns without any information about the number of pattern classes, based on a finite set of patterns in a metric pattern space. Because a metric is not suitable in all the cases to build an equivalence relation, an ultrametric is generated from indexed hierarchies. The contributions presented in this paper consists in the proposal of multilevel parallel algorithms for bottom-up hierarchical clustering, and hence for generating ultra-metrics based on the metrics provided by the user. The algorithms were synthesized and optimized for clusters having the Roadrunner architecture (the first supercomputer that breaks 1PFlops barrier [1]).

1 Introduction

Let E be a set of n patterns with p features from a metric space. The set $N(E)-\{(x,f_x)$ / $x\in E, f_x\in R_+\}$, where f_x are the weights of patterns, it is called cloud of weighted patterns. It is required to determine a partition of M equivalence classes $P_M(E)=\{C_1, C_2, \ldots, C_M\}$ based only on the $N(E)$ and the metric $d{:}E\times E\rightarrow R_+$ defined on the pattern space. The problem of determining a partition of a finite set E is an NP-hard problem. It is called also unsupervised learning, because the system must learn how to classify the patterns without any help from the human expert. The solution to determine the partition $P_M(E)$ consists in identifying groups of patterns in the set E by minimizing a performance criterion generally based on a metric. Various algorithms are proposed to determine a partition in M classes, such as k-means and its variants [2].

Stefan-Gheorghe Pentiuc · Ioan Ungurean
University "Stefan cel Mare" of Suceava, str. Universitatii no.13,
RO-720229 - Suceava, Romania
e-mail: {pentiuc,ioanu}@eed.usv.ro

F.M.T. Brazier et al. (Eds.): Intelligent Distributed Computing V, SCI 382, pp. 71–80.
springerlink.com

A metric does not always ensure the possibility to define an equivalence relation over E, which would allow the building of equivalence classes for partition $P_M(E)$, and ultrametrics are used for this aim [3].

Definition 1. Let E be the set of patterns. It's called ultra-metric distance the application $\delta : E \times E \rightarrow R_+$ that comply with the following axioms:

1. $\forall x \in E$, $\delta(x,x)=0$;
2. $\forall x,y \in E$, $\delta(x,y)=\delta(y,x)$;
3. $\forall x,y,z \in E$, $\delta(x,y) \leq max\{\delta(x,z), \delta(z,y)\}$.

In practice, it is very difficult for the users of pattern recognition systems to define an ultra-metric. For this reason, we seek to build an ultra-metric based on the metric provided by the user. Another practical difficulty for the user is to indicate the number M of clusters of the partition. Therefore, it is more convenient to offer a string of partition of equivalence classes, $P_k(E)=\{C_1, C_2, \ldots, C_k\}$ with k=1,..,n, $P_1(E)=\{E\}$ and $P_n(E)=\{\{x_1\},\ldots,\{x_n\}\}$- the banal partition. From this string of partitions, the user will select the most appropriate partition for his problem. It is preferred to adopt an algorithm for determining a hierarchy of subsets for the set of patterns.

2 Hierarchies and Ultra-Metrics

In order to emphasize the relation between ultrametrics - hierarchies and string of partitions, in this section will provide some definition and an equivalence theorem.

Definition 2. Let $P(E)$ be the set of all equivalence partitions of the set E. The set $H \subset P(E)$ is an hierarchy of E if the following conditions are met:

1. $E \in H$;
2. $\forall x \in E, x \in H$;
3. $\forall h, h' \in H$ with $h \cap h' \neq \emptyset$ then either $h \subset h'$ or $h' \subset h$.

Definition 3. An indexed hierarchy is the couple (H, f), where H is a hierarchy and $f: H \rightarrow R_+$ a map that comply the following axioms:

1. $\forall h \in H$ with $card(h)=1$, $f(h)=0$;
2. $\forall h,h' \in H$ with $h \subset h'$, $f(h)<f(h')$.

An example of indexed hierarchy is presented in fig. 1 (let $h=\{x_4,x_5\}$, $h'=\{x_3,x_4,x_5\}$, is observed that $f(h)=1.4$, $f(h')=2$, and $h \subset h'$).

It can be noticed that an indexed hierarchy contains subtrees whose leaves are patterns belonging to the same cluster (the left subtree h_1 in fig. 1 represents the cluster $C_1=\{x_1,x_2,x_3,x_4,x_5\}$ and right subtree h_2 the cluster $C_2=x_6,x_7,x_8,x_9$).

Given an indexed hierarchy it can be defined an ultra-metric [4].

Lemma 1. *If (H,f) is an indexed hierarchy of E, then the application $\delta: E \times E \rightarrow R_+$ defined as $\delta(x,y)=min\{f(h)/h \in H, x \in h, y \in h\}$ is an ultra-metric.*

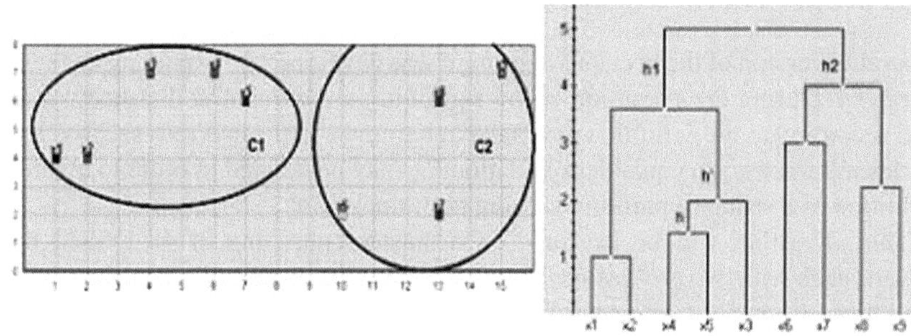

Fig. 1 Example of an indexed hierarchy of patterns

The proof is based on the properties of the indexed hierarchies. Having an ultra-metric, it can be defined an equivalence relation over E, which will allow to obtain a partition of equivalence classes of E. Let be the relation R_α defined as: $\forall(x,y) \in E \times E, xR_\alpha y$ if and only if $\delta(x,y) \leq \alpha$, with $\alpha \in R_+$.

It is noted that the relation R_α is an equivalence relation. The reflexivity and symmetry result from the ultra-metrics properties. If $xR_\alpha z$ and $zR_\alpha y$ then $xR_\alpha y$, will see that $xR_\alpha z$ and $zR_\alpha y$ implies $\delta(x,z) \leq \alpha$ and $\delta(z,y) \leq \alpha$, so max $\delta(x,z), delta(z,y)$ $\leq \alpha$, and from the ultra-metric properties results, $\delta(x,y) \leq \alpha$, so $xR_\alpha y$. It emphasizes that if a metric is used to define the relation R_α, this is not transitive and thus it is not an equivalence relation.

Lemma 2. *The set of all equivalence classes generated by the relation R_α for various values of α forms an indexed hierarchy.*

For proof, check the properties of indexed hierarchy on set *P(E)* obtained with R_α for a series of ascending values of the α [4].

Using these two lemmas the following theorem of equivalence [4] can be proved.

Theorem 1. *Let φ an application that based on an indexed hierarchy an ultrametric distance are built, and ψ an application that based on an ultra metric and a relation R_α leads to a hierarchy, then $\varphi = \psi^{-1}$ and $\psi = \varphi^{-1}$.*

The algorithms used to determine a hierarchy of pattern classes uses two main strategies of partitioning: bottom-up and top-down. The majority uses a metric as in [5] or similarity measures [6] for grouping the patterns. We start from a serial bottom-up algorithm [2] based on a metric supplied by user, and an index of aggregation $f(h \cup h', u)$ used to build the indexed hierarchy. Initially $H=\{h_1,...,h_n\}$, with $h_i=\{x_i\}$, corresponding to $P_1(E)$, and *f(h,h')* is represented by the metric matrix. At each step k, starting from the partition $P_k(E)$, a new partition $P_{k-1}(E)$ is obtained by merging a couple of clusters (h,h') with f(h,h')= minimum. Due to the new cluster $f(h = h \cup h'$ built, the index *f(h,h')* needs to be re-evaluated according to aggregation index. The algorithm continues until the partition $P_1(E)$ is obtained.

3 Parallel Hierarchical Classification Algorithms

A parallel version of the previous algorithm was developed for obtaining an indexed hierarchy. During the execution of the algorithm, an ultrametric distance is generated according to the definition presented in Lemma 1 in the previous section. Based on this ultrametric, an equivalence relation R_α may be defined in order to assure the definition of a string of partition varying the values of α.

This algorithm will be executed on a cluster consisting of 48 QSS22 blade servers, each with two processors PowerXCell 8i. This processor is based on the Cell Broadband Engine Architecture (CBEA) [7], which is a new architecture, derived from 64-bit PowerPC architecture. The Cell Broadband Engine solutions offered by IBM is a combination of processors based on CBEA integrated in IBM BladeCenter servers, and software tools for applications developing. This architecture consists of nine processors integrated on a single chip, interconnected and connected with external devices via a high bandwidth bus, with coherent memory - practically a multi-core processor. The most important feature of the Cell BE processor is that all internal cores share the same memory location for storage (the effective address space that includes main memory). However, the function of internal cores is specialized in two areas, each covered by one type of processor: PowerPC Processor Element (PPE) and Synergistic Processor Element (SPE) [8]. The processor based on CBEA architecture containing a PPE module and eight SPE.

Based on this architecture, the parallelization must be done on two levels: on the first level, the computing tasks are distributed to PPE processors, and on the second level, each processor PPE distributes the sub-tasks to SPE cores. The first step was to implement this algorithm only for PPE processor. In this case, the algorithm is running on a single core without using the SPE cores. After implementation of this variant, the points where the SPE cores can be enabled and used are identified.

The MPI parallelization of this algorithm is achieved by dividing the ultrametrics matrix to the Cell BE processors. The parallel algorithm for the first level is as follows:

*)*Read the input data set from a file*
*)*Allocates memory for the index matrix s[x][y]=d(s,y) where d is the metric provided by the use*
*) *$P_n(E)$ is initialized with clusters h_i=x_i and index 0*
*) *distribute the computing tasks to PowerXCell 8i processors by splitting the index matrix*
*) *compute the part of index matrix which is assigned to current d PowerXCell 8i processor*
*) *MPI_Allreduce() for determining the whole metric matrix*
repeat
  *) *distribute the charges to PowerXCell 8i processors by splitting the index matrix*

*) *determine the minimum value in the parts of index matrix assigned to current d PowerXCell 8i processor*
*) *determine the global minimum value s[h][h'] with* $h \neq h'$, *by MPI_Allreduce() function*
*) *build a new cluster* $h = h \cup h'$ *with index s[h][h']*
*) *change, in the index matrix, the values for column and line assigned to the new cluster h*
*) *remove from the index matrix the line and column assigned to cluster h'*
until *) *the index matrix contains only one element.*

The inputs of the algorithm are the patterns matrix of dimensions *n*x*p* and the metric. The index matrix of dimensions *n*x *n* stores the values *f(x,y)* at each step, initial being equals to the metric matrix. Because the index matrix is symmetric, the algorithm uses only the part above the main diagonal in a linearized form.

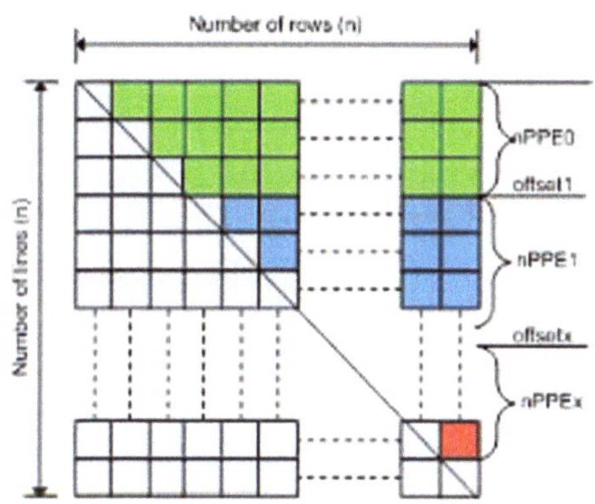

Fig. 2 The allocation of metrics matrix to PowerXCell 8i processors

The distribution of computing tasks to PPE processors is made by the division of the distance matrix to the *noPPE* processors. The problem of balancing the charges is reduced to split equally the area of a triangle (above the main diagonal) into *noPPE* polygons of equal areas (fig. 2).

It must be determined the number of lines associated and the offset of these lines. This allocation is updated at every iteration because a column and a line from the index matrix are removed.

Similarly, it is determined the number of lines assigned to each PPE processor by replacing the value of offset0 with the offset value associated with each processor.

4 Improving the Proposed Algorithm by Using the SPE Cores

In the proposed algorithm it can be identify two points where the SPE cores can be enabled and used: computing the index matrix and searching the closest two cluster (representing the search of minimum value from the index matrix).

The allocation is made taking into account the fact that data blocks used for DMA transfer between main memory and local memory are multiples of 128 bytes. It is considered that, for trapezoid from fig. 3, one of the parallel sides is $n - offsetPPE$ and the other parallel sides are $n - offsetPPE - nl$, so the trapezoid area is $(2n - 2offsetPPE - nl)/2$. Thus, the area associated with each SPE processor is $(2n - 2offsetPPE - nl)/(2 * noSPE)$.

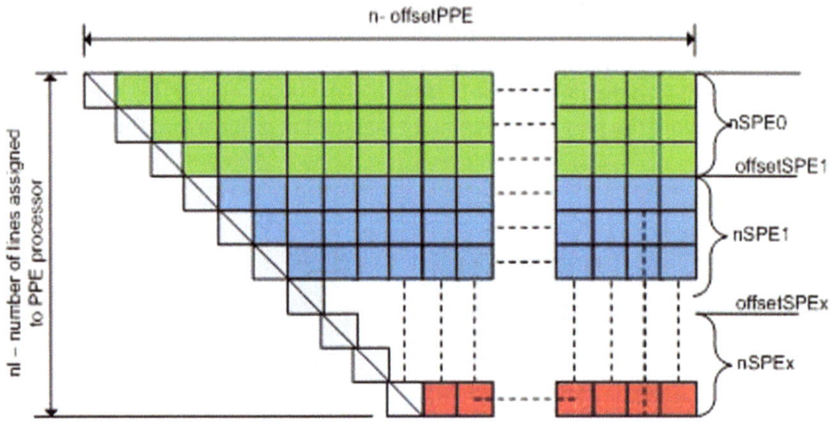

Fig. 3 The allocation of ultrametrics matrix to SPE cores

In order to find the number of lines associated with the first SPE processor (nSPE0), it is solved the following quadratic equation:

$$\frac{(n-offsetPPE-offset0+n-offsetPPE-offset0-nSPE0)*nSPE0}{2} =$$
$$= \frac{2n-2offsetPPE-nl}{2*noSPE}$$

The application is designed so that at its execution, on each SPE processor starts a thread [10]. These threads can perform operations only on data from local memory, whose size is 256KB. For this reason, if it is intended to process data from shared main memory then the data are transferred by DMA to the local memory associated to SPE processor, the desired processing is performed, and the result is transferred to the main memory to be retrieved by the PPU processor [11].

It is very important to synchronize the PPU processor with the SPE cores. Thus, when a thread starts on the SPE cores, he expects through a mailbox commands from PPE processor. The commands that can be received by an SPE core are: end of execution, computing the index matrix associated to the current core, searching the minimum value from the part of index matrix associated to the current core. Thread associated with the SPE processor finishes the execution if it receives the command for the end of execution.

In the other two commands, the index matrix is split equally to each SPE core. The PPU processor determines the number of lines assigned to each SPE cores and it store the data at a default address in memory (address that is sent to SPE thread on its creation). Through a mailbox, it sent to SPE cores the commands to compute the index matrix or to search the minimum in this matrix. Further the PPU processor waits for a mailbox message from all processors SPE in order to know when they completed the required tasks. On reception of these commands, the SPE cores transfers by DMA needed data from shared main memory into local memory, process the data and transfer by DMA the results to main memory. After the data transfer in main memory, the SPE core signals the PPU processor by a message through a mailbox the end of the required task.

The principles outlined so far was implemented taken into account two important aspects: the DMA transfer can be accomplished into and from the main memory, just with data aligned on 128 bits, as well and in data blocks, whose size is multiple of 128 bytes, but not higher than 16KB. Because of this aspect, the memory where each pattern is stored will be aligned to 128 bits (the last 7 bits of the address are of value 0), and the data block size associated to each pattern is approximated to a value multiple of 128 bytes.

5 Experimental Results

In order to perform the performance tests, two input sets were used, provided by the UCI Machine Learning Repository [9]: Bag of Words NIPS full papers, which includes 1500 documents, where the size of dictionary is of 12419 terms, as well as Bag of Words KOS blog entries, which includes 3430 documents and the size of dictionary is of 6906 terms. The first test was to determine the series of partitions using only the PPE cores from PowerXCell 8i processors. The first test was to determine the series of partitions using only the PPE cores from PowerXCell 8i processors. It started with running the application on a single processor, after that the number of processors is doubled until the maximum number (96 PowerXCell 8i processors) is achieved.

In fig. 4 are presented the execution time and communication time obtained by executing the algorithm at the first level (only on PPE cores from PowerXCell 8i processors).

It can be seen that, the communication time between processor increase if the numbers of processors is growing, so after a certain number of processors, the performance improvement by increasing the number of processors cannot be achieved.

In the fig. 5 are presented the execution time and communication time obtained by executing the application using all SPE cores from PowerXCell 8i processors.

In Table 1 and Table 2 are presented the speedups obtained when the number of processors is increasing, with and without activation of the SPE cores. Speedups are

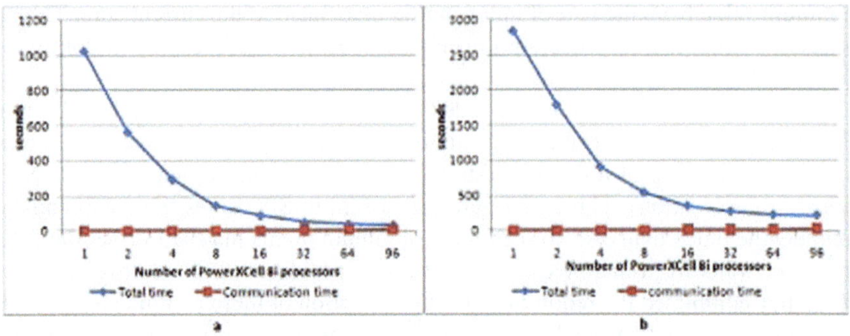

Fig. 4 Execution and communication times for NIPS (a) and KOS (b) datasets

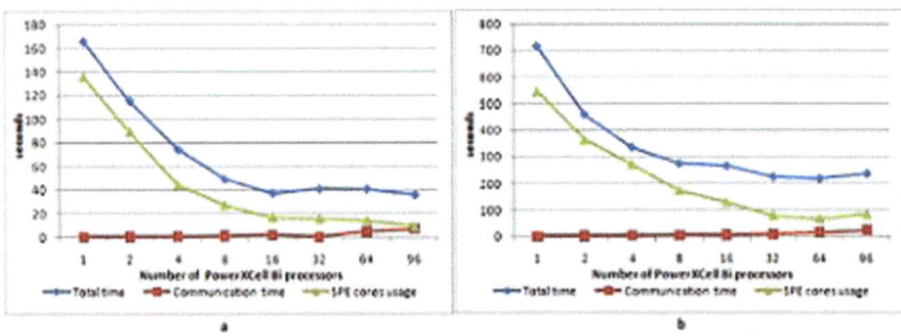

Fig. 5 Execution and SPE cores usage time for NIPS (a) and KOS (b) datasets

Table 1 Speedups obtained for NIPS dataset, as the relation with the execution on a single PowerXCell 8i processor, using only the PPE core

Number of PowerXCell 8i processors	1	2	4	8	16	32	64
Speedup without SPE cores	1.00	1.82	3.46	7.12	11.80	20.62	27.57
Speedup with SPE cores	6.16	8.84	13.70	20.71	27.38	24.82	24.98

measured as the relation with the execution on a single PowerXCell 8i processor, using only the PPE core.

In Table 3 is presented the speedup obtained when the number of processors is increasing, and the SPE cores are activated. Speedup is measured as the relation with the execution on a single PowerXCell 8i processor, using all SPE cores.

It can be observed that the speedup achieves a maximum after that is decreased. The decrease in execution time is obtained when the communication time are more significant than computing time [10, 11]. As a consequence, a decrease of speedup is achieved.

Table 2 Speedups obtained for KOS dataset, as the relation with the execution on a single PowerXCell 8i processor, using only the PPE core

Number of PowerXCell 8i processors	1	2	4	8	16	32	64
Speedup without SPE cores	1.00	1.59	3.15	5.28	8.16	10.68	12.92
Speedup with SPE cores	3.99	6.21	8.49	10.38	10.72	12.66	13.03

Table 3 Speedups obtained, as the relation with the execution on a single PowerXCell 8i processor, using all SPE core

Number of PowerXCell 8i processors	1	2	4	8	16	32	64
Speedup for NIPS dataset	1.00	1.37	2.40	3.88	5.53	7.02	6.17
Speedup for KOS dataset	1.00	1.40	2.43	4.08	5.47	5.95	4.69

6 Conclusions

In this paper is presented a solution for determining an indexed hierarchy for a finite set of patterns consisting in a new multilevel parallel algorithm for hierarchical classification. This algorithm has been optimized for a cluster with Roadrunner architecture. The tests emphasized that the algorithm can be used for very large data sets and performance improvement achieved by parallelization of the algorithm for the cluster with Roadrunner architecture, by using all the cores from PowerXCell 8i processors. From a functional perspective, the proposed variant has the advantage of a solid theoretical background that besides indexed hierarchy of forms provides an ultrametric over a finite space of patterns that may be used for building an equivalence relations that guarantee the partitioning in equivalence classes of patterns.

Acknowledgements. Experiments were carried out in High Performance Computing Lab, from Stefan cel Mare University of Suceava, who has a cluster of 48 blade servers IBM QS22. This system was purchased within project "GRID FOR DEVELOPING PATTERN RECOGNITION AND DISTRIBUTED ARTIFICIAL INTELLIGENCE APPLICATIONS-GRIDNORD", contract 80/13.09.2007, in the frame of PN II, Capabilities Programme.

References

1. Barker, K.J., Davis, K., Hoisie, A., Kerbyson, D.J., Lang, M., Pakin, S., Sancho, J.C.: Entering the petaflop era: The architecture and performance of Roadrunner. High Performance Computing, Networking, Storage and Analysis, 1–11 (2008), doi:10.1109/SC.2008.5217926
2. Dhillon, I.S., Guan, Y., Kulis, B.: Kernel k-means: spectral clustering and normalized cuts. In: Proceedings of the Tenth ACM SIGKDD International Conference on Knowledge Discovery and Data Mining (KDD 2004), pp. 551–556. ACM, New York (2004), doi:10.1145/1014052.1014118

3. Milligan, G.W., Isaac, P.D.: The validation of four ultrametric clustering algorithms. Pattern Recognition 12(2), 41–50 (1980); doi:10.1016/0031-3203(80)90001-1, ISSN 0031-3203
4. Celeux, G., Diday, E., et al.: Classification automatiqur de donees. In: Environement statistique et informatique, Dunod Informatique. Dunod Informatique, Bordas Paris (1989)
5. GuoYan, H., DongMei, Z., Jiadong, R., ChangZhen, H.: A Hierarchical Clustering Algorithm Based on K-Means with Constraints. In: Innovative Computing, Information and Control (ICICIC), pp. 1479–1482 (2009), doi:10.1109/ICICIC.2009.18
6. Jun, Q., Qingshan, J., Fangfei, W., Zhiling, H.: A Hierarchical Clustering Based on Overlap Similarity Measure. Software Engineering, Artificial Intelligence, Networking, and Parallel, Distributed Computing 3, 905–910 (2007) ;doi:10.1109/SNPD.2007.502
7. Sandeep, K.: Practical Computing on the Cell Broadband Engine. Springer, Heidelberg (2009); ISBN: 978-1-4419-0307-5
8. Vogt, J.S., Land, R., Boettiger, H., Krnjajic, Z., Baier, H.: IBM BladeCenter QS22: Design, performance, and utilization in hybrid computing systems. IBM Journal of Research and Development 53(5) (2009); doi:10.1147/JRD.2009.5429069
9. Asuncion, A., Newman, D.J.: UCI Machine Learning Repository. University of California, School of Information and Computer Science, Irvine, CA (2007)
10. Guochun, S., Kindratenko, V., Pratas, F., Trancoso, P., Gschwind, M.: Application Acceleration with the Cell Broadband Engine. Computing in Science & Engineering 12(1), 76–81 (2010); doi:10.1109/MCSE.2010.4
11. Dursun, H., Barker, K.J., Kerbyson, D.J., Pakin, S.: Application profiling on Cell-based clusters. Parallel & Distributed Processing, 1–8 (2009); doi:10.1109/IPDPS.2009.5161092

Dual Manner of Using Neural Networks in a Multiagent System to Solve Inductive Learning Problems and to Learn from Experience

Florin Leon and Andrei Dan Leca

Abstract. Learning can increase the flexibility and adaptability of the agents in a multiagent system. In this paper, we propose an automated system for solving classification and regression problems with the use of neural networks, where agents have three different learning algorithms and try to estimate a good network topology. We establish a competitive behavior between the types of agents, and we address a way in which agents can optimize the utility received for problem solving. Thus, the agents use neural networks to solve "external" problems given by the user, but they can also build their own "internal" problems out of their experience in order to increase their performance. The resulting behavior is an emergent property of the multiagent system.

1 Introduction

Multiagent systems represent an active, promising research direction that can highlight many phenomena typical of biological or social systems. Following the use of adaptive and deliberative agents, learning enables systems to be more flexible and robust, and it makes them better able to deal with uncertainty and changing conditions, which are key factors for intelligent behavior. This is particularly important in multiagent systems, where the task of predicting the results of interactions among agents can be very difficult.

Among the types of learning typically used when agents deal with an unknown environment, reinforcement learning is a convenient way to allow the agents to autonomously explore and learn the best action sequences that maximize their overall value, based on successive rewards received from the environment. However,

Florin Leon · Andrei Dan Leca
Technical University of Iaşi, Bd. Mangeron 27, 700050 Iaşi, Romania
e-mail: fleon@cs.tuiasi.ro, andy_leca@yahoo.com

F.M.T. Brazier et al. (Eds.): Intelligent Distributed Computing V, SCI 382, pp. 81–91.
springerlink.com

a multiagent reinforcement problem adds an additional level of complexity. Since classic algorithms such as *Q-learning* [24] estimate the values for each possible discrete state-action pair, each agent causes an exponential increase of the size of the state-action space. Several learning algorithms have been proposed for multi-agent systems, among which we could mention: *Fictitious Play* [2], *Nash-Q* [10], *AWESOME* [3].

Many multiagent system applications can be found in the literature where agents use neural networks for learning. Such systems were used for intrusion detection in computer networks [9], classification of medical records [16], optimization of multiagent reinforcement learning algorithms [4], market modeling and simulation [23, 8], or for studying phenomena such as the Baldwin effect in evolving populations [18].

The vast majority of works consider the use of neural networks, or other machine learning methods, either to solve problems posed by the user in some form, or to make the agents self-adaptive to the environment. However, a human being uses his/her cognitive abilities to solve both external problems and problems derived from his/her own experience.

The main original contribution of this paper is the combination of these two approaches, i.e. the agents in the system use neural networks to solve "external" problems, but they can also build their own "internal" problems out of their experience in a competitive multiagent environment in order to optimize their performance. An emergent behavior of the system is that a group of agents whose original performance is below that of other groups can greatly improve their operation by learning from their own experience in an autonomous way. Thus, they succeed in outperforming the other groups only by developing an adaptive strategy, not by changing their initial problem solving capabilities.

2 Description of the Task Allocation System

In the following, we describe the way in which individual agents operate when presented with learning tasks. Their goal is to build models based on neural networks for data provided by an external user.

2.1 Learning Problems

In order to study the behavior of the multiagent system where the agents use neural networks to solve classification and regression tasks, we considered 20 typical classification and regression problems.

The problems and-sym, xor-sym define the Boolean logic problems AND and eXclusive OR. sin and max are the corresponding functions of 1 or 2 variables. iris-d3 and iris-d5 are variants of the Iris flower problem [7], where the values are discretized into 3 or 5 intervals, while iris-num uses the original dataset. The drugs problem tries to find the rules to prescribe one of the two classes of drugs, according to the patient's age, blood pressure, and gender. monks{1-3} are the three classical MONK's benchmark problems [22]. The shepard{1,2,4,6} are problems proposed while studying human performance on a category learning task [21]. weather and weather-sym are variants of the golf-playing problem [17] where data values are symbolic or numerical, and only symbolic, respectively. Finally, in order to study the regression capabilities, three problems based on sampled trigonometric functions were devised, assumed difficult to learn because of their shape and large number of instances. A detailed description of all these problems can be found in some previous papers [15, 12].

2.2 *Behavior of Individual Agents*

The multiagent system is implemented using the JADE framework [1] and thus it can be easily distributed. It is comprised of a varying number of agents that apply one of three different algorithms to train their neural networks: *Backpropagation* [20], *Quickprop* [6], and *RProp* [19]. The agents are given a number of tasks, consisting in repeated versions of the 20 classification and regression problems mentioned earlier, problems that appear in a random order. Before testing the overall behavior of the multiagent system, the problems were considered separately, so that their complexity could be estimated in terms of execution time while executing the training algorithms.

Since finding the optimal topology of a neural network is a difficult problem, the agents successively try increasingly complex network configurations. The performance of the algorithms depend on the Mean Square Error (MSE) desired by the user, and also on the number of training epochs. In this scenario, the agents need to find an MSE of 0.001 or lower, and run the algorithms for 500 epochs. The MSE can be higher for classification problems, if the percent of correctly classified instances is 100% for both training and testing.

They first use a network topology of 1 hidden layer with 10 neurons. If this configuration proves to be too simple to learn the model, and the performance criteria are not met, they use a network topology of 1 hidden layer with 20 neurons. The third attempt is a topology of 2 hidden layers with 15 and 5 neurons, respectively. Finally, they use a topology of 2 hidden layers with 30 and 10 neurons, respectively.

These particular values were considered sufficient for most learning problems, based on authors' previous experience with neural network models on real-world problems [5, 13, 14]. Since the process of building a proper neural network is meant to be automated and the agents are also required to solve the problems as quickly as possible, and therefore they do not have the option of an extensive search for the optimal topology, these four variants, although not guaranteed to be optimal, should be representative for the complexity of the learning problems. The ratio between the number of neurons in the first and the second hidden layer follows Kudrycki's heuristic [11], which recommends a ratio of 3:1.

In this approach, the number of epochs is rather small and the target MSE is rather large. However, these values can emphasize the difference in convergence speed of the training algorithms. Fig. 1 displays the average values of 10 tests for each problem and algorithm. The *Backpropagation* (BP) time was taken as a reference, and displayed in the labels on the abscise, with values given in milliseconds. The *Quickprop* (QP) and *RProp* (RP) times are given as percents compared to the reference times. This way of displaying the data was preferred because of the great difference (in orders of magnitude) between the training times for different learning problems.

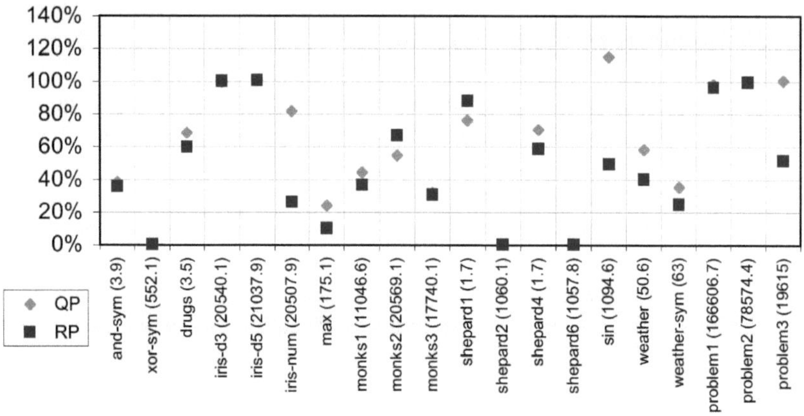

Fig. 1 Average execution time of *QP* and *RP* compared to *BP*

3 Behavior of the Multiagent System

The multiagent system assumes the following rules for task allocation. Tasks are proposed to the agents and they have the freedom of accepting a task or not. Only the available agents, i.e. agents that are not engaged in solving some other task can bid for a new one. The system randomly selects up to 3 agents to solve a task from the bidding agents. This redundancy is useful because the neural network training depends on the initial, random initialization of connection weights. Therefore, a trained network can get stuck in a local minimum. However, the 4 successive attempts to solve the problem in addition to a maximum of 3 agents to solve the same problem should be enough to find a reasonable solution.

If the system has less than 3 available agents, the task can be assigned to only 1 or 2 agents. If there are no agents available, the task allocation system waits until some agents become available again.

Agents do not know the task complexity *a priori*. They can only have descriptive quantitative information such as: the number of attributes, the type (classification or regression), the number of training instances, and the number of testing instances.

Nevertheless, two tasks with similar structures may have completely different complexities, depending on the unknown internal model. Agents also know the number of other available agents in the system, i.e. possible competitors, but they do not know the type of the training algorithm the others use.

The agents receive rewards for solving tasks. However, only the agent that provides the smallest error out of the competitors receives the reward. If more agents provide the same error rate, the network with the smallest topology, and then the agent with the smallest execution time is declared the winner.

The utility that agents receive for solving a task is computed by taking into account the classification accuracy or the MSE of a regression problem.

For classification problems, the utility is given in terms of the percent of incorrectly classified instances (P_{ICI}): $U^C = 100 - P_{ICI}$.

If the problem has both training and testing data, then the utility is given as a weighted sum between the percent of incorrectly classified instances for training and testing, respectively, in order to emphasize the importance of generalization: $U^C = 100 - \left(P_{ICI}^{training} + 2 \cdot P_{ICI}^{testing}\right)/3$.

For regression problems, the received utility takes into account the ratio between the training MSE achieved by the agent and the maximum allowed error, which is 0.001 in our case: $R = MSE_{training}/MSE_{max}$.

The utility decrease U_d follows the shape given in Fig. 2. The utility of a solution for a regression problem is computed by the following equation: $U^R = 100 - U_d$.

Similar to the classification case, if the problem also has testing data, the formula becomes: $U^R = 100 - \left(U_d^{training} + 2 \cdot U_d^{testing}\right)/3$.

4 Case Studies

In this section, we study the overall performance of the agents in terms of received utility. First, we consider the simple case when agents accept tasks automatically. Then, we try to optimize their behavior by allowing them to learn from their previous experience, so that they can only accept tasks that they believe would yield them a greater utility. We emphasize a surprising result of the emergent behavior of the agents. Finally, we analyze the way in which agent performance scales when the number of agents and tasks varies.

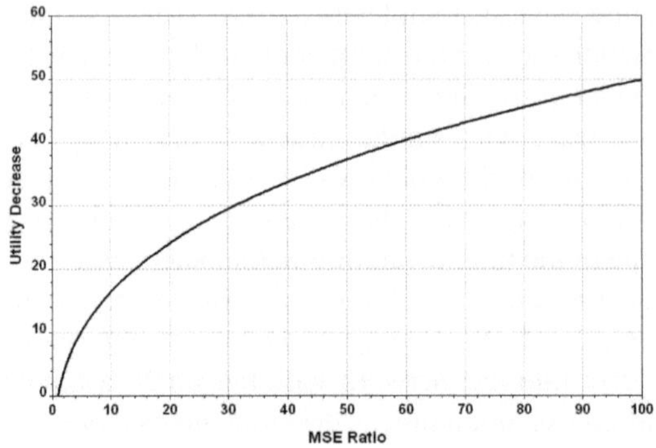

Fig. 2 The decrease in utility as a function of the MSE ratio (R)

4.1 Non-adaptive Agents

In order to analyze and later optimize the utilities received by the agents, we establish a reference result, when all the agents accept any given task, provided that they are available. The total utilities received by the agents are displayed in Fig. 3, as box plots. Box plots display statistics of data by showing the *smallest value*, the *lower quartile* ($Q1$ - 25th percentile), the *average value*, the *upper quartile* ($Q3$ - 75th percentile), and the *largest value*. The utilities received by the three types of agents are the box plots to the left of each section. The three sections refer to the three types of agents: *Backpropagation* (BP), *Quickprop* (QP) and *RProp* (RP). It can be noticed that *Backpropagation* agents perform rather poorly compared to the other agents. *Quickprop* agents have good results and *RProp* agents seem to perform the best.

4.2 Adaptive Backpropagation Agents

Given these results, we aim to improve the performance of *Backpropagation* agents, and analyze how a change in their utility affects the other agents in the system. The strategy that we consider is based on task acceptance. The *Backpropagation* algorithm appears to be slower that the other two. Therefore, it would be logical for *Backpropagation* agents to refuse the most complex, time-consuming tasks, that would take them longer to execute and even then, the error could be worse than that of their competitors and thus no utility would be received. By getting more simple tasks more often, one could hope to improve the total utility. Also, if there are more available agents in the system, there are more chances that they should be *Quickprop* or *RProp* agents, which usually provide better solutions. Thus, if fewer agents are available, there are more chances that a *Backpropagation* agent should be the winner of a task utility. This intuition is later confirmed by the experiments presented in Fig. 5.

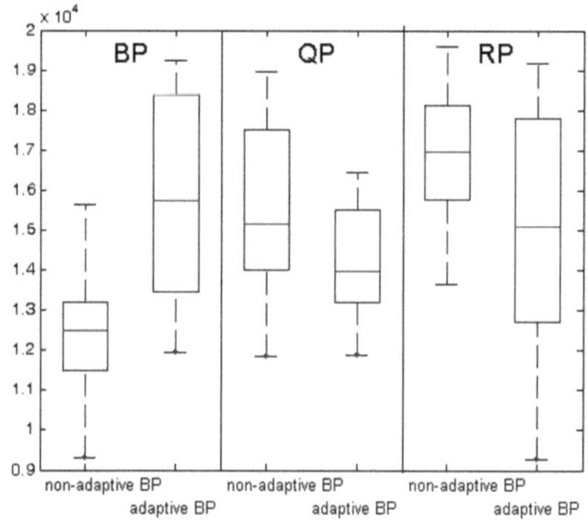

Fig. 3 Box plot summarizing the utilities of agent types when *Backpropagation* agents use ("adaptive BP") or do not use ("non-adaptive BP") a neural network to accept or reject tasks

We thus consider the task information and the final result of accepting it: winning a reward or losing it. This represents another inductive learning problem, built out of the internal experience of the agents and not given by an external user. The attributes taken into account are: the number of attributes, the number of training instances, the number of testing instances, the type of the problem, and the number of competitors (other available agents in the system). The output is given by gaining or losing a reward.

The optimization of the strategy has clear effects on the performance of the agents. The box plots on the right side of each section in Fig. 3 display the resulting behavior of the system when *Backpropagation* agents use the neural network trained based on their execution experience.

It can be clearly seen that the average utility of the *Backpropagation* agents significantly improves compared to that of other agents. This increase in performance also affects the other types of agents in the system: the *Backpropagation* agents receive an average total utility greater than that of *Quickprop* and even *RProp* agents.

4.3 *Scaling*

Finally, we address the way in which the system scales, by analyzing the evolution of the average utility for each type of agents when the number of agents varies and the number of tasks is constant: 5000 (Fig. 4a), and when the number of tasks varies and the number of agents is constant: 30, with 10 agents of each type (Fig. 4b). The figure displays the "adaptive BP" (A) and "non-adaptive BP" (NA) cases for the three types of agents involved.

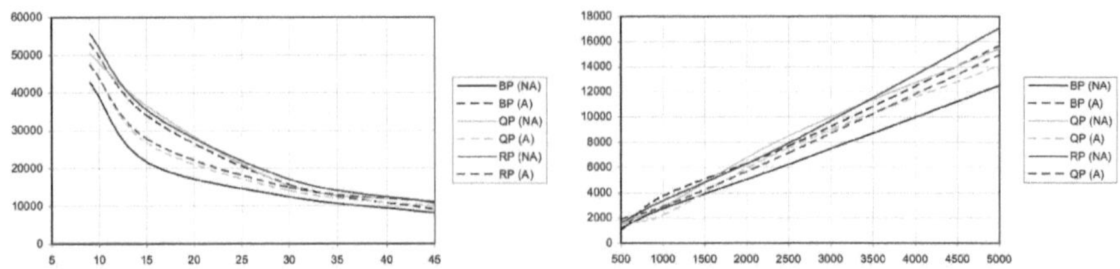

Fig. 4 The evolution of average utilities when: a) the number of agents varies; b) the number of tasks varies

When the number of agents varies, it can be noted that the average utility decreases, because the tasks are divided among more solving entities. When the number of agents grows, the difference between the types of agents becomes less important. When the number of tasks increases, the average utility received by agents increases in an approximately linear manner.

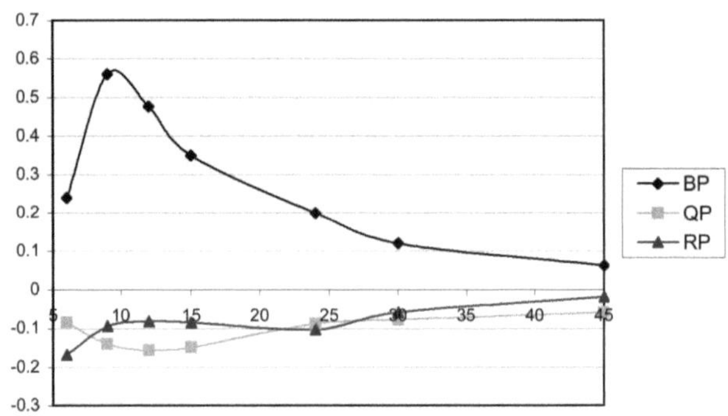

Fig. 5 The percent variation of utilities in adaptive vs. non-adaptive situations

An interesting result appears when studying the percent variation of utility (V) between the utilities in the non-adaptive and the adaptive case, for 1000 tasks, when the number of agents varies: $V = (U_A - U_{NA}) / U_{NA}$.

This situation is reflected in Fig. 5. One can see that there is an optimal number of agents for which the optimization has the greatest effect. For 1000 tasks, this number is around 9 agents (3 of each type). When the number of tasks increases, this optimal number also tends to increase. We can also explain this phenomenon by considering that when the number of agents is small, there is not enough information, or "shared experience", from which *Backpropagation* agents can build an effective learning model. Conversely, when the number of agents increases, the differences between the agent types become less important, as previously suggested by Fig. 4a.

5 Conclusions

By modifying the acceptance strategy of tasks, the overall performance of a group of agents can be improved. With an optimized strategy for task acceptance, the *Backpropagation* agents outperform both *Quickprop* and *RProp* agents. The emergent behavior of the system gives a surprising result, because the performance of the *Backpropagation* algorithm itself remains the same (usually inferior to the other two algorithms). The difference is made by the "intelligence" of accepting only the tasks that seem to fit the reduced capability of *Backpropagation* agents, and this is accomplished by merging the common group experiences, without any other type of direct communication.

Another important feature of the multiagent system is that agents use neural networks to solve problems provided by the user, but also to learn from their own experience and they consequently improve their behavior. Thus, they use the same type of model for both external and internal purposes.

Our analysis shows that for a given total number of tasks there is an optimal number of *Backpropagation* agents where the effect of the optimization is the most significant.

It should be noted that a high classification or regression accuracy *per se* is not the main goal of the paper. Other approaches, such as the Support Vector Machines (SVMs) or boosting techniques (e.g. AdaBoost) have been shown to provide better results than neural networks for many practical problems. The focus of our work is to present a meta-model that can both address user-defined tasks and develop internal strategies for the agents to optimize their behavior in an autonomous way based on their own experience. Any other inductive learning method could be used instead of the neural networks. This feature can prove especially important in real-world, open multiagent systems, where adaptability and simplicity conveyed by the use of a unique, consistent learning model can provide a decisive advantage for the agents involved.

Acknowledgments. This work was supported by CNCSIS-UEFISCSU, project number PNII-IDEI 316/2008, *Behavioural Patterns Library for Intelligent Agents Used in Engineering and Management*.

References

1. Bellifemine, F.L., Caire, G., Greenwood, D.: Developing Multi-Agent Systems with JADE. Wiley Series in Agent Technology (2007)
2. Brown, G.W.: Iterative solutions of games by fictitious play. In: Koopmans, T.C. (ed.) Activitiy Analysis of Production and Allocation. ch. XXIV, pp. 374–376. Wiley, Chichester (1951)
3. Conitzer, V., Sandholm, T.: AWESOME: A general multiagent learning algorithm that converges in self-play and learns a best response against stationary opponents. In: Proceedings of the 20th International Conference on Machine Learning, ICML 2003, Washington, US, pp. 83–90 (2003)

4. Crites, R., Barto, A.G.: Elevator Group Control Using Multiple Reinforcement Learning Agents. Machine Learning 33, 235–262 (1998)
5. Curteanu, S., Leon, F.: Hybrid Neural Network Models Applied to a Free Radical Polymerization Process. In: Polymer-Plastics Technology and Engineering, vol. 45, pp. 1013–1023. Marcel Dekker, USA (2006)
6. Fahlman, S.E.: Faster Learning Variations on BackPropagation: An Empirical Study. In: Proceedings of the 1988 Connectionist Models Summer School, Morgan Kaufmann, Los Altos (1988)
7. Fisher, R.A.: The use of multiple measurements in taxonomic problems. Annual Eugenics, Part II 7, 179–188 (1936)
8. Grothmann R.: Multi-Agent Market Modeling based on Neural Networks. PhD Thesis, Faculty of Economics. University of Bremen, Germany (2002)
9. Herrero, A., Corchado, E., Pellicer, M.A., Abraham, A.: Hybrid Multi Agent-Neural Network Intrusion Detection with Mobile Visualization. Collection of Innovations in Hybrid Intelligent Systems, 320–328 (2008)
10. Hu, J., Wellman, M.P.: Multiagent reinforcement learning: Theoretical framework and an algorithm. In: Proceedings of the 15th International Conference on Machine Learning, ICML 1998, Madison, US, pp. 242–250 (1998)
11. Kudrycki, T. P.: Neural Network Implementation of a Medical Diagnostic Expert System. Master's Thesis, College of Engineering. University of Cincinnati, USA (1988)
12. Leon, F., Leca, A.D., Atanasiu, G.M.: Strategy Management in a Multiagent System Using Neural Networks for Inductive and Experience-based Learning. In: Bratianu, C., Pop, N.A. (eds.) Management & Marketing, Challenges for Knowledge Society, vol. 5(4), pp. 3–28 (2010)
13. Leon, F., Piuleac, C.G., Curteanu, S.: Stacked Neural Network Modeling Applied to the Synthesis of Polyacrylamide Based Multicomponent Hydrogels. In: Macromolecular Reaction Engineering, vol. 4, pp. 591–598. WILEY-VCH Verlag GmbH & Co., Germany (2010)
14. Leon, F., Zaharia, M.H.: Stacked Heterogeneous Neural Networks for Time Series Forecasting. Mathematical Problems in Engineering 2010, Article ID 373648, 20 (2010)
15. Leon, F., Zaharia, M.H., Gâlea, D.: Performance Analysis of Categorization Algorithms. In: Proceedings of the 8th International Symposium on Automatic Control and Computer Science, Iaşi, Romania (2004)
16. Oeda, S., Ichimura, T., Yoshida, K.: Immune Multi Agent Neural Network and Its Application to the Coronary Heart Disease Database. In: Gelbukh, A. (ed.) CICLing 2001. LNCS, vol. 2004, pp. 1097–1105. Springer, Heidelberg (2001)
17. Quinlan, J.R.: Induction of Decision Trees. Machine Learning 1, 81–106 (1986)
18. Red'ko, V.G., Mosalov, O.P., Prokhorov, D.V.: A model of Baldwin effect in populations of self-learning agents. In: Proceedings of the International Joint Conference on Neural Networks, IJCNN 2005, Montreal, Canada, pp. 1355–1360 (2005)
19. Riedmiller, M., Braun, H.: A direct adaptive method for faster backpropagation learning: The RPROP algorithm. In: IEEE Int. Conf. Neur. Net., vol. 1, pp. 586–591 (1993)
20. Rumelhart, D.E., Hinton, G.E., Williams, R.J.: Learning internal representations by error propagation. In: Rumelhart, D.E., McClelland, J.L. (eds.) Parallel Distributed Processing: Explorations in the Microstructure of Cognition, vol. 1, pp. 318–362. The MIT Press, Cambridge (1986)
21. Shepard, R.N., Hovland, C.I., Jenkins, H.M.: Learning and memorization of classifications. Psychological Monographs 75 (1961)

22. Thrun, S. B., et al.: The MONK's Problems - A Performance Comparison of Different Learning Algorithms. Technical Report CS-CMU-91-197, Carnegie Mellon University (1991)
23. Veit, D., Czernohous, C.: Automated Bidding Strategy Adaption using Learning Agents in Many-to-Many e-Markets. In: Proceedings of the Workshop on Agent Mediated Electronic Commerce V, AMEC-V, Melbourne, Australia (2003)
24. Watkins, C.J.C.H.: Learning from Delayed Rewards, PhD Thesis, King's College, Cambridge University (1989)

Optimizing the Semantic Web Service Composition Process Using Cuckoo Search

Viorica Rozina Chifu, Cristina Bianca Pop, Ioan Salomie,
Dumitru Samuel Suia, and Alexandru Nicolae Niculici

Abstract. The behavior of biological individuals which efficiently deal with complex life problems represents an inspiration source in the design of meta-heuristics for solving optimization problems. The Cuckoo Search is such a meta-heuristic inspired by the behavior of cuckoos in search for the appropriate nest where to lay eggs. This paper investigates how the Cuckoo Search meta-heuristic can be adapted and enhanced to solve the problem of selecting the optimal solution in semantic Web service composition. To improve the performance of the cuckoo-inspired algorithm we define a 1-OPT heuristic which expands the search space in a controlled way so as to avoid the stagnation on local optimal solutions. The search space is modeled as an Enhanced Planning Graph, dynamically built for each user request. To identify the optimal solution encoded in the graph we define a fitness function which uses the *QoS* attributes and the semantic quality as selection criteria. The cuckoo-inspired method has been evaluated on a set of scenarios from the trip planning domain.

1 Introduction

Due to the continuous increase of the number of Web services that provide similar functionality, the selection of the optimal solution in semantic service composition is an NP hard problem. Consequently, the design of efficient methods that provide the optimal or a near-optimal solution in a short time and without processing the entire search space has become mandatory. Research studies demonstrated that bio-inspired principles can be successfully applied in solving optimization problems.

This paper investigates how the Cuckoo Search [7] meta-heuristic can be adapted and enhanced to solve the problem of selecting the optimal solution in semantic

Viorica Rozina Chifu · Cristina Bianca Pop · Ioan Salomie · Dumitru Samuel Suia · Alexandru Nicolae Niculici
Department of Computer Science, Technical University of Cluj-Napoca, Romania
e-mail: Viorica.Chifu@cs.utcluj.ro

F.M.T. Brazier et al. (Eds.): Intelligent Distributed Computing V, SCI 382, pp. 93–102.
springerlink.com

Web service composition. To achieve this goal we analyze the biological source of inspiration, and model the biological concepts to fit the problem of selecting the optimal composition solution. Finally, we adapt and enhance the algorithm proposed by the Cuckoo Search meta-heuristic for solving the service selection problem. To improve the performance of the cuckoo-inspired algorithm we define a 1-OPT heuristic which maintains a good balance between exploration and exploitation. By exploration, new areas of the search space are considered while by exploitation, the search space in the neighborhood of an optimal solution is explored. The search space for the cuckoo-inspired method is represented by an Enhanced Planning Graph (EPG) dynamically built for a given user request. To identify the optimal composition solution encoded in the graph we define a fitness function which uses the QoS attributes and the semantic quality as selection criteria. The cuckoo-inspired method has been evaluated on a set of scenarios from the trip planning domain.

The paper is organized as follows. Section 2 presents the EPG. Section 3 presents the cuckoo-inspired selection method, while Section 4 evaluates its performance. Section 5, overviews related work. We end our paper with conclusions.

2 The Enhanced Planning Graph

The Web service composition problem is modeled as an *Enhanced Planning Graph* (EPG) we previously introduced in [3]. The EPG is obtained by mapping the AI planning graph problem [4] to semantic Web service composition and also by enhancing the mapped concepts with the new abstractions of service cluster and parameter cluster. A service cluster groups services which provide the same functionality. The functionalities of the services belonging to the same cluster are annotated only with *is-a* related ontological concepts. A parameter cluster groups input and output service parameters annotated with *is-a* related ontological concepts. The EPG construction is an iterative process which operates at the semantic level. At each step, a new layer consisting of a tuple (A_i, L_i) is added to the graph where A_i represents a set of service clusters and L_i is a set of clusters of service input/output parameters. In the tuple (A_0, L_0) from layer 0, A_0 is an empty set of services and L_0 contains the user provided input parameters. For each layer $i > 0$, A_i is a set of clusters of services for which the input parameters are in L_{i-1}, and L_i is a set of clusters of parameters obtained by adding the outputs of the services in A_i to the set L_{i-1}. The EPG construction ends either when the requested outputs are contained in the current set of parameters or when the sets of service and parameters clusters are the same on two consecutive layers.

We formally define a composition solution encoded in the EPG as follows:

$$sol = \{solElem_1, ..., solElem_n\} \tag{1}$$

where: (*i*) $solElem_i$ is a solution element composed of a set of services, s_{ijk}; s_{ijk} is the k-th service from cluster j on layer i of the EPG (from each cluster only one service is selected as part of the $solElem_i$); (*ii*) n is the number of layers in the EPG.

To evaluate a composition solution, *sol*, we define a fitness function QF as:

$$QF(sol) = \frac{w_{QoS} * QoS(sol) + w_{Sem} * Sem(sol)}{(w_{QoS} + w_{Sem}) * |sol|} \tag{2}$$

where: (*i*) $QoS(sol)$ is the QoS score of *sol* [3]; (*ii*) $Sem(sol)$ is the semantic quality score of *sol* [3]; (*iii*) w_{QoS} and w_{Sem} are the weights corresponding to the relevance of QoS and semantic quality.

3 The Cuckoo-Inspired Selection Method

This section presents the cuckoo-inspired method for identifying the optimal or a near-optimal composition solution encoded in the EPG.

3.1 Problem Formalization

We have mapped the behavior of cuckoos in search for the best nest where to lay their eggs [7] to the problem of selecting the optimal solution in semantic Web service composition as follows: (*i*) a cuckoo becomes an artificial cuckoo, (*ii*) the egg laid by the cuckoo in the host nest becomes a service composition solution encoded in the EPG, (*iii*) the nest with eggs becomes a container for a set of composition solutions, (*iv*) the degree of survivability of a cuckoo egg becomes the quality of a composition solution evaluated with the fitness function QF (see Formula 2), and (*v*) the forest with nests becomes a set of composition solution containers.

We formally define an artificial cuckoo as:

$$cuckoo = (id, sol, gSolRec, container) \tag{3}$$

where: (*i*) *id* is the cuckoo identification number; (*ii*) *sol* is a composition solution associated to the artificial cuckoo; (*iii*) *gSolRec* is the global recommended composition solution, which some of the cuckoos will use to build a new solution; (*iv*) *container* is a randomly chosen container of composition solutions.

A solution container is formally defined as:

$$container = (Sol, cSolRec, score, strength) \tag{4}$$

where: (*i*) *Sol* is the set of composition solutions found at a given time inside a container; (*ii*) *cSolRec* is the container recommended solution established using QF; (*iii*) *score* is the quality of the container recommended composition solution evaluated with QF; (*iv*) *strength* is a container parameter which defines the probability to destroy a container of composition solutions (similar to the destruction of the nest by the host bird). The strength is continuously updated as:

$$strength = strength - \frac{noSc * (s - scoreL)}{1000} \tag{5}$$

where: (*i*) *noSc* is the number of solutions in the container; (*ii*) *scoreL* is the score of the latest solution added to the container evaluated with the fitness function *QF*. The weaker this value is, the more probable the container will be discarded. The strength of a container depends on the number of solutions already in the container and on the *QF* value of the currently added solution. If there are a lot of solutions in the container and a solution with a very low score is added, then the probability to discard the container increases. Initially, the strength is 1.

3.2 The Cuckoo-Inspired Selection Algorithm

The cuckoo-inspired selection algorithm (see Algorithm 1) determines a ranked set *Sol* of composition solutions, where each solution is the best one identified in each iteration. The inputs of the algorithm are: (*i*) the EPG, (*ii*) the weights w_{Sem} and w_{QoS} which state the importance of the associated solution's semantic quality and *QoS*, (*iii*) the number of artificial cuckoos, *noCuckoos*, and (*iv*) the number of containers, *noContainers*, used in the search process. The formula for computing *noCuckoos* is:

$$noCuckoos = Round(\sqrt[n]{noSol}) \tag{6}$$

where: (*i*) *noSol* is the number of composition solutions calculated by multiplying the number of services in each EPG cluster; (*ii*) $n \in N$ is experimentally determined.

Based on experimental results we concluded that for efficiency reasons, in certain situations, *noCuckoos* should increase. However, *noCuckoos* should not grow too much because it would increase the time needed to run a simulation. This constraint is solved by stopping the growth of *noCuckoos* when it reached the root of degree $n-1$ of the search space size. The factor by which *noCuckoos* increases was determined such that *noCuckoos* grows slowly and the high growth variations are eliminated as much as possible (the increase rate was limited to 5%). On the other hand, *noCuckoos* must be kept constant when the algorithm does not stagnate in a local optimum because in such situations there are enough cuckoos to cover wide areas of the search space. The initial number of cuckoos and the number of stagnations should be determined before the beginning of the selection algorithm. To ensure the diversity of the composition solutions, the number of containers (i.e. cuckoo's nests) has to be higher than the number of artificial cuckoos and each artificial cuckoo should be able to place different composition solutions (i.e. eggs) in at least β containers. Therefore, we compute the number of solution containers as:

$$noContainers = \alpha * noCuckoos + \beta \tag{7}$$

where the number of β solution containers are added to ensure the diversity of the composition solutions. The values for α and β should be carefully selected so as not to affect the running time of the algorithm because the more containers are considered, the slower is the algorithm execution time.

Initially, each cuckoo is given a recommended solution depending on its identification number (line 6). In this process, only few cuckoos are given a random recommended solution while the rest of the cuckoos are given recommended solutions

Algorithm 1. Cuckoo_Inspired_Web_Service_Selection

```
1  Input: EPG, w_QoS, w_Sem, noCuckoos, noContainers - the number of containers.
2  Output: Sol
3  Comments: Cuckoos - set of artificial cuckoos, Containers - set of solution containers,
   sol_best - the recommended solution in each algorithm iteration.
4  begin
5     Sol = ∅
6     Cuckoos = Initialize_Cuckoos(noCuckoos)
7     Containers = Initialize_Containers(noContainers)
8     noStag = Initialize_Stagnations()
9     while(!Stopping_Condition()) do
10       Cuckoos = Fly_Cuckoos(Cuckoos,Containers)
11       Containers = Assign_Recommended_Solution(Containers)
12       sol_best = Compute_Best_Solution(Containers)
13       Cuckoos = Assign_Recomended_Solution(sol_best)
14       Containers = Destroy_Containers(Containers)
15       if (Sol_n == Sol_{n-1}) then
16          Do_1_Opt(Cuckoos,Containers)
17          Increase_noCuckoos(Cuckoos,Containers,noStagnations)
18          Containers = Assign_Recommended_Solution(Containers)
19          sol_best = Compute_Best_Solution(Containers)
20       end if
21       Sol = Sol ⋃ sol_best
22    end while
23    return Sol
24 end
```

obtained by performing a greedy search in the EPG. The recommended solutions are established based on the weights w_{QoS} and w_{Sem}. The containers are also given a recommended solution based on the same criteria as in the case of cuckoos (line 7). Actually, the container initialization replicates the natural situation where the host nests initially contain only the eggs belonging to the host bird. After the cuckoo and container initialization, the selection algorithm iteratively searches for the optimal composition solution until a stopping condition is satisfied. In our case, the algorithm stops when the quality of the best solution obtained so far has not improved over the last *noIt* iterations. Within each iteration, the solution associated to each cuckoo is modified and placed in a new container (line 10). After associating a new composition solution to each cuckoo, the best solution in each container is identified and set as the container recommended one (line 11). Next, the best solution of the current iteration (over all containers) is found. This solution is also set as the recommended solution for each cuckoo for the following iterations (line 13). Some of the containers (having low strength) are destroyed as they contain low quality solutions (line 14). In the container destruction process only the best container solution is kept while the others are discarded. If the last two best solutions are identical, then a 1-OPT heuristic (line 16) is applied to the current solutions of the cuckoos and also to the recommended solutions of all the containers. The next step consists of deciding whether to increase the number of cuckoos or not based on the number of stagnations (line 17). Consequently, the best solution in each container is identified and set as the container recommended one (line 18). Finally, the best solution of the current iteration (over all containers) is recomputed (line 19) and added to the set *Sol* (line 21). In the end, the algorithm returns the set *Sol* of the best solutions.

The *Fly_Cuckoos* procedure is of particular interest as it aims to generate new composition solutions. First, the solution of each artificial cuckoo is reset and each cuckoo has to choose a new container where it will place its future solution. The new solution that will be assigned to a cuckoo can be generated by considering the recommended solution of the container, the global best solution, or it can be a brand new one. The optimal number of cuckoos which decide to consider either the recommended or the global best solution was experimentally determined. After a solution is assigned to a cuckoo, the container and its strength value will be updated.

When generating a new solution for an artificial cuckoo, the process of selecting the appropriate service from each of the clusters in every layer plays an important role. This process is performed only by a certain number of cuckoos. The selection can be influenced by the global best solution or by the local best solution of the chosen container. As a result, an artificial cuckoo chooses to add a service to its partial solution with a probability P defined as:

$$P(cServ_i, sol_{part}) = \alpha * \frac{PQ(cServ_i, sol_{part})}{\sum_{k=1}^{n} PQ(cServ_k, sol_{part})} \tag{8}$$

where: (*i*) $cServ_i$ is a candidate Web service which can be added to the partial solution sol_{part}; (*ii*) α is a confidence parameter; (*iii*) n is the number of elements in the set of candidate services $cServ$; (*iv*) PQ evaluates the quality of a partial solution.

The confidence parameter α is used to promote a candidate service $cServ_i \in sol_{rec}$ to become a component of the cuckoos' partial solution:

$$\alpha = \begin{cases} 1, if cServ_i \notin sol_{rec} \\ 2, otherwise \end{cases} \tag{9}$$

By using the confidence parameter it is ensured that the cuckoos' solutions converge to the global optimal one. The function PQ evaluates the quality of a partial solution sol'_{part} by using Formula 2, where sol'_{part} is defined as follows:

$$sol'_{part} = sol_{part} \bigcup cServ_i \tag{10}$$

The candidate services are further sorted according to their probability to be chosen. The service that will be added to the partial solution of the considered cuckoo is selected according to the service probability and a destiny value randomly chosen from the interval [0.5, 1]. The destiny value is introduced to avoid the algorithm stagnancy in a local optimum and to ensure the exploration of more solutions.

4 Experimental Results

We evaluated the cuckoo-inspired selection method on a set of scenarios from the trip planning domain. The services used in our experiments were developed in-house and annotated according to SAWSDL [8]. A user request is specified as ontological concepts semantically describing the provided inputs and requested outputs as well as by weights indicating the relevance of *QoS* compared to the property of semantic quality. Services are semantically described with concepts annotating their functionality, input and output parameters.

In this section we present a scenario from the trip planning domain used in our experiments. Let's assume that the user wants to make the travel arrangements for attending the IDC2011 conference in Delft. First, a flight from Cluj to Delft should be searched for October 3 as well as a flight from Delft to Cluj for October 9. Second, a hotel and a car rental company should be searched in Delft. Third, some companies organizing activities in Delft should be searched. Based on the search results the most convenient flight and hotel should be booked, the most convenient offer for renting a car should be accepted and an appropriate activity should be programmed. Finally, the associated invoices should be issued. Consider that all these subtasks are published by businesses as Web services and our goal is to compose them for scheduling a trip to Delft. Besides the functional requirements, the user also expresses non-functional requirements related to the importance of the *QoS* attributes and semantic quality. We have considered four *QoS* attributes: availability (*Av*), reliability (*Rel*), cost (*Ct*), response time (*Rt*). For the considered scenario, (*i*) the user provided inputs are *SourceCity*, *DestinationCity*, *StartDate*, *EndDate*, *HotelType*, *NumberOfPersons*, *NumberOfRooms*, *CarType* and *ActivityType*; (*ii*) the user requested outputs are *AccommodationInvoice*, *FlightInvoice* and *CarInvoice*; (*iii*) the total *QoS* weight is 0.25 and the individual weights for each *QoS* attribute are Av = 0.15, Rel = 0.25, Ct = 0.45 and Rt = 0.15; (*iv*) the semantic quality weight is 0.75. The EPG built based on the specified user request is organized on 3 layers consisting of 11 clusters that contain in total 51 services.

As a starting point for testing our cuckoo-inspired selection method we generated all the 13996800 possible solutions encoded in the EPG built for the considered user request. For each of these solutions we calculated the quality score (Formula 2) and have identified that the optimal composition solution has the quality score of 0.6947. The score of the optimal composition solution was initially used to adjust the parameters of the cuckoo-inspired algorithm.

The parameters that need to be adjusted for the cuckoo-inspired selection method are the following: *noRec* - the number of artificial cuckoos that will use the recommended solutions, *noDCont* - the number of containers to be destroyed and, *noStag* - the number of stagnations. For establishing the most appropriate values for these parameters we performed approximately 43000 tests with 100 simulations for each combination of parameters having the initial value for $noCuckoos = 61$ and $noContainers = 124$ (for $\alpha = \beta = 2$). The most relevant results are presented in Tables 1 and 2. In Table 1, *FinalnoCuckoos* and *FinalnoCont* represent the values for the number of cuckoos and containers respectively after the algorithm execution completes. Although the first three rows in Table 1 contain parameter configurations for which the algorithm provides good results, the execution time is high. Generally, the algorithm execution time increases when a large number of containers have to be destroyed, due to the overhead induced by searching the destroyable containers.

Table 1 Results obtained for setting the parameters of the cuckoo-inspired selection method

noDCont (%)	noRec (%)	Final noCuckoos	Final noCont	Avg. no. of iterations	Avg. no. of processed solutions	Avg. fitness	Avg. no. of optimal solutions (%)
1	2	166	334	16.32	1519	0.694108	91
1	3	151	304	16.72	1565	0.694036	90
2	2	151	304	16.55	1561	0.693883	86
2	3	126	254	16.81	1617	0.694151	91
3	2	182	366	16.8	1583	0.693884	86
3	4	126	254	16.79	1583	0.69372	86
4	1	182	366	16.85	1597	0.693828	85
4	2	115	232	16.54	1580	0.693552	83
5	1	126	254	16.78	1594	0.694066	88
5	2	151	304	16.65	1546	0.693773	84
6	1	138	278	16.52	1548	0.693884	86
6	3	105	212	16.76	1594	0.693681	84

The number of stagnations, *noStag*, in the same local optimal solution is considered as the stopping criteria. Thus, the initial *noStag* is computed with the formula:

$$noStag = Round(\sqrt[n]{noSol}) \tag{11}$$

The initial value of *noStag* decreases if the number of cuckoos increases above a predefined threshold. We adopted this approach to reduce the search time. In addition, if the number of cuckoos is too big, the chances for stagnating in a local optimum solution are low which means that *noStag* can be lower than the initial value. Table 2 presents some experimental results for identifying the most appropriate number of initial stagnations considered for the selection method. After adjusting the parameters for the cuckoo-inspired method further testing has been performed to assess the method performance according to the criteria: the average percent of explored search space, the average simulation time, and the average number of cases in which the optimal solution has been obtained. By performing 10000 simulations of the cuckoo-inspired method we obtained the following results: the optimal solution has been obtained in an average of 98% of the 10000 simulations by exploring in average 0.013% of the search space in an average time of 7.25 seconds.

Table 2 Results obtained for establishing the initial value for *noStag*

No Stagnations	Final No Cuckoos	Final noCont	noIt	Avg. no. of processed solutions	Avg. fitness	Avg. no. of optimal solutions (%)
6	126	254	14.93	1309	0.69385	87
7	126	254	16.2	1508	0.693908	88
8	138	278	17.74	1766	0.694206	92
9	138	278	18.76	1935	0.694634	98
10	151	304	19.82	2107	0.694618	99
11	166	334	21.34	2383	0.694747	100

5 Related Work

This section reviews bio-inspired methods for selecting the optimal solution in Web service composition. In [5], genetic algorithms are used to find the optimal composition solution. The composition method is based on a given abstract workflow, where each abstract task has a set of concrete services with different *QoS* values associated. Genetic algorithms are used to bind concrete services to the abstract ones aiming at identifying the optimal workflow instance in terms of *QoS* attributes. The genome is encoded as a binary string, where each position is associated to an abstract task and indicates the concrete service which is selected to be used. The approach uses a random mutation operator to generate new workflow instances which are evaluated with a fitness function. Such an approach suffers from a high rate of stagnation in a local optimum. Immune-inspired approaches are an alternative due to their ability of avoiding the stagnation in a local optimum. In [6], an immune-inspired selection algorithm is proposed for Web service composition. The approach uses an abstract composition plan which is mapped to concrete services, resulting in a graph with services in the nodes and *QoS* values on the edges. The immune-inspired algorithm is applied to select the optimal composition solution based on *QoS* attributes. In [2], Particle Swarm Optimization [1] is used to identify the optimal service composition solution. As in [5], the composition approach is based on a predefined workflow associated to concrete services. The particle position is represented by the concrete services mapped on workflow tasks, while particle velocity indicates how the concrete services should be changed. By performing adapted operations of addition, subtraction and multiplication on positions and velocities, the algorithm identifies the optimal composition based on *QoS* attributes.

The differences between our approach and the ones discussed above are: (*i*) our composition method starts from the user request and not from a predefined workflow, (*ii*) our criteria for evaluating the quality of a composition solution includes not only *QoS* attributes but also the property of semantic quality denoting the services degree of match, (*iii*) in our case, the chance of randomly choosing a service s_i to replace a service s_j when improving the current solution is proportional to the quality of the solution obtained by using s_i, (iv) our approach uses a 1-OPT heuristic to explore the search space in a controlled way avoiding local optimum stagnation.

6 Conclusions

This paper proposes a method for selecting the optimal Web service composition which was inspired by the behavior of cuckoos in search for the best nest where to lay their eggs. The method optimizes the selection process without considering the entire search space and avoids the local optimum stagnancy problem. The search space is encoded as an Enhanced Planning Graph dynamically built for each user request. The selection method is evaluated on a set of scenarios from the trip planning domain. First, we performed a set of experiments to adjust the parameters of the method so that it provides the optimal composition in few iterations and without processing the entire search space. Then, using the optimal values established for each adjustable parameter we conducted a set of experiments to analyze the performance of the method according to the following criteria: the average percent of explored search space, the average simulation time, and the average number of cases in which the optimal solution is obtained. As future work, we intend to compare our selection method with other population-based methods on the same test scenarios.

Acknowledgement. This work has been supported by the European Commission within the GAMES project [9] funded under EU FP7.

References

1. Kennedy, J., Eberhart, R.C.: Particle swarm optimization. In: Proc. of the Int. Conference on Neural Networks, USA, pp. 1942–1948 (1995)
2. Ming, C., Zhen-wu, W.: An Approach for Web Services Composition Based on QoS and Discrete Particle Swarm Optimization. In: Proc. of the 8th Int. Conference on Software Engineering, Artificial Intelligence, Networking, and Parallel, Distributed Computing, China, pp. 37–41 (2007)
3. Pop, C.B., Chifu, V.R., Salomie, I., Dinsoreanu, M.: Immune-Inspired Method for Selecting the Optimal Solution in Web Service Composition. In: Lacroix, Z. (ed.) RED 2009. LNCS, vol. 6162, pp. 1–17. Springer, Heidelberg (2010)
4. Russell, S., Norvig, P.: Artificial Intelligence: A Modern Approach. Prentice Hall, Pearson Education, Upper Saddle River, RJ (2003); ISBN: 0137903952
5. Wang, J., Hou, Y.: Optimal Web Service Selection based on Multi-Objective Genetic Algorithm. In: Proc. of the Int. Symposium on Computational Intelligence and Design, China, vol. 1, pp. 553–556 (2008)
6. Xu, J., Reiff-Marganiec, S.: Towards Heuristic Web Services Composition Using Immune Algorithm. In: Proc. of the Int. Conference on Web Services, China, pp. 238–245 (2008)
7. Yang, X.S., Deb, S.: Cuckoo search via Levy flights. In: Proc. of the World Congress on Nature and Biologically Inspired Computing, India, pp. 210–214 (2009)
8. Farrell, J.: Semantic Annotations for WSDL and XML Schema, `http://www.w3.org/2002/ws/sawsdl/spec/`
9. GAMES, `http://www.green-datacenters.eu/`

An Implementation of Evolutionary Computation Operators in OpenCL

István Lőrentz, Răzvan Andonie, and Mihaela Maliţa

Abstract. We discuss the parallel implementation of Genetic Algorithms and Evolution Strategy on General-Purpose Graphical Units, using the OpenCL framework. Multiple evolutionary operators are tested (tournament, roulette wheel selection, uniform and Gaussian mutation, crossover, recombination), as well as different approaches for parallelism, for small and large problem sizes. We use the Island Model of Parallel GA, with random migration. Performance is measured using two graphic cards: NVidia GeForce GTX 560Ti and AMD Radeon 6950. Tests are performed in a distributed grid, using the Java Parallel Processing Framework.

1 Introduction

Evolutionary Algorithms (EA) are a collection of optimization methods inspired from natural evolution [1], [2], [3]. The problem is formulated as finding the minimum value of an evaluation function over a set of parameters defined on a search space. EAs are *meta-heuristic*, as they don't make many assumptions of the function being optimized (for example, they do not require known derivatives). From a meta heuristic point of view, the function to be optimized is a 'black-box', only controlled by the input parameters and the output value. EAs are parallel by their nature. Parallel implementations of optimization algorithms are generally complex problems and this becomes more challenging on fine grained architectures with

István Lőrentz
Electronics and Computers Department, Transylvania University, Braşov, Romania
e-mail: isti_spl@yahoo.com, andonie@cwu.edu

Răzvan Andonie
Computer Science Department, Central Washington University Ellensburg, WA, USA
and
Electronics and Computers Department, Transylvania University, Braşov, Romania
e-mail: andonie@cwu.edu

Mihaela Maliţa
Computer Science Department, Saint Anselm College Manchester, NH, USA
e-mail: mmalita@anselm.edu

F.M.T. Brazier et al. (Eds.): Intelligent Distributed Computing V, SCI 382, pp. 103–113.
springerlink.com

inter-processor communication burdens. We implement two related techniques: Genetic Algorithms (GA) and Evolutionary Strategy (ES). We choose the multi-population (island model) parallel genetic algorithms, from a survey of multiple methods presented in [5].

Essentially, our contribution is the discussion and implementation of several GA and ES models on General-Purpose Graphical Units (GP-GPU), using the OpenCL framework. Using Graphical Processors to parallelize and accelerate genetic algorithms had inspired many authors, for example the site `http://www.gpgpgpu.com/` contains a list of related articles, starting from the year 2005.

OpenCL (Open Computing Language) [10] is an open industry standard parallel computing framework, designed to address a heterogeneous computing environment, across devices containing general-purpose GPUs (Graphics Processing Units) and multi-core CPUs. OpenCL uses a subset of the ISO C99 language, with extensions for parallelism.

An OpenCL platform consists of a *host* computer, which controls one or more OpenCL *devices*. In our scenario, the host computer is an x86-based commodity personal computer, whereas the device is an OpenCL-capable video card.

The application, running on the host computer, controls the execution flow. The host communicates with devices by OpenCL *command queue*s.

While OpenCL supports task-parallelism, the primary model is the data-parallel programming. This model is related to the Stream Processing paradigm, and to a relaxed Single Instruction Multiple Data (SIMD) model.

A device contains one or more *compute units* (CU). A compute unit is composed of one or more *processing elements* (PE), and local memory. Work items are lightweight threads grouped into *work groups*. A work group executes on a single compute unit.

The programmer partitions the data into work-items to be executed in parallel and defines a *kernel function*. The same kernel code will operate on the different data elements. For data partitioning, OpenCL gives the choice of uni-, bi- and three-dimensional blocks called `NDRange`. The OpenCL runtime will launch on the device, for each work-item a thread. Each thread has associated an N (1,2 or 3)-dimensional global index, accessible through the `get_global_id(dimension)` function, inside the kernels.

The OpenCL memory access model is hybrid. Compute Units in a device have shared access to the device memory, thus CUs are programmed using shared-memory parallel techniques, however communication with the host memory or with other devices requires explicit memory block transfers, similar to distributed systems.

We test multiple evolutionary operators: tournament, roulette wheel selection, uniform and Gaussian mutation, crossover, and recombination. We use different approaches for parallelism, for small and large problem sizes. Performance is measured using two graphic cards: NVidia GeForce GTX 560Ti, and AMD Radeon 6950. Tests are performed in a distributed grid, using the Java Parallel Processing Framework (JPPF) [21].

Our present work can be related to [16], where similar models were implemented in CUDA. In addition to [16], we test the code in a distributed framework. We have applied similar parallelization techniques on a different parallel computer architecture in an earlier paper [11].

2 Algorithms and Methods

We first summarize in a unified way, in accordance to the EA general scheme, two standard optimization algorithms: Genetic Algorithms and Evolution Strategy.

In the original introduction of 'Genetic Algorithms' [9], the population of 'chromosomes' is encoded as an array of binary strings. Inspired from biological evolution, every offspring is produced by selecting and combining two parents based on their fitness. The operators of the original Genetic Algorithm (Algorithm 1) are the crossover and single-bit mutation.

Algorithm 1. Genetic Algorithm

Initialize population, as M vectors over the $\{0,1\}$ alphabet, of length N.
repeat
  Create M child vectors, based on:
  1. Select 2 parents, proportionate to their fitness
  2. Crossover the parents, on random positions
  3. Mutate (flip) bits, randomly
  The created M child vectors will form the new population; the old population is discarded.
until The termination criterion is fulfilled (solution found or max number of iterations reached).

Evolution Strategy (ES)[18, 4] is also a population based optimization method, with canonical form written as $(\mu \overset{+}{,} \lambda)$-ES. Here μ denotes the number of parents and λ the number of offsprings created in each iteration. Usually the population is represented by vectors containing the parameters of the functions, encoded as floating point or integer values. The '+' or ',' symbol denotes the variant of the ES algorithm, described below:

- In $(\mu + \lambda)$-ES the μ parents produce λ offsprings and the new generation is selected deterministically from the union of the previous generation and the offsprings. This method guarantees that the best candidates are never replaced (similar to the elite-selection from GA). We use this type of selection in our OpenCL implementation of ES.
- In (μ, λ)-ES, select the best candidates only from the offsprings, similar to the GA.

Algorithm 2. Evolution Strategy algorithm ($\mu \overset{+}{,} \lambda$)-ES

Initialize randomly the population $\mathbf{X}_\mu = \{\mathbf{x}_1, \ldots, \mathbf{x}_\mu\}$. Each individual $\mathbf{x}$ is a a vector of N components, encoding the decision variables (the search space) of the problem.
repeat
  Generate λ offsprings $\tilde{\mathbf{x}}$ forming the offspring population $\{\tilde{\mathbf{x}}_1, \ldots, \tilde{\mathbf{x}}_\lambda\}$ where each offspring $\tilde{\mathbf{x}}$ is generated by:
  1. Select (randomly) parents from $\mathbf{X}$.
  2. Recombine the selected parents to form a recombinant individual $\tilde{\mathbf{x}}$.
  Select new parent population (using deterministic truncation selection) from either:
  - the offspring population $\tilde{\mathbf{X}}$ (usually denoted as (μ, λ)-selection), or
  - the union of the offspring $\tilde{\mathbf{X}}$ and parent $\mathbf{X}$ population (denoted as $(\mu + \lambda)$-selection)
until termination criterion fulfilled (solution found or maximum number of iterations reached).

2.1 Evolutionary Operators in OpenCL

We present the building blocks of an evolutionary algorithm using OpenCL. The control flow of the algorithm is sequential, and the crossover, mutation, evaluation and selection operations are parallelized. We use the parallel prefix-sum (scan) [8] to compute cumulative sums for roulette wheel selection in GA, parallel sum (reduction) [7] for fitness scaling, to compute population fitness average, min, max, and to evaluate the fitness function, and finally the parallel bitonic sort [15] algorithm, to sort population by fitness values. This is required for the fitness-proportionate and rank based GA/ES selection. For tournament based GA selection population, no explicit sorting is performed, resulting a faster kernel.

In our implementation, M denotes population size, G the group (island) size and N the number of variables of the evaluation function. The population is represented as a matrix of M rows and N columns in the global memory of the OpenCL device. Other control parameters are: the algorithm type (GA or ES), selection type (linear fitness proportionate, rank or tournament). recombination type (crossover, arithmetic mean), mutation type (linear or Gaussian) and migration rate.

OpenCL provides a two-level parallel hierarchy. Work items (threads) are combined into work groups. Each work group is submitted for execution on a CU that on some devices might contain a block of low latency local memory (explicit cache). To exploit this feature, we organize our population into islands of sub-populations which fit into the local memory of CUs. Each sub-population is assigned to one OpenCL workgroup, evolving in parallel and independent from other sub-populations.

To minimize the number of transfers between the host and the device, two nested loops are used, as described in Algorithm 3. The outer loop runs on the host machine and invokes the OpenCL kernels (Fig. 1), which in turn execute the inner loop.

Inside the inner loop (Fig. 2), each CU runs independently, accessing only the local memory. We allocate in the local memory a $\mathbf{L}_{2G \times N}$ matrix for the candidate solutions and a $\mathbf{f}_{2G}$ vector for storing the fitness values. The first half of $\mathbf{L}$ stores the

Algorithm 3. All-in-one kernel

```
repeat // outer loop executed on host
    Seed the random number generator (one seed for every work-item)
    Submit OpenCL Kernel for execution: total M work-items in workgroups of size G
    each. A work-item is assigned to one vector in the population.
    for all work-items and work groups do // in parallel on the OpenCL device
        load data from global memory into local memory. The loading address is random-
        ized, providing a random-migration mechanism.
        load random seed from global memory into register
        for all internalIter times do // inner loop
            evolve (according to GA/ES), using local memory
            update local random seed
        end for
        store data from local memory to the global memory corresponding to this popula-
        tion
        store random seed into global memory
    end for
until stopping criterion met
```

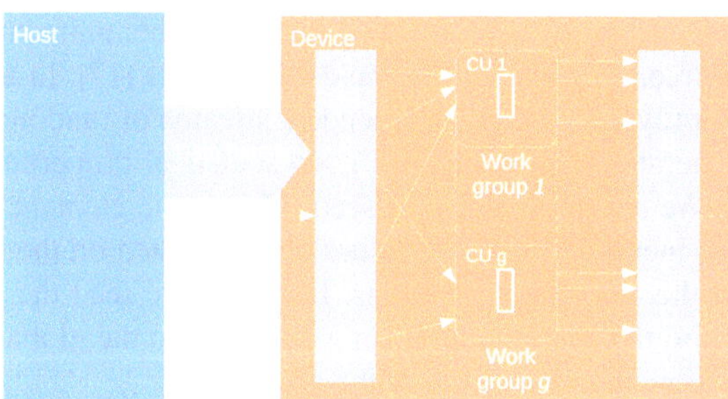

Fig. 1 The outer loop of the algorithm (epochs).

current generation, while the second half stores the intermediate generation (the individuals resulted from the fitness-proportionate selection). Each CU runs *G* threads and each thread is assigned to a corresponding individual in the population. Each thread evaluates the fitness value of the associated individual and no threads cooperate in evaluating a single fitness, whereas the fitness value of an entire group is evaluated in parallel. The inner loop executes Algorithm 1 in case of GA, or Algorithm 2 in case of ES.

Migration between groups is done using the global memory. Before entering the inner loop, the global population is distributed to each of the groups. Inside each group, every thread generates a random address r = [0 ... M-1] and copies the individual from the global population. This addressing scheme ensures that individuals evolved in different groups are mixing together.

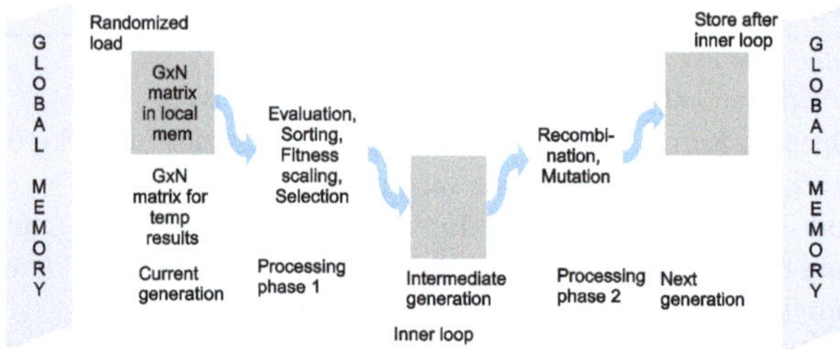

Fig. 2 Processing flow inside a work group (island population). Local memory is divided in two halves to hold intermediate results.

In a second variant we select for migration two candidates randomly from the global memory, and choose the one with best fitness value. We noted a faster convergence on the test problems.

The random load is not an efficient memory access pattern for the graphics devices, however it is executed rarely (compared to the iterations of the inner loop).

An important component of evolutionary algorithms is the pseudo-random number generator. An ideal random number generator should be: uniformly distributed, uncorrelated, cycle-free, satisfy statistical randomness tests [17]. In addition, parallel generators must provide multiple independent streams of random numbers. We use the Xorshift generator [12], which has good statistical properties and requires a few instructions. We assign a vector of seed values, one element for each individual, in the global memory. The seed values are initialized on the host computer at the beginning of the outer-loop iterations. In Linux, we read the special device `/dev/urandom` into the host memory, copy the seeds to the global device memory, and then to local memories of each of the groups.

For problems with large numbers of parameters (where $G \times N$ does not fit into the local memory) we use separate kernels, each specialized to one evolutionary block: one kernel for recombination+mutation, one kernel for evaluation, and another kernel for sorting+selection.

The evaluation of the fitness function is problem dependent and specific optimizations are needed. We implement the Rosenbrock function [19], both as a sequential summing of eq. (1) by each thread, to evaluate its assigned individual from the population. Another choice is to parallelize the evaluation itself, by letting many threads to cooperate in evaluating a single individual. This is possible if the function is decomposable into a data-parallel phase followed by a parallel reduction phase. In our experiments, the first approach gave faster results.

2.2 *Distribution to Multiple OpenCL Devices*

Our initial implementation uses a single OpenCL device. To use multiple devices (either in the same host, or distributed in a network), we need to distribute subpopulations to each device. Again, we use an island model parallel GA, each island corresponding to a computing node containing one OpenCL device.

The host code is written in Java, using the Java Binding for OpenCL (JOCL) API [20]. This gives us flexibility in packaging the code to be easily adaptable in a distributed framework. We use for this purpose the Java Parallel Processing Framework [21], chosen for its simplicity in development and deployment.

We use the master-worker paradigm of the JPPF framework: the master node launches worker tasks on different nodes. Each task runs the evolutionary algorithm over the local population and periodically exchanges some individuals with peer nodes (migration). The "Population" and "Parameters" are serializable Java classes and the JPPF framework takes care of the internals of distributing the data to the worker nodes. After a predefined number of iterations, the master node collects the results from the workers and selects the best solution (Fig. 3).

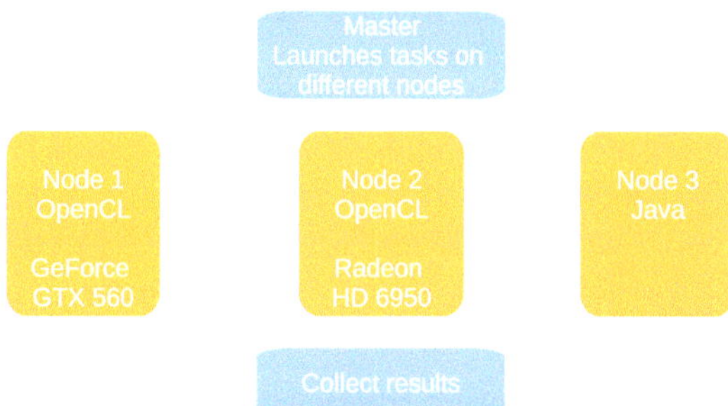

Fig. 3 Flowchart of the distributed algorithm.

The migration method is random-selection based. Each node broadcasts its population to the peers. Each peer makes a random-tournament selection from the merged populations. For each individual, it draws two random candidates and chooses the one with better fitness. This is the same migration method that we used before, between CUs inside a device, but with different migration rates.

3 Results

In our experiments, we use two benchmark problems: the generalized N-dimensional Rosenbrock function (f_1), described in [19, 6], and the Rastrigin function (f_2) described in [13]:

$$f_1(\mathbf{x}) = \sum_{i=1}^{N-1} \left[(1-x_i)^2 + 100(x_{i+1}-x_i^2)^2\right] \quad ; \quad \forall x \in [-2.048, +2.048] \tag{1}$$

$$f_2(\mathbf{x}) = \sum_{i=1}^{N} \left[x_i^2 - A\cos(2\pi x_i)\right] + A \cdot N \quad ; \quad A = 10, \forall x \in [-5.12, +5.12] \tag{2}$$

In all cases, N denotes the number of variables of the problem, G the group (island) population size, and M the global population size. For both functions, we considered:

a small case (N=2), to compare with the results from [16], a medium case (N=8) and a large case (N=256), for testing the outer-loop implementations.

The performance metric is defined as $perf[Mops/sec] = \frac{N \cdot M \cdot I}{T[\mu sec]}$, where I is the number of iterations (generations) and T is the execution time (microseconds).

Two OpenCL-capable graphic cards are used for experiments: the NVidia GeForce GTX 560 Ti and AMD Radeon HD 6950. Both devices fall into the same consumer category and have similar graphics performance and price.

We are not using any vendor-specific build options for compiling the OpenCL sources, neither any device-conditional `#ifdef`-s. The optimization guidelines (using the local memory, avoiding bank conflicts) are taken from the OpenCL Programming Guide for the CUDA Architecture [14]. We develop and profile the code on a GTX560, and test on a HD6950 device as well. The slower measured performance on HD6950 could be caused by the fact that we haven't fully optimized our code and not by the processor's physical performance.

The all-in-one kernel is running on populations that fit entirely in the local-memory block of an OpenCL CU, that is `G*N*sizeof(float)` $\leq$ Local_Mem_Size. Current devices are supporting 32-48 KBytes of local memory per workgroup.

The results for GA with tournament selection, GA with rank-based roulette selection and the $(\mu + \lambda)$ ES are given in Table 1.

Table 1 Results (10^6 operations/sec) for population size M=8192 and island group size G=256

Fitness function	N	Algorithm	Performance GTX 560	Performance HD 6950
Rosenbrock	2	GA Tournament	3230	2200
Rosenbrock	8	GA Tournament	4600	3560
Rosenbrock	2	GA Roulette	370	140
Rosenbrock	8	GA Roulette	1230	210
Rastrigin	2	GA Tournament	2730	2330
Rastrigin	8	GA Tournament	2910	3680
Rastrigin	2	GA Roulette	360	140
Rastrigin	8	GA Roulette	1060	310
Rosenbrock	2	ES $(\mu+\lambda)$	500	270
Rosenbrock	8	ES $(\mu+\lambda)$	1040	470
Rastrigin	2	ES $(\mu+\lambda)$	470	250
Rastrigin	8	ES $(\mu+\lambda)$	971	440

Choosing the appropriate size of the groups (islands) and the number of the groups is critical. We found that a global population of size $M = 8192$ and $G = 256$ gives good results on both devices, so we use these sizes in all tests (Table 2).

The fastest method in our tests is the GA Tournament, because it does not need sorting and computing cumulative fitness sums of the population. Using smaller M or G gives sub-optimal results, due to processor under-utilization. Increasing the sizes

(G=32, ..., 512 and for M=1024, ..., 16384), did not improve performance. However, for future devices, is expected that performance will scale by adding more CUs.

For larger problems, where the all-in-one kernel was not usable, we tested the Rosenbrock and Rastrigin functions with N=256 variables (Table 2). Here, the algorithm's performance might become memory transfer-bounded, since the inner loop requires multiple kernel invocations and accesses to the global memory.

Table 2 Results (10^6 operations/sec) for population size M=8192 and island group size G=256

Fitness function	N	Algorithm	Performance GTX 560	HD 6950
Rosenbrock	256	ES ($\mu+\lambda$)	50	30
Rastrigin	256	ES ($\mu+\lambda$)	48	28

For distributed evolution, we use the layout from Fig. 3, on maximum 6 nodes (2 OpenCL + 4 Java-only). The Java-only nodes run the sequential versions of the algorithms, with the same evolution methods as in the OpenCL nodes. We measure the convergence of the algorithm, in the one node vs. many-node case, in the same time interval (Fig. 3).

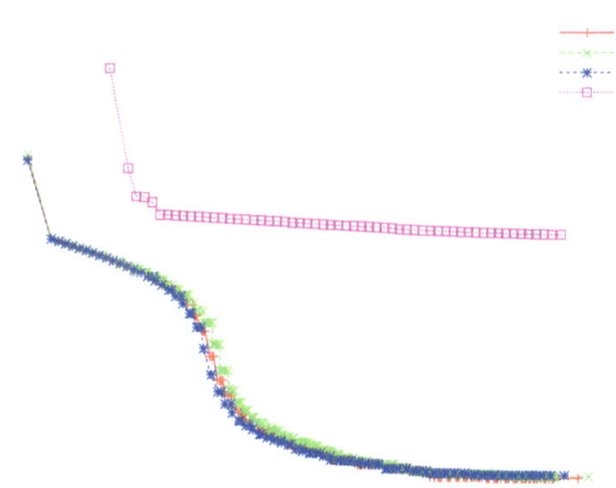

Fig. 4 Results of distributed minimization of the 64-variable Rosenbrock function.

4 Conclusions

Computing the inner loops of the evolutionary algorithms inside the kernel gives the best performance. This is the most suitable strategy for optimizing functions with a relatively small number of parameters that fit into the local memory of the devices.

We notice on graphic cards, the obtained performance is very sensitive to choosing the right workgroup size, number of workgroups, and proper usage of local memory. Small changes in these parameters results in major performance changes.

In the distributed algorithm, we notice overheads at startup and communication. As increasing the number of compute units in a device translates into multiple islands in our implementation, the migration strategy becomes a crucial factor for a good scalability. Further work is to examine the migration strategies, for both the intra-device and distributed islands.

References

1. Bäck, T.: Evolutionary algorithms in theory and practice. Oxford University Press, Oxford (1996)
2. Bäck, T., Fogel, D.B., Michalewicz, Z. (eds.): Handbook of Evolutionary Computation. IOP Publishing Ltd. (1997)
3. Bäck, T., Fogel, D.B., Michalewicz, Z. (eds.): Basic Algorithms and Operators. IOP Publishing Ltd. (1999)
4. Beyer, H.G., Schwefel, H.P.: Evolution strategies A comprehensive introduction. Natural Computing 1, 3–52 (2002)
5. Cantú-Paz, E.: A survey of parallel genetic algorithms. Calculateurs Paralleles 10 (1998)
6. De Jong, K.A.: An analysis of the behavior of a class of genetic adaptive systems. Ph.D. thesis, Ann Arbor, MI, USA (1975)
7. Harris, M.: Optimizing Parallel Reduction in CUDA. NVIDIA Developer Technology (2008)
8. Harris, M., Sengupta, S., Owens, J.D.: Parallel Prefix Sum (Scan) with CUDA. In: Nguyen, H. (ed.) GPU Gems 3, Addison-Wesley, Reading (2007)
9. Holland, J.: Adaptation in natural and artificial systems. University of Michigan Press, Ann Arbor (1975)
10. Khronos OpenCL Working Group: The OpenCL Specification, version 1.1 (2010), `http://www.khronos.org/registry/cl/specs/opencl-1.1.pdf`
11. Lőrentz, I., Malița, M., Andonie, R.: Evolutionary Computation on the Connex Architecture. In: The 22nd Midwest Artificial Intelligence and Cognitive Science Conference, Cincinnati, Ohio, vol. 710, CEUR-WS.org (2011)
12. Marsaglia, G.: Xorshift RNGs. Journal of Statistical Software 8(14), 1–6 (2003)
13. Mühlenbein, H., Schomisch, M., Born, J.: The parallel genetic algorithm as function optimizer. Parallel Comput. 17, 619–632 (1991)
14. NVIDIA: OpenCL Programming Guide for the CUDA Architecture, 3.2 edn. (2010)
15. Pharr, M., Fernando, R.: GPU Gems 2: Programming Techniques for High-Performance Graphics and General-Purpose Computation. chap. 46. Addison-Wesley Professional, Reading (2005)
16. Pospichal, P., Jaros, J., Schwarz, J.: Parallel Genetic Algorithm on the CUDA Architecture. In: Di Chio, C., Cagnoni, S., Cotta, C., Ebner, M., Ekárt, A., Esparcia-Alcazar, A.I., Goh, C.-K., Merelo, J.J., Neri, F., Preuß, M., Togelius, J., Yannakakis, G.N. (eds.) EvoApplicatons 2010. LNCS, vol. 6024, pp. 442–451. Springer, Heidelberg (2010)

17. Quinn, M.J.: Parallel Programming in C with MPI and OpenMP. McGraw-Hill, New York (2003)
18. Rechenberg, I.: Evolutionsstrategie 1994. Frommann-Holzboog (1994)
19. Rosenbrock, H.H.: An Automatic Method for Finding the Greatest or Least Value of a Function. The Computer Journal 3(3), 175–184 (1960)
20. JOCL - Java binding for the OpenCL API, `http://jogamp.org/jocl/www/`
21. JPPF - Java Parallel Processing Framework, `http://www.jppf.org/`

Part IV
Autonomous and Adaptive Distributed Systems

A Hybrid Mapping and Scheduling Algorithm for Distributed Workflow Applications in a Heterogeneous Computing Environment

Mengxia Zhu, Fei Cao, and Jia Mi

Abstract. Computing intensive scientific workflows structured as a directed acyclic graph (DAG) are widely applied to various distributed science and engineering applications to enable efficient knowledge discovery by automated data processing. Effective mapping and scheduling the workflow modules to the underlying distributed computing environment with heterogeneous resources for optimal network performance has remained as a challenge and attracted research efforts with many simulations and real experiments carried out in the grid and cloud infrastructures. Due to the computing intractability of this type of optimization problem, heuristic algorithms are commonly proposed to achieve the minimum end-to-end delay (EED) or other objectives such as maximum reliability and stability. In this paper, a Hybrid mapping algorithm combining Recursive Critical Path search and layer-based Priority techniques (HRCPP) is designed and developed to achieve the minimum EED. Four representative mapping and scheduling algorithms for minimum EED are compared with HRCPP. Our simulation results illustrate that HRCPP consistently achieves the smallest EED with a low algorithm running time observed from many different scales of simulated test cases.

1 Introduction

Many large-scale scientific applications executed on network resource are represented as complex workflow structure defining a set of interdependent tasks. Fast and reliable execution of entire workflow across heterogeneous computer

Mengxia Zhu · Fei Cao · Jia Mi
Computer Science Department, Southern Illinois University, Carbondale, Illinois 62901
e-mail: {mzhu, fcao, jiami}@cs.siu.edu

F.M.T. Brazier et al. (Eds.): Intelligent Distributed Computing V, SCI 382, pp. 117–127.
springerlink.com

environments is desired. An efficient mapping and scheduling algorithm is critical in minimizing the end-to-end delay experienced by the users.

A distributed computing task usually starts with a graph/linear structured workflow to produce the time-varied simulation data and preprocess the raw data to be visualized and analyzed by end users. Users always would like to have an interactive experience with prompt response from remote processing servers. In order to address this problem, the mathematical model and algorithm design need to be investigated for workflows with different dependency structures which could be as simple as a linear pipeline or as complicated as a DAG, which models computing module as vertices and data transfer as directed edges. Then, algorithm can be designed to tackle the various optimization objectives among which the minimum EED is usually the most important criteria.

In this paper, we propose the HRCPP mapping algorithm which integrate the idea of critical path module mapping [2] with priority-based non-critical path modules mapping. Namely, modules from CP always get a high priority value and other modules are assigned appropriate priority values based on its computing and communication loads. A topological sorting is exploited to group modules into layers and mapping are conducted from layer to layer in a top down fashion. Among all possible mapping nodes for a particular module, the one which leads to the smallest execution time from the source module to the current module will be selected. After each round of complete mapping, a new critical path can be calculated and the same mapping process is repeated. We also consider resource sharing for nodes and links to improve the accuracy. When more than one module or one edge is mapped to the same node or network link respectively. A fair-sharing mode is assumed during the period when computing and transfer jobs are concurrently being executed.

Simulation results using different scales of workflows and underlying computer network have demonstrated that our HRCPP consistently achieved the lowest EED with comparative low algorithm running time compared with some popular mapping algorithms such as RCP [2], Greedy A* [3], Streamline [4] and Naive Greedy.

The rest of paper is organized as follows. Related works are presented in Section 2. Analytical models are given in Section 3. The HRCPP algorithm is elaborated in Section 4. Experiments results are shown in Section 5. Finally, conclusion can be found in Section 6.

2 Related Works

There are many research works on the mapping and scheduling of general computing tasks with a DAG structure with minimum EED in distributed systems [5, 1, 6, 7, 8]. The general DAG mapping problem is known to be NP-complete [4, 9] even on two processors without any topology or connectivity restrictions [10]. These approaches can be approximately classified into several categories [11] such as subgraph isomorphism methods [12], the critical path method [13], clustering algorithms [14] etc. The optimization objectives are to minimize either communication costs, processor costs, EED or a combination of these factors on network environments that usually have certain constraint assumptions such as homogeneous processors and computing environment and universal resource access etc.

Kwok et al. proposed a Dynamic Critical-Path (DCP) scheduling algorithm to map task graphs with arbitrary computation and communication costs to a multiprocessor system with unlimited and identical processors in a fully connected graph. Each node's absolute earliest start time (AEST) and absolute latest start time (ALST) were used to determine the nodes along the dynamic critical path [1]. The Recursive Critical Path (RCP) algorithm explored dynamic programming and the recursive critical path selection to iteratively minimize the EED [2]. Streamline, proposed in [4], is a workflow scheduling scheme for streaming data which places a coarse-grained dataflow graph on available grid resources. Sekhar et al. proposed Greedy A* [3] algorithm for similar mapping problems in sensor network area. Most of the algorithms that have been discussed so far are primarily targeted towards static homogeneous environments with fully connected networks, while the proposed work investigates the dynamic mapping and scheduling in heterogenous environments with arbitrary network topology.

3 Analytical Models

We construct analytical cost model for computing modules in task graphs and network nodes and communication links in computer networks to facilitate a mathematical formulation of the application. A computing workflow/ task graph is represented as a directed acyclic weighted graph $G_w = (V_w, E_w)$ with $|V_w| = m$. Vertices represent computing modules: $w_0, w_1, ...w_{m-1}$, starting from w_0 and ending at w_{m-1}.

The network graph is modeled as a directed arbitrary weighted graph $G_c = (V_c, E_c)$ with $|V_c| = n$. Details about the notions used are shown in Table 1. We estimate the computing/execution time of module w_i running on node v_j to be $T_{comp}(w_i, w_j) = \frac{f_{w_i}(z_{w_i})}{p_j}$ and the data transfer time of a message of size z over a network link $l_{i,j}$ to be $T_{tran}(z, l_{i,j}) = \frac{z}{b_{i,j}} + d_{i,j}$.

4 Algorithm Design

Our hybrid algorithm seamlessly integrates the techniques of critical path finding and topological sorting with module priority assignment to achieve boosted network performance with low algorithm complexity. The details of these techniques are given in the following two subsections.

Table 1 Parameter used in the mathmetical models

Parameters	Definitions
$G_w = (V_w, E_w)$	computation workflow
m	number of modules in the workflow
w_i	the i-th computing module
$e_{i,j}$	dependency edge from modle w_i to w_j
$z_{i,j}$	data size transferred over dependency edge $e_{i,j}$
z_{w_i}	aggregated input data size of module w_i
$f_{w_i}(\cdot)$	computational complexity of module w_i
$G_c = (V_c, E_c)$	computer network
n	number of nodes in the network
v_i	the i-th computer node
v_s	source node
v_d	destination node
p_i	computing power of node v_i
$l_{i,j}$	network link between nodes v_i and v_j
$b_{i,j}$	bandwidth of link $l_{i,j}$
$d_{i,j}$	minimum link delay of link $l_{i,j}$

Algorithm 1. HCPT$(G_w, G_c, w_0, w_{m-1}, v_s, v_d)$

```
Create G*_w by assuming resource homogeneity and connectivity completeness in G_w;
Calculate time cost components C_{G*_w} based on G*_w;
call FindCriticalPath(C_{G*_w}, w_0, w_{m-1});
call ModulesMapping(G_w, G_c, w_0, w_{m-1});
calculate Exact EED_0 based on the current mapping;
t = 1;
while |EED_t - EED_{t-1}| >= Threshold do
    calculate new time cost components C^t_{G_w} based on G_w;
    call FindCriticalPath(C^t_{G_w}, w_0, w_{m-1});
    call ModulesMapping(G_w, G_c, w_0, w_{m-1});
    calculate Exact EED_t based on the current mapping;
    t = t + 1;
end while
return EED_t
```

4.1 Recursive Critical Path-Based Mapping Algorithm

The NP-completeness of the DAG workflow scheduling problem rules out any polynomial optimal solutions [15]. Initially, a homogeneous computing environment with identical computing nodes and network links G_m^* is assumed to find the initial critical path. Subsequently, the initial computing and transport time cost for each workflow component $C_{G_m^*}$ can be easily determined. Modules along the critical path have the same earliest start time (EST) as their corresponding latest start time (LST).

Function of *ModulesMapping* is later called to map all modules to the real computer network G_c by removing the assumption on connectivity completeness. Based on the current mapping, the new critical path as well as the initial EED_0 can be calculated. We keep computing the new critical path based on current round of mapping strategy and iteration terminates when the reduction of EED value reaches certain convergency point as shown in Algorithm 1.

The *ModulesMapping* function is responsible for assigning all modules onto appropriate computer nodes after a critical path is identified. The topological sorting is first executed to sort modules into different layers starting from layer 0. Each module will be given a priority number depending on their computing and communication demands. In other words, the higher the load, the higher the priority value for this module. The CP module from a topological layer will be given the highest priority value without considering its workloads. After modules are assigned priority values, we start to resolve the mapping strategy for all modules from each layer in a top down approach, namely layer 0 is considered before layer 1 is considered .

The critical path in left of Fig. 1 shown in shaded modules are given the highest priority in each level. The order to map those modules could be, for instance $w_0, w_2, w_1, ..., w_s, w_t, w_l, w_{m-1}$ as indicated by the shallow arrows. Modules with heavier computing and communication load will be mapped before modules with lighter loads. So they have privileges to utilize better resources. After each mapping round, a new critical path will be computed and the priority assignment and module mapping procedures will be repeated. our algorithm will iteratively improve the mapping strategy until the difference between the current EED and previous EED is below certain threshold. The pseudocode is given in Algorithm 1.

The complexity of this algorithm is $O(k \cdot (\mathrm{m{\cdot}n}+|E_w|) \cdot PL)$, where k is the number of iterations, m and n represents the number of modules in the task graph and nodes in the computer graph respectively, E_w denote the number of links in task graph, and PL represent the longest path length in the computer graph.

4.2 Prioritized Modules Mapping

A topological sort of a directed acyclic graph is a liner ordering of its vertices constraint by the edge dependence. By applying topological sorting to DAG-structured workflow, we can separate modules into different layers with happen-before dependencies. Modules from the same layer can be executed simultaneously. There will be no more than one module from the critical path in the same layer. The module which has more unfinished workload and earlier start time will get a higher priority. The priority for module w_i is computed as: $Priority = UF_{m_i}$ $-PEED_{m_i} +$ $(MinPEED_{m_i}$ $-CurPEED_{m_i})$, where UF_{m_i} represents the unfinished EED from m_i to destination module from previous mapping, $PEED_{m_i}$ denotes the partial EED from source module to m_i, $(MinPEED_{m_i} - CurrPEED_{m_i})$ is the difference between the minimum partial EED from current mapping and the partial EED from previous mapping.

Algorithm 2. ModulesMapping($G_w, G_c, w_0, w_{m-1}, v_s, v_d$)

```
for all w_j ∈ CP do
    set w_j.flag = 1;
end for
calculate computing requirement for each module;
sort all modules in a topological order;
if w_i.flag == 1 then
    assign w_i highest priority;
else
    sort modules in current layer in a decreasing order based on their computing require-
    ments;
end if
MaxLayer = the number of maximum layers in G_w;
l = 0;
while l ≤ MaxLayer do
    for all all modules in layer l do
        case i: w_i is mapped to the same node onto which one of its predecessor module is
        mapped as long as this node also has direct network link with the nodes onto which
        w_i's other predecessor modules are mapped;
        case ii: w_i is mapped to the node which has direct link with all nodes onto which
        module w_i's all predecessor modules are mapped. Since we map modules in a top
        down approach, modules's predecessor modules are already mapped in the previous
        round;
        MNList(w_i)= all possible mapping nodes for w_i;
        for each mapping node v ∈ MNList(w_i) do
            call ResourceSharingStatus(w_0, w_i, G_w, G_c);
            calculate partial EED from source module w_0 to w_i;
            mapping w_i to node v which achieve lowest partial EED;
        end for
        i = i + 1;
    end for
    initialize ResourceSharingStatus();
    l = l + 1;
end while
```

As shown in Algorithm 2, our objective is to choose the best mapping node for module w_i in order to achieve the lowest partial EED from the source module w_0 to w_i. Among all possible mapping choices:

Case I: the new unmapped module w_i is mapped to the same node onto which one of its predecessor module is mapped as long as this node also has direct network link with the nodes onto which w_i's other predecessor modules are mapped.

Case II: the new unmapped module w_i is mapped to the node which happens to have direct link with all nodes onto which module w_i's all predecessor modules are mapped. Since we map modules in a top down approach, modules's predecessor modules are already mapped in the previous round.

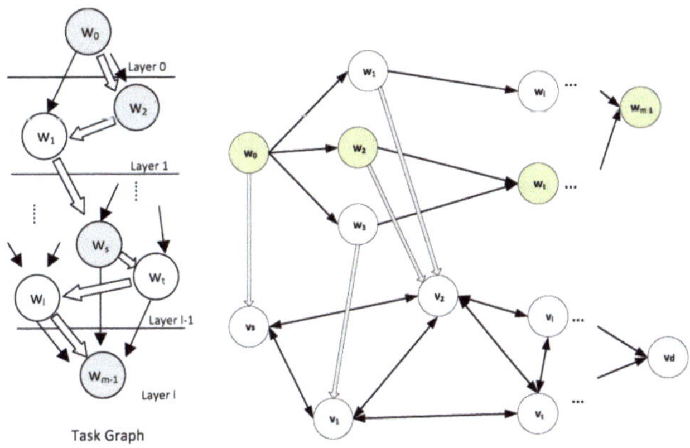

Fig. 1 Left figure: Task Graph Model with shaded modules on the identified critical path. Right figure: An example on how to conduct module layer mapping under all possible cases.

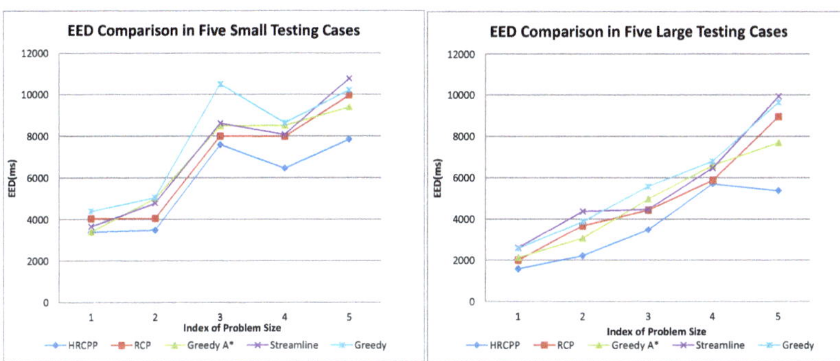

Fig. 2 Comparison EED of HRCPP, RCP, Greedy A*, Streamline, and Greedy.

We use right of Fig. 1 to demonstrate an example for possible mapping choices. The upper figure represents the DAG workflow with shaded modules along the CP. The lower figure shows the network graph. After topological sorting, w_0 belongs to layer 0; w_1, w_2 and w_3 belongs to layer 1. Modules from layer 0 will be mapped to v_0 first, then modules from layer 1 and so on. For example, m_t has its predecessor modules as w_2 and w_3, which are mapped to v_2 and v_1 respectively. For case I, w_t could be mapped either on v_1 or v_2 since v_1 and v_2 are directly connected. For case II, w_t can only be mapped to v_t since only v_t has direct link with both v_1 and v_2.

The mapping strategy that leads to the minimum partial EED is chosen for that module. We also assume that the data movement time between two modules that are mapped to the same node is negligible.

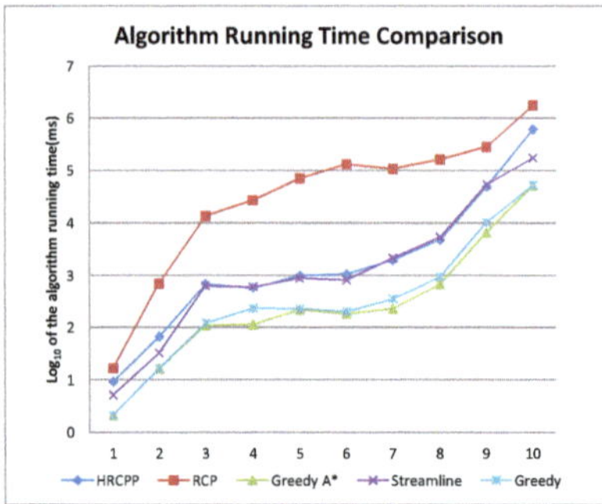

Fig. 3 Comparison of logarithmic algorithm running time.

5 Performance Evaluation

For performance comparison purposes, we implement five DAG mapping algorithms, namely, purposed HRCPP algorithm, RCP, Greedy A*, Streamline and Naive Greedy under the same computing environment. We conduct a set of experiments using some workflow and network graphs generated by randomly varying the following attributes: (i) the number of modules and the complexity of each module, (ii) the number of inter-module communications and the data size between modules, (iii) the number of nodes and the processing power of each node, and (iv) the number of links and the BW and MLD of each link.

The total end-to-end delay of these testing cases are computed and compared in Fig. 2. We also compared the algorithm's running time in Fig. 3 to demonstrate the algorithm efficiency since as the problem size rapidly scales up, only an efficient algorithm can give solutions in a reasonable amount of time.

5.1 Performance Comparison

The set of experiments in left figure of Fig. 2 consists of 5 workflow graphs with 6, 10, 13, 14, and 15 modules and 10, 18, 24, 28, and 30 dependency edges respectively. The corresponding network graphs contain 15, 15, 18, 18, and 28 nodes and 207, 207, 288, 288, and 376 network links respectively. The set of experiments in the right figure of Fig. 2 are comparably bigger in scales and contain 5 workflow graphs with 19, 22, 30, 40, and 45 modules and 36, 44, 62, 75, and 96 dependency edges respectively. The corresponding network graphs have 28, 31, 40, 60, and 63 nodes and 753, 927, 1558, 4205, 3520 network links respectively.

It can be observed that HRCPP achieves the minimum EED for all test cases arranging from small scale to bigger scales (shown in Fig. 2). Such performance improvement becomes more significant as the problem size goes up. The advantages of our approach are not significant when it is applied to small scale problems.

In addition, we also found out that as the size of problem grows, it is harder for the RCP algorithm to come up with a feasible mapping solution with a failed output. It is probably because only the connectivity along the critical path is guaranteed [17]. Greedy A*. and Streamline algorithms are designed for complete graphs without considering network topology restrictions. Some mapping alternatives may be missed when two connected modules are mapped to two nodes without direct link. The same disconnected problem may also occur in Greedy. In comparison, our HRCPP approach considers not only the network topology restrictions but also the importance of differently prioritized modules to search a much wider solution space for better performance.

In Fig. 3, the running time of the five algorithm executing on 10 mapping cases are illustrated. Generally speaking, the running time of RCP is about 10 to 100 folds of other approaches although RCP has the closest performance result to our approach. It will not serve as a feasible approach when the problem size reaches a high level. Among these five approaches, Greedy, Streamline and Greedy A* have a relatively lower running time cost with a relatively poor performance. Our HRCPP maintains a relatively low time cost across different sizes of test cases.

6 Conclusion and Future Work

A hybrid mapping algorithm for DAG structured workflow to underlying heterogeneous distributed computing environments is proposed. Critical path finding and topological sorting are seamlessly integrated in a two-step approach to iteratively reduce the end-to-end execution delay for the entire workflow by adjust mapping strategies from previous round.

Our HRCPP gives modules along the CP the highest priority among modules in the same topological sorting layer, thus powerful resources can be assigned to them in order to achieve fast response time. Then topologically sort is applied to the DAG and then assign each module a priority value based on a combined consideration of the computation and communication loads of this module and current round improvement. Each unmapped module will have multiple mapping choices based on the module dependencies, network connectivity and current mapping scheme. The computer node, which leads to the minimum partial EED from the source module will be chosen for the current module. After all modules are mapped, a new CP can be determined and the previous mapping steps are repeated until the total EED reaches certain convergency point. Our simulation results demonstrate that our hybrid approach consistently achieves the lowest total execution time with comparatively low algorithm running time compared with some existing representative approaches. We plan to investigate new strategies to further reduce algorithm complexity as well as the end-to-end delay from various aspects.

References

1. Kwok, Y., Ahmad, I.: Dynamic critical-path scheduling: An effective technique for allocating task graph to multiprocessors. IEEE Trans. on Parallel and Distributed Systems 7(5) (May 1996)
2. Wu, Q., Gu, Y.: Supporting distributed application workflows in heterogeneous computing environments. In: Proc. of the 14th IEEE Int. Conf. on Parallel and Distributed Systems, Melbourne, Australia, pp. 3–10 (December 2008)
3. Sekhar, A., Manoj, B., Murthy, C.: A state-space search approach for optimizing reliability and cost of execution in distributed sensor networks. In: Proc. of Int. Workshop on Distributed Computing, pp. 63–74 (2005)
4. Agarwalla, B., Ahmed, N., Hilley, D., Ramachandran, U.: Streamline: a scheduling heuristic for streaming application on the grid. In: The 13th Multimedia Computing and Networking Conf., San Jose, CA (2006)
5. Wu, M.Y., Gajski, D.D.: Hypertool: A programming aid for message-passing systems. IEEE Trans. on Parallel and Distributed Systems 1(3), 330–343 (1990)
6. Wang, L., Siegel, H.J., Roychowdhury, V.P., Maciejewski, A.A.: Task matching and scheduling in heterogeneous computing environment using a genetic-algorithm-based approach. Journal Parallel and Distributed Computing 4, 175–187 (1997)
7. Carter, B.R., Watson, D.W., Freund, R.F., Keith, E., Mirabile, F., Siegel, H.J.: Generational scheduling for dynamic task management in heterogeneous computing systems. Information Science 106(3-4), 219–236 (1998)
8. Ma, T., Buyya, R.: Critical-path and priority based algorithms for scheduling workflows with parameter sweep tasks on global grids. In: Proc. of the 17th Int. Symp. on Computer Architecture on High Performance Computing, pp. 251–258 (2005)
9. Kwok, Y.K., Ahmad, I.: Static scheduling algorithms for allocating directed task graphs to multiprocessors. ACM Computing Surveys 31(4), 406–471 (1999)
10. Afrati, F.N., Papadimitriou, C.H., Papageorgiou, G.: Scheduling dags to minimize time and communication. In: Proc. of the 3rd Aegean Workshop on Computing: VLSI Algorithms and Architectures, pp. 134–138. Springer, Heidelberg (1988)
11. Topcuoglu, H., Hariri, S., Wu, M.: Performance effective and lowcomplexity task scheduling for heterogeneous computing. IEEE Trans. on Parallel and Distributed Systems 13(3) (2002)
12. Cordella, L., Foggia, P., Sansone, C., Vento, M.: An improved algorithm for matching large graphs. In: Proc. of the 3rd IAPR-TC-15 Int. Workshop on Graph-based Representations, Italy (2001)
13. Rahman, M., Venugopal, S., Buyya, R.: Ant colony system: A dynamic critical path algorithm for scheduling scientific workflow applications on global grids. In: Proc. of the 3rd IEEE Int. Conf. on e-Science and Grid Computing, pp. 35–42 (2007)
14. Boeres, C., Filho, J., Rebello, V.: A cluster-based strategy for scheduling task on heterogeneous processors. In: in Proc. of 16th Symp. on Computer Architecture and High Performance Computing, pp. 214–221 (2004)
15. Rahman, M., Venugopal, S., Buyya, R.: A dynamic critical path algorithm for scheduling scientific workflow applications on global grids. In: In Proc. of the 3rd IEEE Int. Conf. on e-Science and Grid Computing, pp. 35–42 (2007)

16. Wu, Q., Gu, Y., Lin, Y., Rao, N.S.V.: Latency Modeling and Minimization for Large-scale Scientific Workflows in Distributed Network Environments. In: Proc. of the 44th Annual Simulation Symposium (ANSS 2011), Boston, MA, USA, April 4-7 (2011)
17. Zhu, M., Wu, Q., Rao, N.S.V., Iyengar, S.S.: Adaptive visualization pipeline decomposition and mapping onto computer networks. In: Proc. of the IEEE Internatioal Conference on Image and Graphics, Hong Kong, China, December 18-20, pp. 402–405 (2004)

Towards an Adaptive Supervision of Distributed Systems

Cédric Herpson, Vincent Corruble, and Amal El Fallah Seghrouchni

Abstract. The traditional, centralized, approach to supervision is challenged when communications between supervision and supervised systems become either slow, disrupted or too costly. To obtain a supervision system that is able to dynamically adapt itself to the communications' state, we propose to distribute the supervision process through several autonomous agents. To evaluate our approach, we made experiments on a simulator for distributed systems using three different supervision approaches. Results show that our agent's decision model does lead to a relevant autonomous supervision in distributed systems where a short response time prevails over a limited repair extra-cost.

1 Introduction

The advent of physically distributed systems and the need to minimize the down-time of services and production processes call for more efficient supervision systems. Indeed, as the complexity of systems increase, humans can no longer process the flow of information arriving at each instant. So, the improvement of system effectiveness requires the delegation of more tasks to the supervision system, and more automation of its decisions/actions.

The supervision of distributed systems such as ones found in services, production, larges infrastructures and organizations is so far mainly centralized. As a result of which, even though a number of malfunctions have predefined repairs, the unbounded communication time between the supervision system and the geographically distributed regions of the supervised system delay the repairs. This – sometimes unnecessary – communication time reduces the effectiveness of the supervision and so increases the supervised system's down-time.

Cédric Herpson · Vincent Corruble · Amal El Fallah Seghrouchni
LIP6, Université Pierre et Marie Curie, 75225 Paris, France
e-mail: {FirstName.LastName}@lip6.fr

F.M.T. Brazier et al. (Eds.): Intelligent Distributed Computing V, SCI 382, pp. 129–139.
springerlink.com

Many distributed approaches are proposed in the literature to address this challenge. However, they mainly focus on the phases of monitoring and diagnosis [14, 3, 7] and do not integrate a repair phase. Thus, the triggering of a repair remains centralized, which lead the supervision system to waste the time saved by the distribution of the diagnosis phase. To our knowledge, [12] is the only work that addresses the distribution of both the diagnosis and repair phases. However, it remains at an abstract and descriptive level and furthermore assumes that communication links are reliable.

Our objective is to go beyond the descriptive level and to propose a concrete supervision mechanism that is effective even in the context of unreliable communications. Our proposal relies on a multi-agent architecture where each supervision agent handles both diagnosis and repair on a given location. We identified three key points to get an anytime supervisory architecture : (1) the agent's decision model, (2) the diagnosis inference, and (3) the maintenance of a consistent view of the system state by the multi-agent system. This paper deals with the first point.

In the following, section 2 introduces our fault-and-repair model and section 3 describes in detail the agent's decision model. We then present and discuss the preliminary evaluation of our approach in section 4. Next, we briefly introduce relevant literature in section 5 before concluding with some perspectives.

2 A MultiAgent Architecture for Distributed Systems Supervision

The supervision process is distributed among several autonomous agents having each a local view of the system to be supervised, and endowed with diagnosis and repair capabilities. The supervised system is divided into regions, each one is supervised by one agent. Our architecture comes within the scope of model-free approaches with spatially distributed knowledge [13].

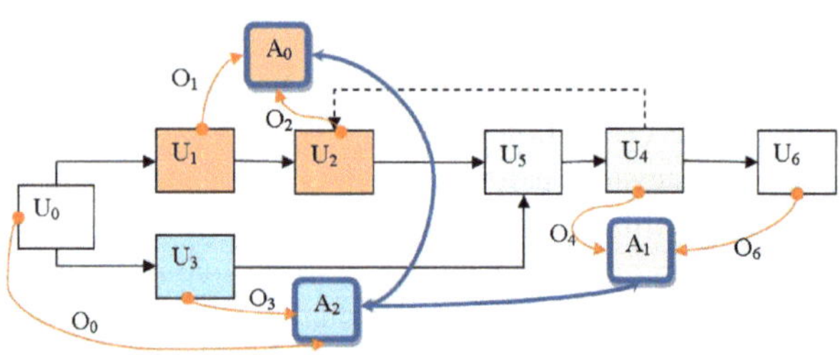

Fig. 1 The supervision agents (A_i) exchange information in order to establish a diagnosis and a repair consistent with the observations (O_j) they get on the various units of the supervised system (U_k).

2.1 Assumptions

We consider that communications are asynchronous and that there is no upper bound on transmission delay. We assume that the messages exchanged between units may be lost or corrupted, and that some units are not supervised (*e.g.* unit U_5 on Figure 1). This last assumption relies on the fact that industrial processes usually involve several parties (e.g subcontractors) that do not share their supervision information. Moreover, we assume in this paper that the observations and the messages between agents are uncorrupted.

2.2 Fault Model and Repair Plan

Let F be the set of known faults of a system S and R be the set of repair plans. The signature of a fault f is a sequence of observable events generated by the occurrence of f. The set of signatures of a given fault f is $Sig(f)$. If a signature lets one identify a fault with certainty, then it is called a *characteristic signature*. Each fault is associated with a t-temporised Petri nets that represents its temporal dynamics (Figure 2). Each fault is supposed to be repairable, it is to say that there exists at least one partially ordered sequence of atomic repairs r_k that repairs it (a repair plan).

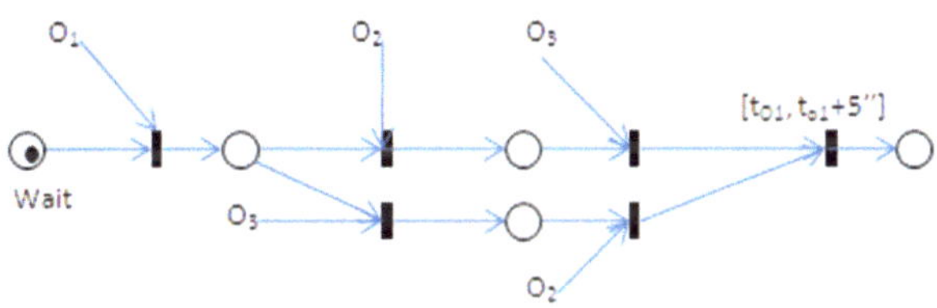

Fig. 2 Example of fault model, $Sig(f) = \{o_1\{o_2,o_3\}^{[t_{o_1},t_{o_1}+5'']}\}$. The o_i are the events observed by the agents. The t_{o_i} indicate the temporal constraints. Thus, $[t_{o_1},t_{o_1}+5'']$ constrains the observations to appear under the 5 time units that follow the occurrence of o_1 for f to be recognized.

2.3 Diagnosis and Multiple Faults

During the time period $[t-\Delta_t,t]$, agent A_i collects a sequence of observations $seqObs_{\Delta_t}(t,A_i)$ generated by the occurrence of faults in the system. However agent A_i does not know which faults have occurred. It thus analyses *seqObs* in order to determine $fp_t(A_i)$, the set of all faults whose signatures partially or totally match elements of *seqObs*. We call a diagnosis (dg), a subset of $fp_t(A_i)$ whose elements are sufficient to explain *seqObs*. Dg is the set of all possible diagnoses of *seqObs*.

Each agent knows the models of the faults that may occur in the region it oversees (A^{rg_i}). Faults may also cover different regions. In that case, agents know the part of the model that concern their region and the other regions concerned. For example:

$$Sig(f) = o_1^{rg_1} o_2^{rg_1} o_3^{rg_2} o_4^{rg_2} \Longrightarrow \begin{cases} Sig_{A^{rg_1}}(f) = o_1 o_2 A^{rg_2} \\ Sig_{A^{rg_2}}(f) = A^{rg_1} o_3 o_4 \end{cases} \quad (1)$$

Note: the global precedence relation between the events that occur within the supervised system is realised by means of a distributed stamping mechanism.

2.4 Fault Cost and Repair Cost

Finally, each fault f (respectively each repair plan $rp(f)$) is associated to a cost of dysfunction $Ct_{dysf}(f,t)$ (respectively a cost of execution $Ct_{Ex}(rp(f))$). The cost of a set of fault (a diagnosis dg_i) for the supervision system is the result of the aggregation of the respective costs of the faults that compose it. Similarly, the execution cost of a repair plan rp associated to a given diagnosis depends on the respective costs of the repairs that compose it.

3 Agent's Decision Model

We are interested in highly dynamic systems. Consequently, information available to an agent at a given time can be insufficient to determine with certainty which action to select. An agent must be able to assess the desirability of immediately triggering the plan made under uncertainty, compared with the option of waiting and communicating with other supervisor agents in order to increase the reliability of its decision. A supervision agent has so to determine the optimal decision (D_{opt}) choosing between an immediate repair action ($Drep_{opt}$) and a repair action considered in k time steps ($Dcom_{opt}$). This waiting time can yield information that reduces uncertainty and thereby improve decision-making. The counterpart is that the elapsed time may have a significant negative impact on the system. The *expected* potential gain in terms of accuracy must be balanced with the risks taken.

Algorithm 1. Online Decision

---

while true **do**

1. Observation gathering
2. Computation of the different sets of faults that can explain the current observations : Dg (set of diagnosis)
3. Sorting of the different diagnoses appearing in Dg based on available information and on the constraints we chose to focus on (Most Probable Explanation-*MPE*, Law of parsimony, Worst case,...)
4. Determination of the immediate repair $Drep_{opt}$ and computation of its estimated cost $Ct(Drep_{opt})$
5. Computation of the waiting cost $Ct_{wait}(k)$, determination of the delayed repair $Dcom_{opt}$ and of the associated cost $Ct(Dcom_{opt})$.
6. Choice between the immediate repair $Drep_{opt}$ and the delayed repair one $Dcom_{opt}$
7. Execution of the decision retained D_{opt}

end while

Let $Ct(x)$ the cost of an action x and $Ct_{wait}(k)$ the cost related to the extra time k before selecting a repair plan. The decision-making process of each supervision agent is described in the following algorithm:

Algorithm 1 determines all the candidate diagnoses and then establishes the various decisions before choosing the one that achieves the best balance between risk and cost. It is executed by each agent when faults occur. We will detail in the following subsections the steps 4 and 5 relative to the determination of the immediate and delayed repair and of their respectives costs.

3.1 Determination of the Immediate Repair $Drep_{opt}$

As presented in introduction, we do not address the diagnosis inference problem in this paper. We therefore assume here that the potential explanations associated to a given sequence of observations are available. The knowledge of the different signatures of faults allows us to establish a list of potential diagnoses Dg. We sort these explanations according to available information and to the constraints we chose to focus on (e.g MPE). The first element of Dg is then the diagnosis considered as relevant. It is then necessary to establish its cost.

The computation of the cost of the immediate repair $Ct(Drep_{opt})$ must take into acount the execution cost Ct_{Ex} of the repair plan associated to the diagnosis retained as well as a cost representative of the uncertainty of the decision, Ct_{Unc}. Indeed, if the only cost considered in the immediate action is Ct_{Ex}, the final decision will always be to act immediately due to the additional waiting cost of the delayed action. In the general case, we have:

$$Ct(rp(dg_i)) = Ct_{Ex}(rp(dg_i)) + Ct_{Unc}(Dg) \tag{2}$$

In the case of Ct_{Unc}, we do not know – for each repair and for all possible execution contexts – the cost of performing a wrong repair (due to misdiagnosis). The consequences of triggering a wrong repair plan are thus approximated by the costs of the repairs remaining to be triggered in case of failure of the first repair. If $|Dg| >= 2$, we have:

$$Ct_{Unc}(Dg) - \sum_{n=2}^{|Dg|} \left[W * (Ct_{ExecTime}(Rep_{|Dg|-(n-1)}) + Ct_{Ex}(Rep_{|Dg|-(n-2)})) \right] \tag{3}$$

with :

$$W = \prod_{i=n}^{|Dg|} (1 - P(dg_{|Dg|-(i-1)} | Dg \backslash \{dg_j | j \leq i-1\}))$$

where Rep_{α} represents the repair plan associated with the α-th diagnosis of the sorted list Dg, and $Ct_{ExecTime}$ represents the consequences of an inefficient repair on the system.

Thus, the cost of the uncertainty of a decision is the sum of the costs of the remaining repair plans plus the waiting cost consequent to the execution of the previous (wrong) repair; the whole weighted by the probability of error of each decision.

3.2 Determination of the Delayed Repair $Dcom_{opt}$

The computation of the delayed repair relies on the capability that each agent have to determine the set of observables (O_{exp}) that can be expected during the next k time steps from the set of all potential faults at time t, $fp_t(A)$ using its knowledge of the faults signatures. From this point an agent is then able to evaluate, for each possible sequence of observables, the potential reduction of the search space induced. The agent then compute for each of these possible futures the optimal immediate decision. The determination of the optimal delayed decision $Dcom_{opt}$ then comes down to choose between the different possible situations by ordering them using the same criteria as for sorting Dg to identify the immediate repair in 3.1.

Note that in a multiple-fault context, the size of the tree of possibilities to explore is $O(2^{O_{exp}})$. Associated with the limited resources of the agents, this complexity does not allow to completely explore the search space and requires adapting the windows' size (k) to the agents computational capabilities.

Once the delayed decision is identified, its cost $Ct(Dcom_{opt})$ is established using Equation (2). We then have to add to this cost the waiting cost Ct_{wait}. This waiting cost depends on the respective costs of the malfunctions associated to the remaining diagnoses and of the elapsed time.

3.3 A Robust Approach

In cases where a significant amount of data (or a prior model) is available, probabilistic models are one possible approach to the forecasting of supervised systems's behaviour [2]. The major drawback of such models is their propensity to retain only the *most probable* hypothesis at the expense of the other *possible* hypotheses to get the optimal decision. While such behaviour may be satisfactory from a macroscopic point of view, the consequences of highly sub-optimal decisions – even if they are not very likely – may be unacceptable. So, it seems essential to consider the utility of a decision according to the cost of the occurrence of a given set of faults rather than its occurrence probability. This reasoning led us to favor a robust criterion [10] for the determination of the cost and decision functions. The decisions taken by the agents will therefore rely on the worst case hypothesis. Thus, in the algorithm 1, the waiting cost Ct_{wait}, the execution cost Ct_{Ex} and the cost of the consequences of an inefficient repair on the system $Ct_{ExecTime}$ are equal to the worst case scenario using the max function. Similarly, $Drep_{opt}$, $Dcom_{opt}$ and D_{opt} will intend to minimize the cost of repair in case of misdiagnosis.

4 Experimental Evaluation

In order to evaluate our approach, we developed a simulator for distributed systems. Based on the JADE multi-agent platform, our environment allows us to model both physical units and communications links, and to simulate the occurrence of failures in it. Our supervision system is deployed on this simulator. When running the simulation, some faults trigger the sending of an alarm message to the agent responsible for the site where they appear. These messages are the observations from which the agents will determine the appropriate behaviour.

In order to obtain generic results, we used twenty-one different experimental environments composed of 7 nodes (see Fig. 1 for an example). One is based on a real infrastructure, and the others are created using a random generator of precedence graphs. With a given number of nodes, the precedence graph generator creates a complete graph with random weights on the edges. It then finds the minimum spanning tree using the Prim's algorithm, randomly chooses a node as the ending one, transforms the graph to a directed one and then adds and connects a master node to existing root nodes in order to obtain a precedence graph.

We fixed *a priori* the locations and responsibilities (the regions) of the supervision agents according to the geographical location of the units that compose the supervised process. The topology variations of the twenty-one environments will therefore have an impact on the agent's messages' transmission delay. The upper bound k that represents is set to 5 units of time. Moreover, we assume in these experiments that the respective costs of the faults that compose a diagnosis are additives.

In order to have benchmarks for the evaluation of the principles underlying our architecture ($RDS2$), we also implemented two additional supervision systems : The first is centralized (*SC*) and the second (*SDI*) considers each unit as an independent system supervised by one agent that only has the information transmitted from the unit. From the set of fault presented on the table 1 below, we established three different scenario. They vary from the case with only local faults to ones with regional and global faults.

- Scenario 1: two faults f_1 and f_2, independent from each other, locally repairable, occur on the system.
- Scenario 2: f_3 which involves all the supervision agents (sites U_0, U_3 and U_1) occurs.
- Scenario 3: three faults, f_1,f_2 and f_4 occur.

Table 1 Faults considered in the experiments.

Fault	Sig	Rp	$Ct_{Ex}(rp)$	Occuring time (ut)	$ExTime(rp)$	$Ct(f)$
f_1	a^{U_0}	r_1	1	10	1	2*t
f_2	b^{U_4}	r_2	1	20	1	2^t
f_3	$a^{U_0}c^{U_3}[2']d^{U_1}[2']$	r_3	2	15	1	t^3
f_4	$a^{U_0}c^{U_3}[2']e^{U_4}$	r_4	4	20	1	$7*ln(t)$
f_5	$a^{U_0}c^{U_3}[2']$	r_5	3		1	3

Each scenario is run 20 times on the twenty-one environments without failure of communications (transmission delay is set to 1 unit of time per edge crossed) and then with a communication cut of 30 unit of time (ut) starting at time t=15ut isolating unit U_4 and the agent A_2 from the rest of the network. We consider in these experiments that the cost of a repair is constant, even if the diagnosis is false. The consequence will appear in the total cost of the repairs performed on a scenario.

The performance evaluation is based on three criteria: (1) The mean time between the first failure of a scenario and the last repair. (2) The number of supervision messages exchanged. (3) The total cost of repairs made.

4.1 Results

The effectiveness of a supervisor varies with the topological structure of the supervised network, the supervisor architecture and the considered fault scenario. Due to lack a space, we only present the aggregated results (Figure 3). If the values changes from one environment to another, average results are here representative of observed behaviours.

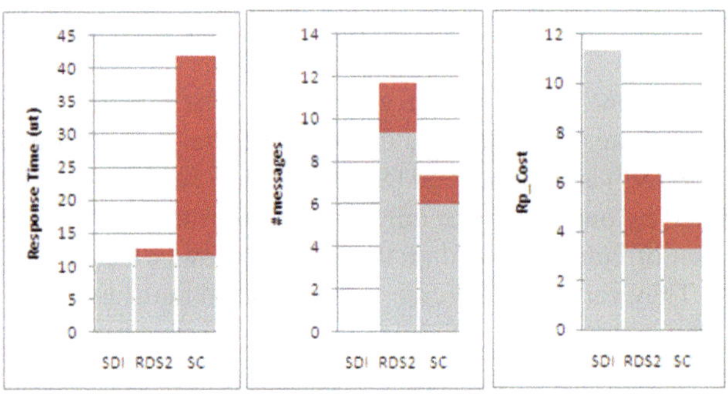

Fig. 3 Each diagram present the aggregated results of the different supervision systems for three scenario over the 21 environments. In gray without communication interruption, in reddish with.

According to the first evaluation criterion, the average response time of the different supervision systems indicate that *SDI* is faster. However, this result is due to the fact that in these experiments, a wrong repair has no impact on the system. The different agents of *SDI* can consequently successfully repair the faults without having enough information by trial-and-error. This behaviour is clearly highlighted by the results of the different supervision systems over the third criterion where the average repair cost of *RDS*2 and *SC* are equal to 3,33 when communication are reliable whereas *SDI* reaches 11,3. When communications are reliable, the fact that *RDS*2' agents need to communicate to establish a diagnosis gives an advantage to *SC* which reaches the same conclusion without extra-cost.

The occurrence of communication failures has no impact on *SDI*. Within *RDS*2, an isolated agent immediately react to the lack of response from other agents in executing the local repair. At the same time, other agents form (if necessary) a coalition to determine candidate diagnoses. As the coalitions do not necessarily have enough information to identify with certainty the faults, the agents may need more exchanges to reach a consensus on a given diagnosis (increasing the number of messages). Moreover, the lack of information sometimes lead the agents to trigger wrong repairs (extra-cost of 3). However, as the decision criterion is robust, the number of wrong decisions stays low and the system is able to quickly return to normalcy. In the case of *SC*, the communication and repair cost over-cost are due to the fact that monitoring information are returned to a central server. As *SC* received a set of events before the breakdown and no more information after, it therefore triggers a repair according to this partial knowledge. The failure of this repair pushes the detection and the triggering of the right repair of a fault until the restoration of communications (extra-time of 30).

Based on these experiments, we can conclude that the proposed decision model lets the agents adapt dynamically and effectively their behaviours to the current state of the supervised system. Considering the reactivity of *RDS*2 and the limited repair extra-cost it generates in case of communication failures, the communication extra-cost can be deemed as an acceptable consequence compared with a total-absence of supervision (*SDI* and *SC*).

5 Related Work

Several approaches for the distributed supervision of distributed systems have been proposed in the literature.

In the literature related to diagnosis, the concept of distribution appears only occasionally. The presentation of different approaches is therefore based on a dichotomy between model-based [11] and non-model-based [1] approaches.
In the work conducted by researchers from areas related to distributed systems, emphasis is usually placed on the distribution of available knowledge. [4] and [13] have addressed the question of the ability of a set of agents to determine an overall diagnosis according to the shape of this distribution (spatial or semantical).

In all these approaches, the correction phase is considered separately and usually left to humans (which requires to centralize the diagnosis result on a given site). This latter point clearly highlights the contradiction between the need of distribution and the reluctance to give to the system the means to realize what it was designed for. In order to benefit from the advantages of distribution, the supervision system must indeed be able to perform diagnosis and repair, its degree of autonomy conditioning its performances. Our proposal is a first attempt in this direction.

The problem of online decision-making under uncertainty is the central point of work by [8], [5] on the control of anytime algorithms. The first distinction between this work and ours is that theirs requires knowing the optimal solution and being able to determine dynamically the distance between the current solution and the optimal one. In our work, talking about the quality of a solution (i.e a diagnosis) is meaningless insofar as a diagnosis is right or wrong, and its "value" is only known a posteriori. The second point of divergence is that we try to select a candidate (a diagnosis or a repair) among a set of potential solutions. The complexity of the task is therefore increased. Facing this problem, [9] recently developed a promising framework to find the best time to determine the winning candidate among a set of potential solutions.

6 Conclusion and Perspectives

In this paper we presented an agent-based approach to guarantee the anytime supervision of distributed systems in a context of unreliable or costly communications. Each agent can autonomously perform both diagnosis and repair, and is able to dynamically adapt its behaviour. Agents are able to find a balance between a quick local diagnosis-and-repair process and a delayed and collaborative one using an adaptive criterion for decision-making. Experiments comparing differents supervision approaches show that our proposal does lead to an efficient adaptive supervision for systems where a short response time is critical and prevails over a limited extra-cost for the repair.

The results obtained in the various environment topologies presented in this paper combined with the evaluation of our approach facing more complex faults scenario using real data [6] will lead us to study in the short-term the behaviour of our agent's decision model as a function of the communications costs. This will require that the agents deal with the problem of the maintenance of a consistent view of the supervised system's state over time, a challenge identified in the introduction as the third key-point of our architecture.

References

1. Betta, G., Pietrosanto, A.: Instrument fault detection and isolation: state of the art and new research trends (1998)
2. Dagum, P., Galper, A., Horvitz, E.: Dynamic network models for forecasting. In: Proceedings of the Eighth Workshop on Uncertainty in Artificial Intelligence, pp. 41–48 (1992)
3. Fabre, E., Benveniste, A., Jard, C.: Distributed diagnosis for large discrete event dynamic systems. In: IRISA (2003)
4. Fröhlich, P., Nejdl, W.: Resolving conflicts in distributed diagnosis. In: ECAI Workshop on Modelling Conflicts in AI (1996)
5. Hansen, E.A., Zilberstein, S.: Monitoring and control of anytime algorithms: A dynamic programming approach. Artificial Intelligence 126, 139–157 (2001)
6. Herpson, C., El Fallah Seghrouchni, A., Corruble, V.: An agent decision model for an adaptive supervision of distributed systems. In: Proc. of EPIA 2011 (2011)
7. Horling, B., Lesser, V., Vincent, R., Bazzan, A., Xuan, P.: Diagnosis as an integral part of multi-agent adaptability. Technical report (2000)
8. Horvitz, E., Rutledge, G.: Time-dependent utility and action under uncertainty. In: Proc. of UAI, pp. 151–158 (1991)
9. Kalech, M., Pfeffer, A.: Decision making with dynamically arriving information. In: Proceddings of AAMAS (2010)
10. Kouvelis, P., Yu, G.: Robust discrete optimization and its applications. Kluwer Academic Pub., Dordrecht (1997)

11. Lafortune, S., Teneketzis, D., Sampath, M., Sengupta, R., Sinnamohideen, K.R.: Failure diagnosis of dynamic systems: an approach based on discrete event systems. In: Proc. American Control Conference, June 25-27, vol. 3, pp. 2058–2071 (2001)
12. Nejdl, W., Werner, M.: Distributed intelligent agents for control, diagnosis and repair. RWTH Aachen, Informatik, Tech. Rep. (1994)
13. Roos, N., Teije, A.T., Bos, A., Witteveen, C.: An analysis of multi-agent diagnosis. In: AAMAS, pp. 986–987 (2002)
14. Wörn, H., Längle, T., Albert, M., Kazi, A., Brighenti, A., Revuelta Seijo, S., Senior, C., Bobi, M.A.S., Collado, J.V.: Diamond: distributed multi-agent architecture for monitoring and diagnosis. Production Planning & Control 15, 189–200 (2004)

Addressing Challenges of Distributed Systems Using Active Components

Lars Braubach and Alexander Pokahr

Abstract. The importance of distributed applications is constantly rising due to technological trends such as the widespread usage of smart phones and the increasing internetworking of all kinds of devices. In addition to classical application scenarios with a rather static structure these trends push forward dynamic settings, in which service providers may continuously vanish and newly appear. In this paper categories of distributed applications are identified and analyzed with respect to their most important development challenges. In order to tackle these problems already on a conceptual level the active component paradigm is proposed, bringing together ideas from agents, services and components using a common conceptual perspective. It is highlighted how active components help addressing the initially posed challenges by presenting an example of an implemented application.

1 Introduction

Technological trends like the widespread usage of smart phones and the increased internetworking of all kinds of devices lead to new application areas for distributed systems and pose new challenges for their design and implementation. These challenges encompass the typical *software engineering* challenges for standard applications and new aspects, summarized in this paper as *distribution*, *concurrency*, and *non-functional properties* (cf. [9]).

Distribution itself requires an underlying communication mechanism based on asynchronous message passing and in this way introduces a new potential error source due to network partitions or breakdowns of nodes. Concurrent computations are an inherent property of distributed systems because each node can potentially act in parallel to all other nodes. In addition, also on one computer true hardware

Lars Braubach · Alexander Pokahr
Distributed Systems and Information Systems Group, University of Hamburg
e-mail: {braubach,pokahr}@informatik.uni-hamburg.de

F.M.T. Brazier et al. (Eds.): Intelligent Distributed Computing V, SCI 382, pp. 141–151.
springerlink.com

concurrency is more and more available by the advent of multi-core processors. This concurrency is needed in order to exploit the available computational resources and build efficient solutions. Non-functional aspects are important for the efficient execution of distributed applications and include aspects like scalability and robustness.

In order to tackle these challenges different *software or programming paradigms* have been proposed for distributed systems. A paradigm represents a specific worldview for software development and thus defines conceptual entities and their interaction means. It supports developers by constraining their design choices to the intended worldview. In this paper *object*, *component*, *service* and *agent orientation* are discussed as they represent successful paradigms for the construction of real world distributed applications. Nonetheless, it is argued that none of these paradigms is able to adequately describe all kinds of distributed systems and inherent conceptual limitations exist. Building on experiences from these established paradigms in this paper the active components approach is presented, which aims to create a unified conceptual model from agent, service and component concepts and helps modeling a greater set of distributed system categories.

The next section presents classes of distributed applications and challenges for developing systems of these classes. Thereafter, the new active components approach is introduced in Section 3. An example application is presented in Section 4. Section 5 discusses related work and Section 6 concludes the paper.

2 Challenges of Distributed Applications

To investigate general advantages and limitations of existing development paradigms for distributed systems, several different classes of distributed applications and their main challenges are discussed in the following. In Fig. 1 theses application

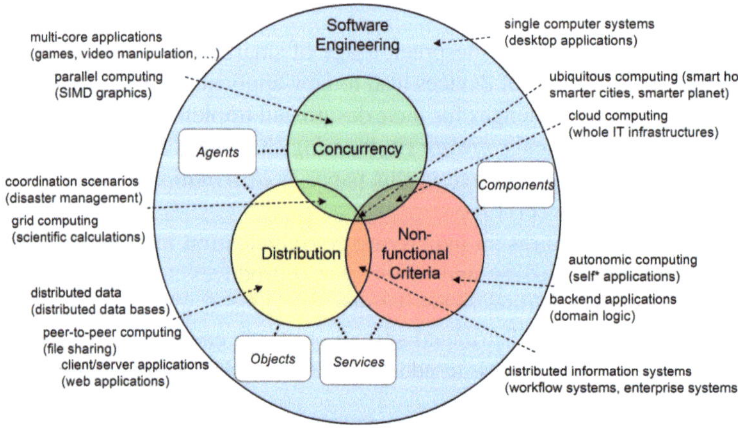

Fig. 1 Applications and paradigms for distributed systems

classes as well as their relationship to the already introduced criteria of software engineering, concurrency, distribution and non-functional aspects are shown. The classes are not meant to be exhaustive, but help illustrating the diversity of scenarios and their characteristics.

Software Engineering: In the past, one primary focus of software development was laid on *single computer systems* in order to deliver typical desktop applications such as office or entertainment programs. Challenges of these applications mainly concern the functional dimension, i.e. how the overall application requirements can be decomposed into software entities in a way that good software engineering principles such as modular design, extensibility, maintainability etc. are preserved.

Concurrency: In case of resource hungry applications with a need for extraordinary computational power, concurrency is a promising solution path that is also pushed forward by hardware advances like multi-core processors and graphic cards with parallel processing capabilities. Corresponding *multi-core* and *parallel computing application* classes include games and video manipulation tools. Challenges of concurrency mainly concern preservation of state consistency, dead- and livelock avoidance as well as prevention of race condition dependent behavior.

Distribution: Different classes of naturally distributed applications exist depending on whether data, users or computation are distributed. Example application classes include *client/server* as well as *peer-to-peer computing* applications. Challenges of distribution are manifold. One central theme always is distribution transparency in order to hide complexities of the underlying dispersed system structure. Other topics are openness for future extensions as well as interoperability that is often hindered by heterogeneous infrastructure components. In addition, today's application scenarios are getting more and more dynamic with a flexible set of interacting components.

Non-functional Criteria: Application classes requiring especially non-functional characteristics are e.g. centralized *backend applications* as well as *autonomic computing* systems. The first category typically has to guarantee secure, robust and scalable business operation, while the latter is concerned with providing self-* properties like self-configuration and self-healing. Non-functional characteristics are particularly demanding challenges, because they are often cross-cutting concerns affecting various components of a system. Hence, they cannot be built into one central place but abilities are needed to configure a system according to non-functional criteria.

Combined Challenges: Today more and more new application classes arise that exhibit increased complexity by concerning more than one fundamental challenge. *Coordination scenarios* like disaster management or *grid computing* applications like scientific calculations are examples for categories related to concurrency and distribution. *Cloud computing* subsumes a category of applications similar to grid computing but fostering a more centralized approach for the user. Additionally, in cloud computing non-functional aspects like service level agreements and accountability play an important role. *Distributed information systems* are an example class

Challenge / Paradigm	Software Engineering	Concurrency	Distribution	Non-functional Criteria
Objects	intuitive abstraction for real-world objects	-	RMI, ORBs	-
Components	reusable building blocks	-	-	external configuration, management infrastructure
Services	entities that realize business activities	-	service registries, dynamic binding	SLAs, standards (e.g. security)
Agents	entities that act based on local objectives	agents as autonomous actors, message-based coordination	agents perceive and react to a changing environment	-

Fig. 2 Contributions of paradigms

containing e.g. workflow management software, concerned with distribution and non-functional aspects. Finally, categories like *ubiquitous computing* are extraordinary difficult to realize due to substantial connections to all three challenges.

Fig. 2 highlights which challenges a paradigm conceptually supports. Object orientation has been conceived for typical desktop applications to mimic real world scenarios using objects (and interfaces) as primary concept and has been supplemented with remote method invocation (RMI) to transfer the programming model to distributed systems. Component orientation extends object oriented ideas by introducing self-contained business entities with clear-cut definitions of what they offer and provide for increased modularity and reusability. Furthermore, component models often allow non-functional aspects being configured from the outside of a component. The service oriented architecture (SOA) attempts an integration of the business and technical perspectives. Here, workflows represent business processes and invoke services for realizing activity behavior. In concert with SOA many web service standards have emerged contributing to the interoperability of such systems. In contrast, agent orientation is a paradigm that proposes agents as main conceptual abstractions for autonomously operating entities with full control about state and execution. Using agents especially intelligent behavior control and coordination involving multiple actors can be tackled.

Yet, none of the introduced paradigms is capable of supporting concurrency, distribution and non-functional aspects at once, leading to difficulties when applications should be realized that stem from intersection categories (cf. Fig. 1). In order to alleviate these problems already on a conceptual level the active component paradigm is proposed in the following.

3 Active Components Paradigm

The active component paradigm brings together agents, services and components in order to build a worldview that is able to naturally map all existing distributed system classes to a unified conceptual representation [8]. Recently, with the service component architecture (SCA) [6] a new software engineering approach has been proposed

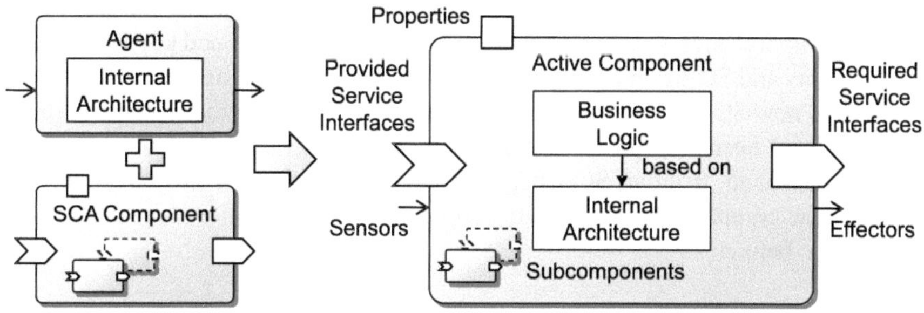

Fig. 3 Active component structure

by several major industry vendors including IBM, Oracle and TIBCO. SCA combines in a natural way the service oriented architecture (SOA) with component orientation by introducing SCA components communicating via services. Active components build on SCA and extend it in the direction of sofware agents. The general idea is to transform passive SCA components into autonomously acting service providers and consumers in order to better reflect real world scenarios which are composed of various active stakeholders. In Fig. 3 an overview of the synthesis of SCA and agents to active components is shown. In the following subsections the implications of this synthesis regarding structure, behavior and composition are explained.

3.1 Active Component Structure

In Fig. 3 (right hand side) the structure of an active component is depicted. It yields from conceptually merging an agent with an SCA component (shown at the left hand side). An agent is considered here as an autonomous entity that is perceiving its environment using sensors and can influence it by its effectors. The behavior of the agent depends on its internal reasoning capabilities ranging from rather simple reflex to intelligent goal-directed decision procedures. The underlying reasoning mechanism of an agent is described as an agent architecture and determines also the way an agent is programmed. On the other side an SCA component is a passive entity that has clearly defined dependencies with its environment. Similar to other component models these dependencies are described using required and provided services, i.e. services that a component needs to consume from other components for its functioning and services that it provides to others. Furthermore, the SCA component model is hierarchical meaning that a component can be composed of an arbitrary number of subcomponents. Connections between subcomponents and a parent component are established by service relationships, i.e. connection their required and provided service ports. Configuration of SCA components is done using so called properties, which allow values being provided at startup of components for predefined component attributes. The synthesis of both conceptual approaches is done by keeping all of the aforementioned key characteristics of agents and SCA components. On the one hand, from an agent-oriented point of view the new SCA properties lead to enhanced software engineering capabilities as hierarchical agent composition and service based interactions become possible. On the other hand, from an SCA perspective internal agent architectures enhance the way how component functionality can be described and allow reactive as well as proactive behavior.

3.2 Behavior

The behavior specification of an active component consists of two parts: service and component functionalities. Services consist of a service interface and a service implementation. The service implementation contains the business logic for realizing the semantics of the service interface specification. In addition, a component may expose further reactive and proactive behavior in terms of its internal behavior definition, e.g. it might want to react to specific messages or pursue some individual goals.

Due to these two kinds of behavior and their possible semantic interferences the service call semantics have to be clearly defined. In contrast to normal SCA components or SOA services, which are purely service providers, agents have an increased degree of autonomy and may want to postpone or completely refuse executing a service call at a specific moment in time, e.g. if other calls of higher priority have arrived or all resources are needed to execute the internal behavior. Thus, active components have to establish a balance between the commonly used service provider model of SCA and SOA and the enhanced agent action model. This is achieved by assuming that in default cases service invocations work as expected and the active component will serve them in the same way as a normal component. If advanced reasoning about service calls is necessary these calls can be intercepted before execution and the active component can trigger some internal architecture dependent deliberation mechanism. For example a belief desire intention (BDI) agent could trigger a specific goal to decide about the service execution.

To allow this kind service call reasoning service processing follows a completely asynchronous invocation scheme based on futures. The service client accesses a method of the provided service interface and synchronously gets back a future representing a placeholder for the asynchronous result. In addition, a service action is created for the call at the receivers side and executed on the service's component as soon as the interpreter selects that action. The result of this computation is subsequently placed in the future and the client is notified that the result is available via a callback.

In the business logic of an agent, i.e. in a service implementation or in its internal behavior, often required services need to be invoked. The execution model assures that operations on required services are appropriately routed to available service providers (i.e. other active components) according to a corresponding binding. The mechanisms for specifying and managing such bindings are part of the active component composition as described next.

3.3 Composition

One advantage of components compared to agents is the software engineering perspective of components with clear-cut interfaces and explicit usage dependencies. In purely message-based agent systems, the supported interactions are usually not visible to the outside and thus have to be documented separately. The active components model supports the declaration of provided and required services and advocates using this well-defined interaction model as it directly offers a descriptive representation of the intended software architecture. Only for complex interactions, such as flexible negotiation protocols, which do not map well to service-based interactions, a more complicated and error-prone message-based interaction needs to be employed.

The composition model of active components thus augments the existing coupling techniques in agent systems (e.g. using a yellow page service or a broker) and can make use of the explicit service definitions. For each required service of a component, the developer needs to answer the question, how to obtain a matching provided service of a possibly different component. This question can be answered at design or deployment time using a hard-wiring of components in corresponding component or deployment descriptors. Yet, many real world scenarios represent open systems, where service providers enter and leave the system dynamically at runtime [4]. Therefore, the active components approach supports besides a static wiring (called *instance* binding) also a *creation* and a *search* binding (cf. [8]). The *search* binding facilities simplified specification and dynamic composition as the system will search at runtime for components that provide a service matching the required service. The creation binding is useful as a fallback to increase system robustness, e.g. when some important service becomes unavailable.

4 Example Application

The usefulness of active components is shown by describing an example system from the disaster management area, implemented using the Jadex framework.[1] The general idea of the *helpline* system, currently implemented as a small prototype, is supporting relatives with information about missing family members affected by a disaster. For information requests the helpline system contacts available helpline providers (hosted by organizations like hospitals or fire departments) and integrate their results for the user. The information is collected by rescue forces using

[1] `http://jadex.sourceforge.net`

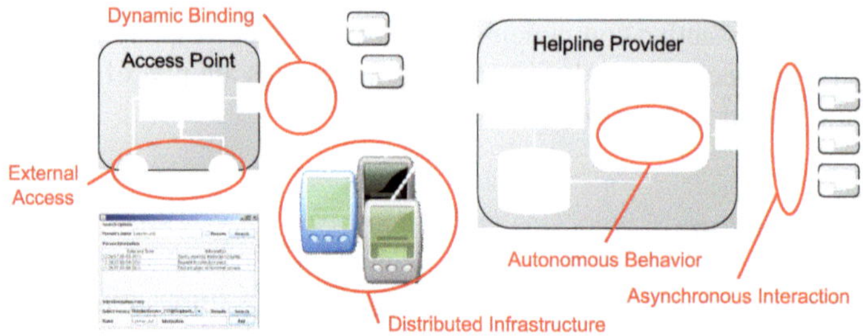

Fig. 4 Helpline system architecture

mobile devices directly at the disaster site. The system can be classified as ubiquitous computing application and exhibits challenges from all areas identified in Section 2.

The main functionality of the system is implemented in the decentralized *helpline provider* components (cf. Fig. 4). Each component offers the *helpline service* allowing to access information as well as adding new information about victims into a database. To improve data quality, helpline providers perform autonomous *information processing*, by querying other helpline providers and integrating information about victims. This behavior is realized using the micro agent internal architecture, which includes a scheduler for triggering agent behavior based on agent state and external events. Simplified versions of the helpline provider components are installed in PDAs used by rescue forces, which synchronize newly added data with backend helpline providers, when a connection is available. To process user requests, *access points* issue information queries to all available helpline providers and present the collected information to the users.

The red markers in Fig. 4 highlight advantages of active components. They provide a natural metaphor for conceptually simplifying system development by supporting *autonomous behavior* and *asynchronous interaction*, which effectively hide details of concurrency and communication and thus reduce the risk of errors related to race conditions or deadlocks. The composition model allows for *dynamic binding* making it especially well suited for open systems in dynamic environments. On a technical level, the *distributed infrastructure* for active components provides a unified runtime environment with awareness features to discover newly available nodes that can also span mobile devices like android phones. Moreover, active components offer *external access* interfaces for easy integration with 3rd-party code, e.g. for integration in a web application or desktop user interface.

5 Related Work

In the literature many approaches can be found that intend combining features from the agent with the component, object or service paradigm. Fig. 5 classifies integration proposals according to the paradigms involved.

Fig. 5 Paradigm integration approaches

In the area of agents and objects especially concurrency and distribution has been subject of research. One example is the active object pattern, which represents an object that conceptually runs on its own thread and provides an asynchronous execution of method invocations by using future return values [11]. It can thus be understood as a higher-level concept for concurrency in OO systems. In addition, also language level extensions for concurrency and distribution have been proposed. One influential proposal much ahead of its time was Eiffel [7], in which as a new concept the virtual processor is introduced for capturing execution control.

Also in the area of agents and components some combination proposals can be found. SoSAA [2] and AgentComponents [5] try to extend agents with component ideas. The SoSAA architecture consists of a base layer with some standard component system and a superordinated agent layer that has control over the base layer, e.g. for performing reconfigurations. In AgentComponents, agents are slightly componentified by wiring them together using slots with predefined communication partners. In addition, also typical component frameworks like Fractal have been extended in the direction of agents e.g. in the ProActive [1] project by incorporating active object ideas.

One active area, is the combination of agents with SOA [10]. On the one hand, conceptual and technical integration approaches of services or workflows with agents have been put forward. Examples are agent-based service invocations from agents using WSIG (cf. JADE[2]) and workflow approaches like WADE (cf. JADE) or JBees [3]. On the other hand, agents are considered useful for realizing flexible and adaptive workflows especially by using dynamic composition techniques based on semantic service descriptions, negotiations and planning techniques.

The discussion of related works shows that the complementary advantages of the different paradigms have led to a number of approaches that aim at combining ideas from different paradigms. Most of the approaches focus on a technical integration that achieves interoperability between implementations of different paradigms (e.g. FIPA agents and W3C web services). In contrast, this paper presented a unified conceptual model that combines the characteristics of services, components and agents.

6 Conclusions and Outlook

In this paper it has been argued that different classes of distributed systems exist that pose challenges with respect to distribution, concurrency, and non-functional properties for software development paradigms. Although, it is always possible to build distributed systems using the existing software paradigms, none of these offers a comprehensive worldview that fits for all these classes. Hence, developers are forced to choose among different options with different trade-offs and cannot follow a common guiding metaphor. From a comparison of existing paradigms the active component approach has been developed as an integrated worldview from component, service and agent orientation. The active component approach has been realized in the Jadex platform, which includes modeling and runtime tools for developing active component applications. The usefulness of active components has been further illustrated by an application from the disaster management domain.

As one important part of future work the enhanced support of non-functional properties for active components will be tackled. In this respect it will be analyzed if SCA concepts like wire properties (transactional, persistent) can be reused for active components. Furthermore, currently a company project in the area of data integration for business intelligence is set up, which will enable an evaluation of active components in a larger real-world setting.

References

1. Baude, F., Caromel, D., Morel, M.: From distributed objects to hierarchical grid components. In: Chung, S., Schmidt, D.C. (eds.) CoopIS 2003, DOA 2003, and ODBASE 2003. LNCS, vol. 2888, pp. 1226–1242. Springer, Heidelberg (2003)
2. Dragone, M., Lillis, D., Collier, R., O'Hare, G.: Sosaa: A framework for integrating components & agents. In: Symp. on Applied Computing. ACM Press, New York (2009)

[2] `http://jade.tilab.com`

3. Ehrler, L., Fleurke, M., Purvis, M., Tony, B., Savarimuthu, R.: AgentBased Workflow Management Systems. Inf. Syst. E-Bus Manage 4(1), 5–23 (2005)
4. Jezek, P., Bures, T., Hnetynka, P.: Supporting real-life applications in hierarchical component systems. In: 7th ACIS Int. Conf. on Software Engineering Research, Management and Applications (SERA 2009), pp. 107–118. Springer, Heidelberg (2009)
5. Krutisch, R., Meier, P., Wirsing, M.: The agentComponent approach, combining agents, and components. In: Schillo, M., Klusch, M., Müller, J., Tianfield, H. (eds.) MATES 2003. LNCS (LNAI), vol. 2831, pp. 1–12. Springer, Heidelberg (2003)
6. Marino, J., Rowley, M.: Understanding SCA (Service Component Architecture), 1st edn. Addison-Wesley Professional, Reading (2009)
7. Meyer, B.: Systematic concurrent object-oriented programming. Commun. ACM 36(9), 56–80 (1993)
8. Pokahr, A., Braubach, L.: Active Components: A Software Paradigm for Distributed Systems. In: Proceedings of the 2011 IEEE/WIC/ACM International Conference on Intelligent Agent Technology (IAT 2011). IEEE Computer Society Press, Los Alamitos (2011)
9. Pokahr, A., Braubach, L., Jander, K.: Unifying Agent and Component Concepts. In: Dix, J., Witteveen, C. (eds.) MATES 2010. LNCS, vol. 6251, pp. 100–112. Springer, Heidelberg (2010)
10. Singh, M., Huhns, M.: Service-Oriented Computing. Semantics, Processes, Agents. Wiley, Chichester (2005)
11. Sutter, H., Larus, J.: Software and the concurrency revolution. ACM Queue 3(7), 54–62 (2005)

An Architectural Model for Building Distributed Adaptation Systems

Mohamed Zouari, Maria-Teresa Segarra, Françoise André, and André Thépaut

Abstract. Dynamic adaptation allows the modification of an application configuration at runtime, according to changes in the environment and/or in users' requirements. The case of adaptive distributed applications has not been substantially addressed. In particular, the distribution of the adaptation system itself has been rarely considered. We address this issue by proposing an architectural model of distributed adaptation systems. Our model allows dynamic adaptation management in a distributed and coordinated manner and expresses variation points of the system. In this paper, we present our model and its use to build distributed adaptation systems. We have applied our results to build an adaptive distributed data replication system used in a medical environment dedicated to remote health care delivery for patients at home.

1 Introduction

Dynamic adaptation [4] is a fundamental requirement of applications running in fluctuating and heterogeneous environments. An adaptation engine generally monitors the execution context in order to trigger dynamic adaptation whenever it detects significant variations. It makes decisions regarding the adaptation and controls the modification of the application in order to achieve the appropriate configuration. When a decentralized application is running in heterogeneous environments, distributed adaptation mechanisms are needed to improve their quality such as efficiency, robustness, and scalability. The distributed management of adaptation leads to the concurrent execution of multiple adaptation processes performed by several

Mohamed Zouari · Françoise André
INRIA/IRISA, Campus de Beaulieu, 35042, Rennes, France
e-mail: {mohamed.zouari,francoise.andre}@irisa.fr

Maria-Teresa Segarra · André Thépaut
Télécom Bretagne, Technopôle Brest-Iroise, 29238, Brest, France
e-mail: {mt.segarra,andre.thepaut}@telecom-bretagne.eu

F.M.T. Brazier et al. (Eds.): Intelligent Distributed Computing V, SCI 382, pp. 153–158.
springerlink.com

engines. Conflicting decisions and inconsistent modifications may occur if the interdependences among the application components are not considered. Therefore, coordinating the activities of multiple adaptation engines is critical in order to preserve the overall consistent state of the application. From this point of view, the development of a distributed adaptation system is challenging.

In this paper, we propose an approach to build a component-based software architecture supporting the distributed and coordinated management of dynamic adaptation. We define an architectural model of adaptation systems that allows the variability of system configuration and provides coordination mechanisms. This model abstracts the adaptation mechanisms in a modular way and allows their specialization so as to facilitate the development process.

The remainder of the paper is organized as follows. Section 2 analyses related work. We describe the general principles underlying our approach for building self-adaptive applications in Section 3. Next, Section 4 presents our architectural model of distributed adaptation systems. Finally, we conclude and discuss future work in Section 5.

2 Related Work

A lot of research has been done in the area of dynamic adaptation [4]. However, we distinguish a limited number of works interested in coordinating activities of multiple adaptation engines [1, 3, 5]. They enable the engines to cooperate for decision-making and/or the control of adaptation actions execution. These works, nevertheless, do not propose reusable and flexible coordination mechanisms. Regarding the decision making, no work proposes mechanisms for distributed and parallel decision making, each based on a local view of the system, in the same adaptation process. To the best of our knowledge, the problem of resolving conflicts among the decisions of several engines has not been addressed in previous works. Concerning the reconfiguration mechanisms, only Cactus [3] proposes a protocol that provides distributed control over modifications. However, it lacks flexibility as the type of modifications (component's implementation), the steps of their achievement, and the participants in the coordination process are predefined. Regarding the facilities for setting up the mechanisms for distributed adaptation management, only [5] provides a solution to customize the adaptation system behaviour. Moreover, no tools are proposed to customize the structure or the distribution of an adaptation system. In order to facilitate the task of the adaptation system designer, it is important to propose adaptation mechanisms that are reusable, easily specializable, and supporting the coordination of multiple adaptation engines' activities when necessary. In addition, tools and methods are required in order to make the specialization tasks easier.

3 Principles for Self-adaptive Applications Building

Our approach aims at facilitating the component-based development of distributed self-adaptive applications. We define a *software component* as a reusable software entity that provides and requires functional interfaces (related to business logic) and may provide a set of control interfaces. The provided interfaces are called *server interfaces* and the required interfaces are the *client interfaces*. The control interfaces allow to manage the non-functional aspects such as suspend/resume the component execution and connect/disconnect components.

We externalize the mechanisms of adaptation control for reusability. Some application components provide control interfaces that define primitive operations to monitor and reconfigure them. The adaptation system is connected to the application trough these interfaces to produce a *self-adaptive application*. We design the adaptation mechanisms in a modular way and we offer two tools, an architectural model and a factory, to facilitate their specialization according to the target application. As shown in Figure 1, an expert provides an architecture description for his system to the factory. The *architecture description* specifies the components that compose the adaptation system, the connections among them, the values of configuration parameters, and the connections with the application components. The *factory* verifies that the constraints of the architectural model are satisfied, deploys the adaptation system and connects it with the application.

The *architectural model* specifies the type of components composing the system, variation points, and constraints. We extend the standard modelling language UML 2.3 to model components and their assembly. This language does not provide the concept of component type. We define a graphical notation and its semantics for modelling such a type (Figure 2) and expressing the variability concerns. The stereotype *type* is used to denote a type of component. We specify below the stereotype the type's name. We represent on the left server interfaces types, on the right the client interfaces types, and at the top the control interfaces types. A multiplicity is indicated for each interface type as well as for the component type. The

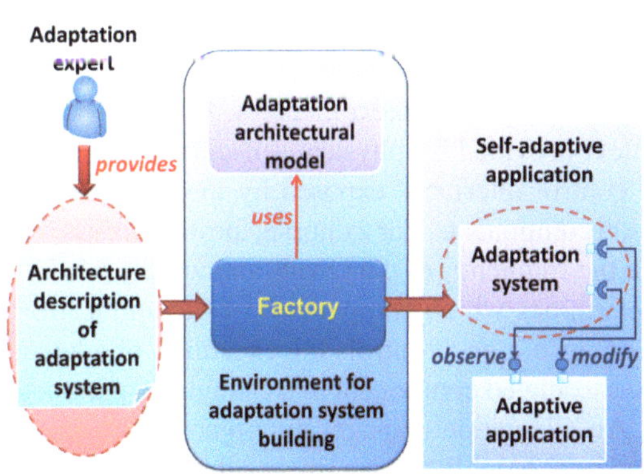

Fig. 1 Building a self-adaptive application

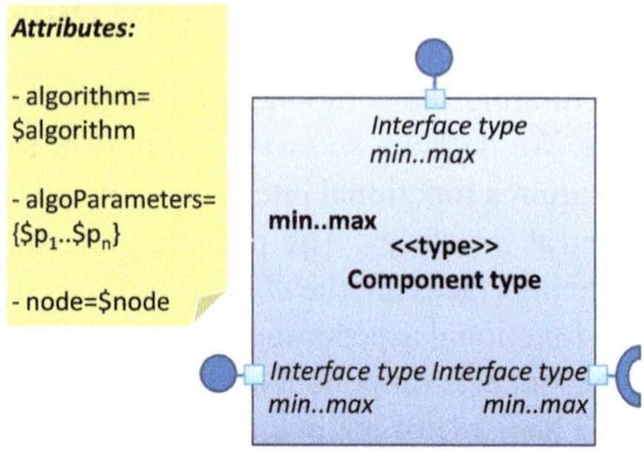

Fig. 2 Graphical notation of a component type

multiplicity of a component type refers to an interval that specifies the allowable number of component instances for a concrete system. The multiplicity of a server interface represents the allowable number of client interfaces that can use it. The multiplicity of client interface indicates the allowable number of the interface instances. Moreover, a list of the component's attributes, which values can be fixed at specialization time, is represented by an annotation. We distinguish three kinds of possible attributes: the algorithm, its configuration parameters, and the node that hosts the component instance. Concerning a *composite component type*, it is defined by the component types that compose it, the possible connections among them, and the type of its external interfaces. Variation points in our model correspond to the number of component and interface instances, the values of the attributes, and the topology to connect components together.

4 Architectural Model of Distributed Adaptation Systems

Our architectural model specifies two composite component types *ContextManager* and *AdaptationManager* to perform the dynamic adaptation control. An adaptation system is composed of several components of each type (the two types have multiplicity 1..*). A component type *ContextManager* collects, interprets, aggregates contextual data and detects the context changes. It collects the information about the execution context provided by physical sensors (e.g., a camera to locate a patient at home) or control interfaces exposed by logical sensors that allow adding probes to application components. For example, the data access requests can be intercepted to calculate the frequency of data reads and writes. A component type *AdaptationManager* determines which components of the application must be adapted and the means to achieve it and controls the execution of adaptation actions. Such actions modify the associated component like changing a parameter value or removing a connection. An adaptation manager may reconfigure an arbitrary number of application components. A publish/subscribe mechanism provides asynchronous interaction between both types of components to trigger adaptation. Moreover, an adaptation manager can query a context manager for specific contextual information in request/response mode. Each adaptation manager may collaborate with one or several context managers depending on their spheres of action. In the following we will give only an overview of the adaptation manager architecture.

Our architectural model defines a component type *AdaptationManager* as a composite that contains three mandatory component types *DecisionMaker*, *Planner*, and *Executor* (Figure 3). An adaptation manager contains a single instance of each type (multiplicity *1*). The component type *DecisionMaker* is responsible for decision making. Such a decision results in an *adaptation strategy* that specifies the appropriate changes to the actual application configuration (e.g., changing the replica placement algorithm). The component type *Planner* identifies a sequence of adaptation actions as a *plan* to apply the strategy chosen by the decision maker. The application of this adaptation plan is controlled by the component type *Executor*.

Our architectural model supports the coordination of adaptation managers' activities. First, the coordination of decision making aims at enabling a group of adaptation managers to make a collective decision in order to ensure non-conflicting and complementary decisions. A component type *DecisionMaker* can assign the task of applying a specific strategy or notify its adopted strategy to an arbitrary number of decision makers using interfaces type *CoordinateItf*. The optional component type *Negotiator* enables a manager to take part in a strategy negotiation process. Second, the plans execution coordination addresses the problem of applying several plans in parallel by a group of adaptation managers when plans contain dependent actions. In such a case, the execution control of these plans must be coordinated in the sense that a specific sequence of their actions execution must be performed to achieve a

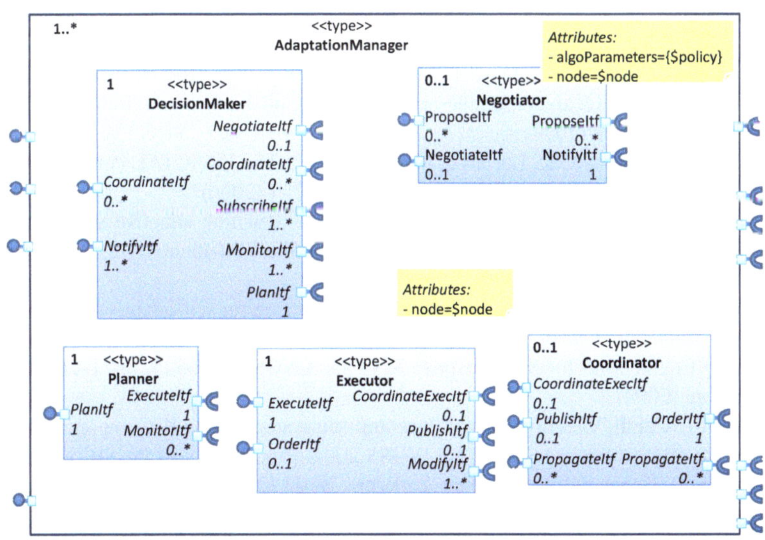

Fig. 3 Architectural model of an adaptation manager

consistent configuration of the application. For this purpose, each component type *Executor* can delegate the control to the optional component type *Coordinator* which interacts with other coordinators. In our previous paper [6], we discussed the steps required to negotiate a strategy and apply adaptation plans in a coordinated manner. Each of these primitive component types has one or two attributes to specialize the behaviour (attributes *policy*) and the distribution (attributes *node*) of each adaptation manager. The value of each attribute *policy* references an external policy describing the adaptation logic.

5 Conclusion

In this paper, we have presented the main concepts related to our approach for building self-adaptive distributed applications. We are particularly interested in the definition and use of an architectural model that specifies the structure and semantics of a software architecture for distributed adaptation management and imposes explicit constraints when building the system. The flexibility of our model allows customizing the behaviour, the structure, and the distribution of adaptation systems using our factory. We have implemented a prototype based on the Fractal component model [2], which is a hierarchical, reflexive and extensible model. We have applied our results to build an adaptive distributed data replication system used in a medical environment dedicated to remote health care delivery for patients at home [6]. As future work, we intend to automate some specialization tasks and make the adaptation system self-adaptive.

References

1. Birman, K.P., Constable, R., Hayden, M., Hickey, J., Kreitz, C., Renesse, R.V., Rodeh, O., Vogels, W.: The horus and ensemble projects: Accomplishments and limitations. Technical report, New York (1999)
2. Bruneton, E., Coupaye, T., Leclercq, M., Quéma, V.: The FRACTAL component model and its support in java. Softw, Pract. Exper., 1257–1284 (2006)
3. Chen, W.K., Hiltunen, M.A., Schlichting, R.D.: Constructing adaptive software in distributed systems. In: ICDCS 2001: Proceedings of the 21st Inter. Conf. on Distributed Computing Systems, Washington, pp. 635–643 (2001)
4. Cheng, B.H., Lemos, R., Giese, H.: Software engineering for self-adaptive systems: A research roadmap. In: Cheng, B.H.C., de Lemos, R., Giese, H., Inverardi, P., Magee, J. (eds.) Software Engineering for Self-Adaptive Systems. LNCS, vol. 5525, pp. 1–26. Springer, Heidelberg (2009)
5. Dowling, J., Cahill, V.: Self-managed decentralised systems using K-Components and collaborative reinforcement learning. In: WOSS 2004: Proceedings of the SIGSOFT Workshop on Self-Managed Systems, New York, pp. 39–43 (2004)
6. Zouari, M., Segarra, M.T., André, F.: A framework for distributed management of dynamic self-adaptation in heterogeneous environments. In: 10th IEEE Inter. Conf. on Computer and Information Technology, Bradford, pp. 265–272 (2010)

Part V
Pervasive and Cloud Computing

TinyMAPS: A Lightweight Java-Based Mobile Agent System for Wireless Sensor Networks

Francesco Aiello, Giancarlo Fortino, Stefano Galzarano, and Antonio Vittorioso

Abstract. In the context of the development of wireless sensor network (WSN) applications, effective programming frameworks and middlewares for rapid and efficient prototyping of resource-constrained applications are highly required. Mobile agents are an effective distributed programming paradigm that is being used for WSN programming. However its diffusion is limited mainly due to the scarce availability of usable mobile agent systems for WSNs. This paper proposes TinyMAPS, a mobile agent system for programming WSNs based on the Sentilla sensor platform. TinyMAPS derives from MAPS (Mobile Agent Platform for Sun SPOT) and is specifically tailored for sensors more constrained than the Sun SPOTs. After providing a description of TinyMAPS and its comparison with MAPS, a simple yet effective case study implemented with TinyMAPS and concerning a real-time WBSN-based system for human activity monitoring is described.

Keywords: Mobile agent platforms, wireless sensor networks, Sentilla JCreates, Java Sun SPOT.

1 Introduction

Wireless Sensor Networks (WSNs) are currently one of the most promising technologies for supporting next generation ubiquitous and pervasive systems. They are suitable for many different high-impact applications such as disaster/crime prevention, emergence management, military applications, environment and infrastructures monitoring, health-care applications, and smart spaces automation. Unfortunately,

Francesco Aiello · Giancarlo Fortino · Stefano Galzarano · Antonio Vittorioso
DEIS, Universitá della Calabria, Via P. Bucci cubo 41c, 87036 Rende (CS), Italy
e-mail: {faiello,avittorioso}@si.deis.unical.it,
g.fortino@unical.it, sgalzarano@deis.unical.it

F.M.T. Brazier et al. (Eds.): Intelligent Distributed Computing V, SCI 382, pp. 161–170.
springerlink.com

developing WSN applications is a complex task due to the constrained resources of the sensor platforms and the lack of general-purpose efficient middleware. Moreover, WSN programming is generally application-specific so that the developed software is tightly dependent on the specific application and scarcely reusable. Thus, to support rapid development and deployment of applications, effective and flexible WSN programming paradigms are needed. Among the paradigms proposed so far [1], the mobile agent paradigm can effectively cope with the programming issues that WSNs have raised. In particular, in [2] authors argue several good reasons for exploiting the benefits of mobile agents in highly dynamic distributed environments, like WSNs. Important network issues like load reduction, latency overcoming, robustness and fault-tolerance can be achieved thanks to the distinctive properties of mobile agents: mobility, dynamic adaptation to environment changes and asynchronous and autonomous execution. Moreover, they fulfill the need for heterogeneity by acting as wrappers for allowing the interaction among systems based on different hardware and software, or by encapsulating properly communication protocols for supporting different networks without any regard for standardization matter.

Mobile agent systems (MASs) are environments for supporting the development and the execution of mobile agent applications. A great variety of agent systems have been to date developed atop conventional distributed systems but only few platforms have been to date proposed and actually implemented for WSN because of the complexity in developing them on very resource-constrained sensor nodes. Agilla [3] is an agent-based middleware developed on TinyOS [4] and based on tuplespace and neighbors list. ActorNet [5], which is specifically designed for TinyOS sensor nodes too, exposes services like virtual memory, context switching and multitasking for providing a uniform computing environment based on the actor model of computation. AFME [6] is an open-source J2ME-CLDC compliant agent platform, whose agents are strongly based on the Belief-Desire-Intention (BDI) paradigm. MAPS [7, 8], which has been conceived for Sun SPOT [9], offers a set of services to mobile agents including message transmission, agent creation, agent cloning, agent migration, timer handling, and easy access to the sensor node resources, whereas the dynamic behavior of agents is modeled as an ECA-based automaton. A comparison among the aforementioned agent systems can be found in [10].

In this paper, a new Java-based mobile agent system, TinyMAPS, is proposed. It is specifically developed for Sentilla JCreate sensor platform [11] and derived from MAPS. The rest of this paper is organized as follows. In Sect. 2, TinyMAPS is described and then compared to MAPS with respect to their architecture, programming model, performance and limitations. Sect. 3 presents a real-time system for human activity monitoring based on TinyMAPS at the sensor-side. Finally, conclusions are drawn and on-going research efforts delineated.

2 TinyMAPS Mobile Agent System

TinyMAPS is a Java-based mobile agent system specifically developed atop the Java-based Sentilla technology [11] and having an architecture and an agent programming methodology directly derived from MAPS. Indeed, TinyMAPS can be considered as a porting of MAPS onto the Sentilla JCreate sensor platform. However, Sentilla sensors are much more resource-constrained than Sun SPOT sensors (as shown in Table 1) so both the mobile system architecture and the agent architecture of TinyMAPS have been purposely tailored to be actually implemented.

Table 1 Comparison between Sentilla JCreate and Sun SPOT

	Sun SPOT	Sentilla JCreate
CPU	ARM920T 180MHz 32bit (floating point)	TI MSP430 8MHz 16bit (NO floating point)
RAM	512KB	10KB
EPROM	4MB	48KB
Virtual Machine	Squawk JVM	custom JVM CLDC compliant
Onboard Sensors	3-axes accelerometer, light and temperature	3-axes accelerometer
Input/Output	6 analogical input/USB port	2 ADC input/8 pinouts

The TinyMAPS architecture (see Fig. 1), similarly to MAPS, is based on components interacting through events but offering a more limited set of core-services (agent creation, migration, communication and sensor resource access) to mobile agents. Agent cloning and timer handling are not provided. In particular, the main components are:

- *Mobile Agent (Agent)*, which is the basic component defined by user. It is designed as a simple class that encloses within the behavior (is possible to define only single-plane agents).
- *Mobile Agent Execution Engine (MAEE)*, which manages the execution of MAs by means of a thread that schedules local or remote events according to a FIFO policy. The MAEE also encapsulates others functions:
  - it uses serialization for sending remote events through the inner component *Mobile Agent Event Sender (MAES)*;
  - it implements a system of naming (*Mobile Agent Naming* component, *MAN*) which keeps the WSN nodes along with the active agents running on them;
  - it interacts with the MAER through the *Mobile Agent Migration Manager (MAMM)* for providing a mechanism of migration for mobile agents.
- *Mobile Agent Event Receiver (MAER)*, which is developed as an independent thread that waits for receiving events from remote MAs. After event reception, it delegates the delivering of event to the MAEE.
- *Resource Manager (RM)*, which allows accessing to the resources of the Sentilla node, i.e. a 3-axial accelerometer and LEDs.

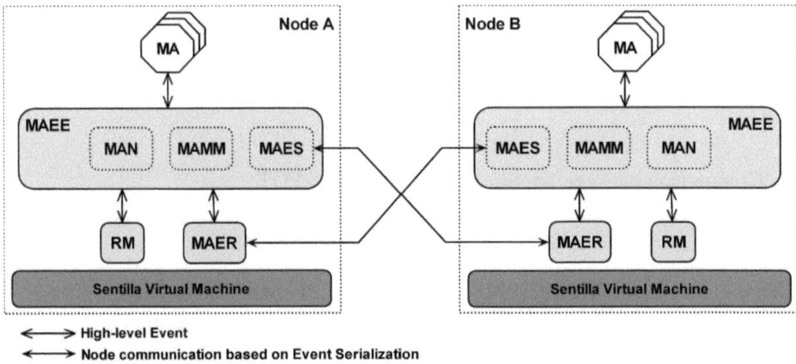

Fig. 1 TinyMAPS architecture

The dynamic behavior of a mobile agent (MA) is modeled through a flat state machine. In particular, it is composed of local variables, local functions, and an automaton whose transitions are labeled by Event-Condition-Action (ECA) rules *E[C]/A*, where *E* is the event name, *[C]* is a boolean expression based on the global and local variables, and A is the atomic action. Thus, agents interact through events, which are asynchronously delivered and managed by the MAEE component. It is worth noting that the state machine-based agent behavior programming allows exploiting the benefits deriving from three paradigms for WSN programming: event-driven programming, state-based programming and mobile agent-based programming.

In the following a comparison between TinyMAPS and MAPS with respect to their architecture, programming model, performance and limitations, is reported.

Both TinyMAPS and MAPS offer similar services for developing WSN agent-based applications. They use state machines to model the agent behavior and directly the Java language to program guards and actions. Moreover, differently from TinyMAPS, MAPS is more powerful and fully exploits the release 5.0 red of the Sun SPOT library to provide advanced functionality of communication, migration, sensing/actuation, timing, and flash memory storage. In MAPS, the implementation of mobile agents is based on the Isolate components defined by the Sun SPOT library. Each Isolate represents "process-like" unit of computation isolated from other instances of Isolate and their migration mechanism is directly offered by the SPOT Squawk JVM. The concept of Isolates within Sentilla technology is different; they are used as system mechanisms together with the concept of Binary when the code is deployed on the motes [11]. TinyMAPS supports migration by simply sending an event that contains agent status information and data (which are hibernated and encapsulated inside the event); the agent needs to restart its execution on the remote node. In any case, both platforms suffer from the current limitation of the Sentilla JCreate and the Sun SPOT that do not allow dynamic class loading, so preventing from the possibility to support code migration (i.e. any class required by the agent must be already present at the destination node). Finally, both MAPS and TinyMAPS allow developers to program agent-based applications in Java according to its rules so no translator and/or interpreter need to be developed and no new language has to be learnt.

To evaluate and compare the performance of TinyMAPS and MAPS two benchmarks have been defined according to [12] for the following mechanisms:

- *Agent communication.* The agent communication time is computed for two agents running onto different nodes and communicating in a client/server fashion (request/reply). Two different request/reply schemes are used: (i) *data Back and Forward (B&F)*, in which both request and reply contain the same amount of data; (ii) *data B*, in which only the reply contains data. Communication Sentilla technology imposes a maximum message payload size of 78 bytes. Comparison results are shown in Fig. 2. MAPS performs better than TinyMAPS; moreover as agent data payload increases, communication time for MAPS is not affected.
- *Agent migration.* The agent migration time is calculated for agent ping-pong among two single-hop-distant sensor nodes. Migration times are computed by varying the data cargo of the ping-pong agent. The obtained migration times are high due to the slowness of the JVM operations supporting the migration process as well as the communication time between two nodes. Comparison results are shown in Fig. 3. For agents with low data payload TinyMAPS performs better than MAPS; however, when the migration event data payload is more than 58 bytes MAPS migration mechanism starts performing better.

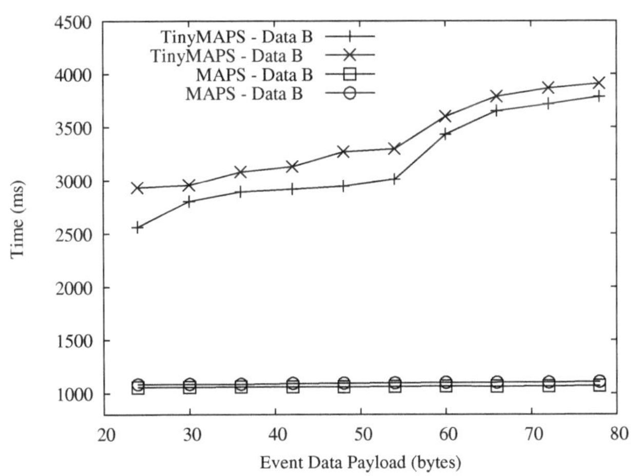

Fig. 2 Agent communication time comparison

3 A Case Study: Real-Time Human Activity Monitoring

The effectiveness of TinyMAPS to support WSN applications is demonstrated by the development of a real-time human activity recognition system that recognizes postures (e.g. lying down, sitting or standing still) and movements (e.g. walking).

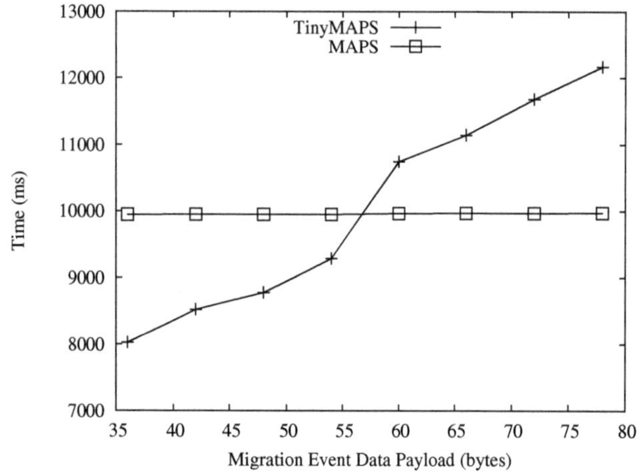

Fig. 3 Agent migration time comparison

Such system has great potential to enable a broad variety of assisted living applications in many context domains like health care, e-fitness, social interaction, and highly interactive games, among the others. The architecture of the system, shown in Fig. 4, is a typical star-based Wireless Body Sensor Network (WBSN) composed of a base station (or coordinator) and two wireless sensor nodes.

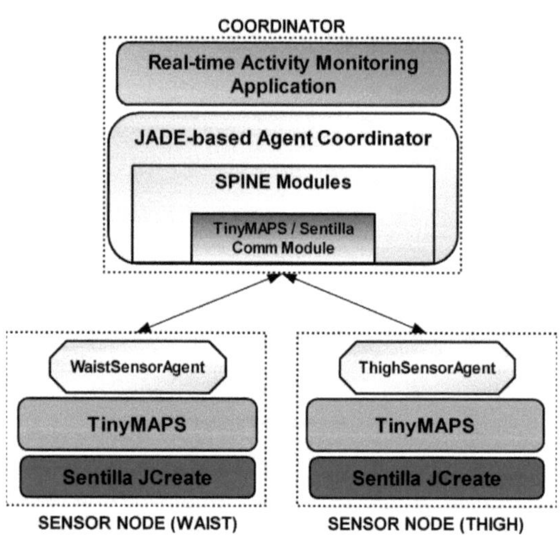

Fig. 4 Architecture of the agent-based real-time activity recognition system

The coordinator is developed through a JADE-based agent that incorporates SPINE-based modules. SPINE is a domain-specific framework for collaborative WBSNs [13] that provides effective APIs (libraries of protocols, utilities and data processing functions) and tools (remote configuration of sensors, data gathering and visualization) for signal processing-based applications for the analysis and the classification of sensor data. SPINE is currently developed for TinyOS and Z-Stack node-side environments [14]. Furthermore, the activity recognition high-level application is the one built in the context of the SPINE project [15].

The JADE agent coordinator is used for issuing commands to the WBSN and is responsible of capturing low-level messages and node events. It relies on the TinyMAPS/Sentilla communication module, specifically developed for supporting Sentilla JCreate sensor nodes. It is composed of a send/receive interface and some components that implement such interface according to the specific sensor platform. Such module translates high-level SPINE messages, formatted according to the SPINE OTA (over-the-air) protocol, into lower-level TinyMAPS/Sentilla messages (named SentillaMessage) and vice versa. It also integrates an application-specific logic for synchronizing the two sensor nodes.

The Sentilla sensor nodes are located on the waist and on the thigh of the monitored person, and TinyMAPS is deployed for supporting the execution of two agents: the WaistSensorAgent and the ThighSensorAgent. They have similar behavior modeled by the FSM shown in Fig. 5.

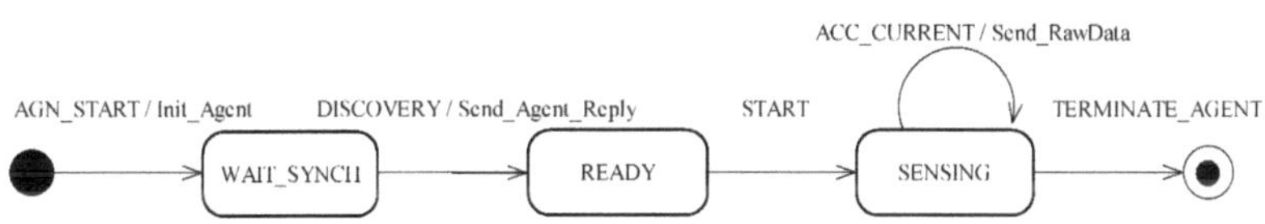

Fig. 5 FSM-based behavior of the Waist/ThighSensorAgent

The interaction diagram depicted in Fig. 6 shows the interaction among the three agents constituting the real-time system: CoordinatorAgent, WaistSensorAgent and ThighSensorAgent. After start-up, each sensor agent is in the WAIT_SYNCH state waiting for synchronization with the CoordinatorAgent. In particular, the CoordinatorAgent broadcasts the DISCOVERY event to check the presence of both nodes; if it detects them, it broadcasts the START event to start the sensing activity. Each sensor agent periodically sends the RAW_DATA event, containing the accelerometer data, to the CoordinatorAgent that, in turn, properly processes such data.

The information sent by the sensor agents are not preprocessed on the nodes because the available memory does not allow for coding particular functions/algorithms along with the TinyMAPS agent execution engine.

The activity monitoring system integrates a classifier based on the K-Nearest Neighbor algorithm which is capable of recognizing postures and movements previously defined in a training phase. The test phase is carried out by considering the pre-defined sequence of postures/movements represented by the state machine

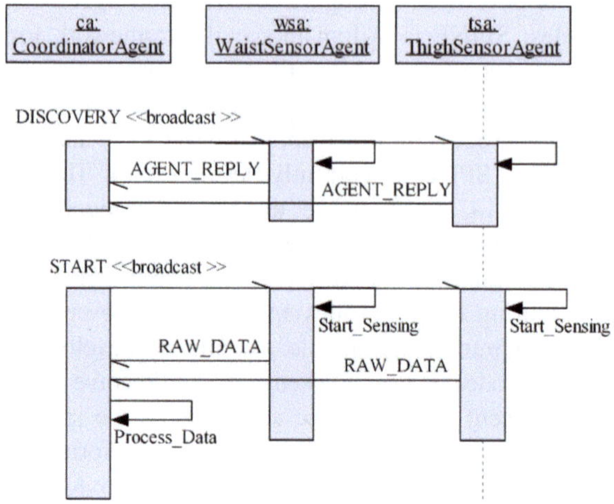

Fig. 6 Agent interaction in the real-time activity monitoring system

reported in Fig. 7. Accordingly, the obtained classification accuracy results are reported in Table 2. The results are good considering that the performance of the adopted sensor platforms is very poor and no processing tasks are performed locally on the nodes before sensing data transmission.

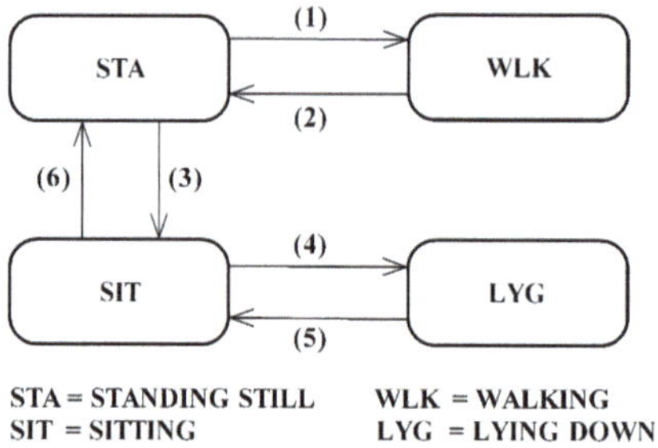

Fig. 7 State machine of the pre-defined sequence of postures/movements

Table 2 Classification accuracy

	Walking	Sitting	Standing still	Lying down
Accuracy (%)	78	97	97	99

The recognized activities transitions are reported in Fig. 8. As can be noted, after a transitory period of 14s from one state to another, all the postures/movements are recognized with an accuracy of 100%. The state transitions more difficult to recognize are $STA \rightarrow WLK$ and $WLK \rightarrow STA$, whereas the transition $LYG \rightarrow SIT$ and $SIT \rightarrow STA$ are recognized as soon as they occur.

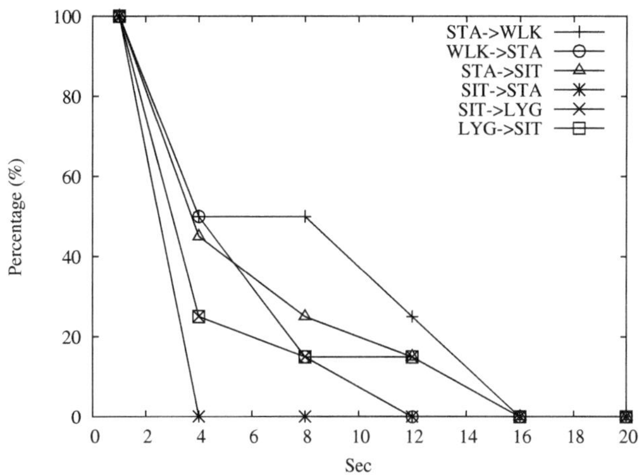

Fig. 8 State machine of the pre-defined sequence of postures/movements

4 Conclusions

Programming WSN applications is a complex task that requires suitable programming paradigms and frameworks to cope with the WSN-specific characteristics. Among them, mobile agent-based programming, which has been formerly introduced for conventional distributed systems, can be more effectively exploited in the context of WSNs. In this paper we have proposed TinyMAPS, a Java-based framework for the development of agent-based applications for Sentilla sensor platforms. By using TinyMAPS, a WSN application can be structured as a set of stationary and mobile agents distributed on sensor nodes supported by a component-based agent execution engine that provides basic services such as message transmission, agent creation, agent migration and easy access to the sensor node resources. Moreover, a complete case study concerning the development and testing of a real-time human activity monitoring system based on WBSNs has been described. It is emblematic of the effectiveness and suitability of TinyMAPS to deal with the programming of complex WSN applications. On-going research efforts are being devoted to optimizing the communication and migration mechanisms of TinyMAPS and using TinyMAPS into other application areas where mobility plays a crucial role.

Acknowledgements. This work has been partially supported by CONET, the Cooperating Objects Network of Excellence, funded by the European Commission under FP7 with contract number FP7-2007-2-224053.

References

1. Yoneki, E., Bacon, J.: A survey of Wireless Sensor Network technologies: research trends and middlewares role. Technical Report UCAM-CL-TR-646, University of Cambridge, UK (September 2005)
2. Lange, D.B., Oshima, M.: Seven Good Reasons for Mobile Agents. Communications of the ACM 42(3) (1999)
3. Fok, C.-L., Roman, G.-C., Lu, C.: Agilla: A Mobile Agent Middleware for Sensor Networks. Accepted to ACM Transactions on Autonomous and Adaptive Systems Special Issue (2011)
4. TinyOS: documentation and software (March 2011), `http://www.tinyos.net`
5. Kwon, Y., Sundresh, S., Mechitov, K., Agha, G.: ActorNet: An Actor Platform for Wireless Sensor Networks. In: 5th Intl. Joint Conference on Autonomous Agents and Multiagent Systems (AAMAS), pp. 1297–1300. ACM, New York (2006)
6. Muldoon, C., O' Hare, G.M.P., Collier, R.W., O' Grady, M.J.: Agent Factory Micro Edition: A Framework for Ambient Applications. In: Alexandrov, V.N., van Albada, G.D., Sloot, P.M.A., Dongarra, J. (eds.) ICCS 2006, vol. 3993, pp. 727–734. Springer, Heidelberg (2006)
7. Aiello, F., Fortino, G., Gravina, R., Guerrieri, A.: A Java-based Agent Platform for Programming Wireless Sensor Networks. The Computer Journal 54(3), 439–454 (2011)
8. Mobile Agent Platform for Sun SPOT (MAPS), documentation and software (April 2011), `http://maps.deis.unical.it`
9. Sun Small Programmable Object Technology (Sun SPOT) (March 2011), `http://www.sunspotworld.com`
10. Aiello, F., Carbone, A., Fortino, G., Galzarano, S.: Java-based Mobile Agent Platforms for Wireless Sensor Networks. In: Workshop on Agent Based Computing: from Model to Implementation VII (ABC:MI 2010). IEEE Press, Poland (2010)
11. Sentilla Developer Community (March 2011), `http://www.sentilla.com/developer.html`
12. Dikaiakos, M., Kyriakou, M., and Samaras, G.: Performance evaluation of mobile-agent middleware: A hierarchical approach. In: Picco, G.P. (ed.) MA 2001. LNCS, vol. 2240, pp. 244–259. Springer, Heidelberg (2001)
13. Bellifemine, F., Fortino, G., Giannantonio, R., Gravina, R., Guerrieri, A., Sgroi, M.: SPINE: A domain-specific framework for rapid prototyping of WBSN applications. Software-Practice & Experiences 41(3), 237–265 (2011)
14. Bellifemine, F., Fortino, G., Giannantonio, R., Guerrieri, A.: Platform-independent development of collaborative WBSN applications: SPINE2. In: IEEE International Conference on Systems, Man, and Cybernetics (SMC), San Antonio, TX, USA, October 11-14 (2009)
15. Signal Processin In Node Environment (SPINE) (March 2011), `http://spine.tilab.com`
16. Maurer,U., Smailagic, A., Siewiorek, D. P., Deisher, M.: Activity recognition and monitoring using multiple sensors on different body positions. In: International Workshop on Wearable and Implantable Body Sensor Networks, pp. 113–116. IEEE Computer Society, USA (2006)

A New Model for Energy-Efficient All-Wireless Networks

Doina Bein and S.Q. Zheng

Abstract. We extend the work of energy-efficient wireless communication initiated by Wieselthier *et al.* in a new direction. Instead of minimizing the total energy consumption, our goal is to minimize the maximum energy consumption among all nodes, thereby achieving the longest network lifetime. We formulate a new model for energy-efficient communication in wireless networks, called the MinMax energy model, that distributes the energy consumption among nodes evenly by constructing appropriate multicast/broadcast trees which may not consume minimum total energy. We propose a few fundamental computational problems related to our model and study their complexities.

1 Introduction

As pointed out in the work of Wieselthier *et al.*[1], a crucial issue in battery operated wireless networks is that there is a trade-off between reaching more nodes with higher energy consumption and interference and reaching fewer nodes with lower energy consumption and interference. This trade-off should be considered in energy-efficient communications, especially for multicast and broadcast communications. To address this issue, in [1] an approach was proposed that involves both controlling transmission energy (network connectivity, a physical layer function) and multicast/broadcast tree (or a path for unicast) formation (a routing function in the network layer). The goal of the joint decisions on connectivity and routing is to achieve energy efficiency, which is measured by the total energy consumption of network nodes. Several heuristic algorithms for constructing minimum total energy

Doina Bein
The Pennsylvania State University, Applied Research Laboratory
e-mail: siona@psu.edu

S. Q. Zheng
University of Texas at Dallas, US Department of Computer Science
e-mail: sizheng@utdallas.edu

F.M.T. Brazier et al. (Eds.): Intelligent Distributed Computing V, SCI 382, pp. 171–181.
springerlink.com

multicast/broadcast trees were proposed in [1]. It was shown in [2, 3] that the problem of constructing multicast/broadcast trees of minimum total energy is NP-hard. Since then, constructing multicast/broadcast trees of minimum total energy has received significant attention and a many heuristics have been proposed (i.e. [2–11]).

We extend the work of energy-efficient wireless communications initiated by Wieselthier *et al.* [1] in a new direction. Instead of minimizing the total energy consumption, our goal is to minimize the maximum energy consumption of all nodes, thereby achieving the longest network lifetime. We propose a new model for energy-efficient communications in wireless networks, namely the MinMax energy model, that distributes energy consumption among nodes evenly by constructing appropriate multicast/broadcast trees which may not consume minimum total energy. The argument for supporting this model is that a network that uses the minimum total energy may become in short time disconnected because of one or more nodes running out of battery power much earlier than other nodes. Balancing the energy consumption can eliminate such a "power bottleneck." [1]

The MinMax model ties the design and use of an energy-efficient wireless network together so that the actual network operation directly follows its intended, optimized operation manner. This connection is missing in all previous related work. The MinMax model makes wireless networks adaptive with respect to changes in network traffic and residual power supplies at nodes. When needed, re-balancing the energy consumption is possible by reconstructing multicast/broadcast trees. The MinMax model is also useful for incremental network redesign. We propose a few fundamental computational problems related to the MinMax model and we study their complexities. Simulation results show that the MinMax energy model is a better alternative for characterizing energy-efficient wireless networks.

The paper is organized as follows. In Section 2 we describe our wireless communication model under simplified assumptions, the proposed MinMax model, and other models derived from it. We also show that the MinMax model, as an optimization problem, is NP-complete. In Section 3 we present some approximation algorithms for finding the minimum-energy multicast trees in a wireless network where all nodes start with the same amount of energy under the MinMax model as well as the other associated models. We conducted extensive simulations and the results are given in Section 4. We conclude in Section 5.

2 MinMax Energy Model

We consider that a wireless network in which the node locations are fixed, nodes are battery operated and equipped with omnidirectional antennas. Every transmission by a node u can be received by all nodes situated within its communication range (the so called *Wireless Multicast Advantage (WMA)*). We also assume that ample bandwidth is available to accommodate all transmission requests. The lifetime of a wireless network is the time between the network starts to operate for its intended application to the time one of its node runs out of its battery power. Our objective

[1] We use the terms power consumption and energy consumption interchangeably.

is to design and operate a wireless network that achieves the longest lifetime. We describe our wireless communication model under simplified assumptions. These assumptions are used in [1] and most papers on the same subject.

Assumption 1. We consider an interference-free model. The energy used by a node for sending a message that will successfully reach another node at distance d is $p(d) = d^{\alpha} + c_e$, where α is a constant parameter depending on the characteristics of the communication medium (typically with a value between 2 and 4), and c_e is a parameter representing an overhead due to signal processing. Moreover, we assume that $p(d)$ is the energy for transmitting one bit of information at distance d. Since the packets sent may have different sizes, the cost of sending one bit of information at distance d would be $p(d)/packet\_size$, but for uniformity reasons we consider $p(d)$ to be that cost instead.

Assumption 2. The energy spent for receiving a bit of information is a constant and is neglected in energy consumption analysis (it is considered to be 0).

Assumption 3. Each node can choose its energy transmission value, not to exceed some maximum value p_{max}, and can continuously adjust it.

The connectivity of a wireless network is represented by an *energy reachability graph* $G = (V,E)$ that is a weighted directed graph where V is the set of nodes and E is the cost function. An edge $u \to v$ exists if node u can send packets to v and the cost of it, $E(u \to v)$, is the energy for sending a bit from u to v. For practical purpose, energy reachability graphs are based on the energy function defined on the metric distance in (2- or 3-dimensional) Euclidean space; we call such a graph an *Euclidean energy reachability graph*. For theoretical investigations, arbitrary graphs with arbitrary non-negative energy weights have also been used as energy reachability graphs, and we call such a graph a *general energy reachability graph*. This study covers both Euclidean and general energy reachability graphs.

This simplified wireless communication model can be generalized to cover more complex features when needed. For example, the impact of node mobility can be incorporated into our model because the energy for transmissions can be adjusted to accommodate the new node's locations.

2.1 MinMax Model: A Formulation for Energy-Efficient Networks

The traffic of a given wireless network $G = (V,E)$, $|V| = n$, can be specified by a set of minimal and distinct communication patterns $P = \{P_1, P_2, \cdots, P_m\}$, where each traffic pattern $P_k = (s_k, D_k)$ has the sender node s_k, the destination nodes D_k specified, $1 \le k \le m$, and can be one of the following three types: *(i)* unicast, when D_k is a single node $D_k = \{d_k\}$, *(ii)* broadcast, when D_k is the set of all other nodes $D_k = V - \{s_k\}$, and *(iii)* multicast, when D_k is a proper subset of all other nodes $D_k \subseteq V - \{s_k\}$. We note that $m \le n$. As an example of traffic, all-to-all broadcast (i.e. each node broadcast a message to all other nodes) is a set of n distinct broadcasts $AA = \{B_1, B_2, \cdots, B_n\}$, one from each node to all other nodes in the network,

$B_k = (s_k, V - \{s_k\})$, $1 \le k \le n$. As another example of traffic, a many-to-one communication pattern can be decomposed into a set of at most n unicasts with the same destination node.

Each traffic P_k involves transmitting a number of bits by the sender node that need to reach the destination nodes; let $Traf_k$ denote this number. Thus, $Traf_k$s define the traffic profile of the network. Two nodes i and j in the network may be part of several traffic patterns that are part of P. Thus, for any two nodes i and j in the network, we define an $n \times n$ traffic matrix $Traf[i,j] = \sum_{\substack{1 \le k \le m \\ s_k = i, j \in D_k}} Traf_k$ that measures how many bits need to be transmitted between i and j to support the traffic P.

Each $P_k = (s_k, D_k)$ can be implemented by a Steiner tree T_k in $G = (V,E)$; in such a tree, nodes not in D_k are used only for forwarding packets. Note that directed paths in G, which may or may not contain nodes other than s_k and d_k, and directed spanning trees of G, which contains all nodes of V, are special cases of Steiner trees [12]. Let T_k be a Steiner tree for P_k in $G = (V,E)$. Based on WMA, define $E_{T_k}(u)$ as the energy used by node u to transmit a bit to its child nodes in T_k; i.e. $E_{T_k}(u) = \begin{cases} 0, \text{if } u \text{ is a leaf of } T_k \\ \max\{E(u \to v) | u \to v \in T_k\}, \text{otherwise} \end{cases}$

Let $E_a(u)$ be the energy available at node u (i.e. the energy stored in the battery of node u) and $E_{min} = \min_{u \in V} \{E_a(u)\}$ be the minimum energy available at any node. We define $r_{E_a}(u) = \frac{E_{min}}{E_a(u)}$ to measure how rich in energy is a node compared with the "poorest" node (lower the ratio, richer the node). If $r_{E_a}(u) = 1$ for all nodes u, then all nodes have the same available energy.

Given P_ks, $Traf_k$s and $E_a(u)$s, our goal is to find T_k^* for P_k, $1 \le k \le m$, such that richer nodes support more traffic, i.e. the value

$$\max_{u \in V} \{ \sum_{1 \le k \le m} r_{E_a}(u) \cdot Traf_k \cdot E_{T_k^*}(u) \} \tag{1}$$

is minimized. We denote this model as the *MinMax Model*. In this model, a node u with smaller $r_{E_a}(u)$ (richer node) will transmit more data. For a given traffic defined by P_ks and $Traf_k$s, the MinMax model balances the energy consumptions at nodes to maximize the network lifetime.

The MinMax model can be formulated as an optimization problem, which we call *optimal MinMax energy routing* (MMER) problem. We show that MMER is NP-hard by considering the following simplified decision problem, called MMER-D problem, and we show that MMER-D is NP-complete.

MMER-D Problem: Given an energy reachability graph $G = (V,E)$, n unicast source-destination pairs (s_i, d_i) with $Traf_i$ in G, and a constant E^*, are there n routes T_i^* for (s_i, d_i) pairs such that $\max_{u \in V} \{ \sum_{1 \le i \le n} Traf_i \cdot E_{T_i^*}(u) \} \le E^*$?

We show that MMER-D is NP-complete by reducing the following NP-complete Bin Packing (BP) problem to MMER-D:

BP Problem: Given a finite set U of n items, a size $s(w_i) \in Z^+$ for each $w_i \in U$, a positive integer bin capacity B, and a positive integer k, is there a partition of U into disjoint sets $U_1, U_2, \cdots, U_k$ such that the sum of the sizes of items in each U_i is $\leq B$?

Proof sketch. It is easy to see that MMER-D is in NP. Given a BP instance, we construct an energy reachability graph of $2(n+k)$ nodes as shown in Fig. 1, where ε is an arbitrary small energy value, and 1 represents the unit energy value.

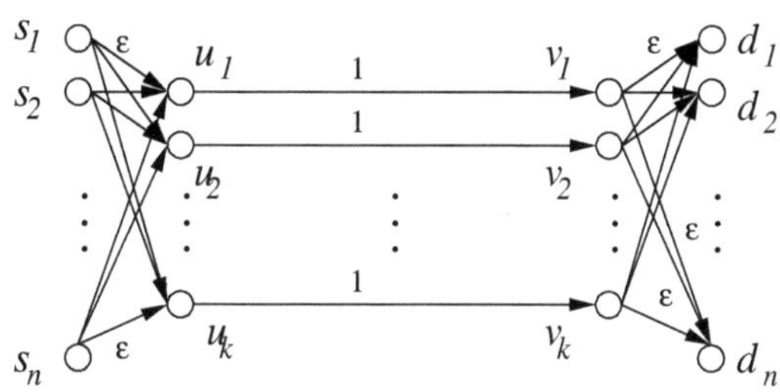

Fig. 1 An MMER-D instance.

Let $Traf_i = s(w_i)$. Clearly, BP has a "Yes" answer if and only if its corresponding MMER-D has a "Yes" answer for $E^* = B$. We also note that it is easy to enforce the relative energy values ε and 1 in the Euclidean space by proper scaling. Then, we have the following claim.

Theorem 1. *MMER problem is NP-hard for general energy reachability graphs and Euclidean energy reachability graphs.*

3 Approximating T_k^*s

As shown in Section 2, the problem of computing optimal T_k^*s is NP-hard. One possible approach for approximation is to find locally optimal or near-optimal T_ks (which may not constitute a set of optimal T_k^*s) iteratively. In this approach, T_ks are found one at a time in a sequential fashion. This approach allows approximating an optimal set of T_k^*s by a set of locally optimal or near-optimal T_ks. This approach also opens a way for iteratively improving the quality of an existent solution using a rip up and reroute technique by adding appropriate constraints, which is an essential element for many powerful optimization meta-heuristics such as simulated annealing and genetic algorithms.

For a given network G, what is a locally optimal T_k for P_k? Previous work implies that T_k is a solution for P_k with minimum total energy consumption. If P_k is a unicast, then finding a path from s_k to $D_k = \{d_k\}$ of minimum total energy is polynomial-time solvable. But if P_k is a multicast or broadcast, then finding a minimum total energy multicast trees and finding a minimum total energy broadcast trees are NP-hard. Despite the computational intractability of these problems, solutions of these problems actually may not lead to a good global solution according to the MinMax model. For example, to broadcast a message from node 1 to all other nodes in the network in Fig. 2, the minimum total energy is used if the message is sent from node 1 to node 2 and node 2 broadcasts the message to all remaining nodes (Fig. $2(a)$). But in this way both nodes 1 and 2 deplete fast, while the other nodes spend no energy. The network will have a longer lifetime if each node (starting with node 1) sends the message to its closest neighbor (indicated by a dotted curve) as shown in Fig. $2(b)$) and so on until node 2 is reached.

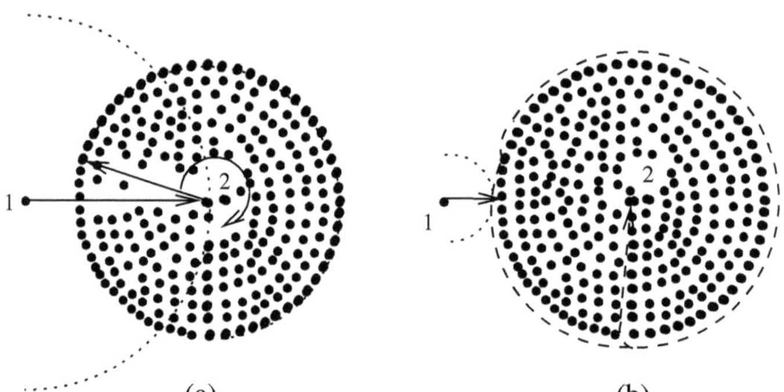

Fig. 2 Communication that extends the life of the network

3.1 Finding MinMax Energy T_ks

We focus first on finding the *MinMax energy tree* for P_k which is the tree $T_k^\star$ such that $\max_{e\in T_k^\star}\{E(e)\} \leq \max_{e\in T_k}\{E(e)\}$, where T_k is an arbitrary tree for P_k. Unless otherwise specified, we assume that $r_{E_a}(u) = 1$ for all nodes u. If this is not true, then we can replace the energy cost $E(u \to v)$ of edge $u \to v$ in energy reachability graph G with $r_{E_a}(u) \cdot E(u \to v)$, and G with such energy costs is equivalent to $r_{E_a}(u) = 1$ for all nodes u. It turns out that the MinMax energy $T_k^\star$ for P_k, which is a unicast, multicast or broadcast, can be found using a modified Dijkstra algorithm, which was designed for finding minimum-cost paths from a single source s_k to all other nodes in G. We name this algorithm for computing $T_k^\star$ with MinMax energy as *MinMax Energy*.

The Dijkstra's algorithm maintains a min-priority queue Q of nodes, based on previously computed costs of paths from s_k to all nodes. The algorithm proceeds in iterations in a prioritized breath-first search. Each node u keeps an upper bound $d[u]$ for the total cost of any path from s_k to u. In each iteration, the node u with the smallest cost in Q is selected, all edges $u \to v$ are examined, and $d[v]$s are reduced if possible by a procedure *Relax*. (Details of Dijkstra's algorithm and its associated procedure *Relax* are given in [13].) In our modified Dijkstra's algorithm, in Procedure *Relax* below, instead of using the total cost estimate $d[u]$, we use a min-max edge cost $mmE[u]$ with initial value ∞.

```
procedure Relax(u,v,E)
  If mmE[v] > max{mmE[u], E(u → v)} then
    mmE[v] := max{mmE[u], E(u → v)};
    p[v] := u;
```

When the algorithm terminates, a tree including all nodes in V is computed, with pointers $p[u]$ of all non-root nodes u pointing at their parent nodes. By reversing the directions of this tree, we obtain a directed spanning tree T'_k rooted at s_k which needs to be processed further. If $P_k = (s_k, D_k) = (s_k, \{d_k\})$, i.e. P_k is a unicast, then we take the path from s_k to d_k in T'_k as $T^\star_k$. If $P_k = (s_k, V - \{s_k\})$, i.e. P_k is a broadcast, we modify T'_k to get a WMA-conforming $T^\star_k$. Let $CHILDREN[u]$ be the set of child nodes of u in T'_k and Q be an FIFO queue. The procedure *Collapse* below transforms T'_k into $T^\star_k$ by a breath-first search.

```
Procedure Collapse(T'_k, E)
  ENQUEUE(Q, s_k); S := V − {s_k}; T*_k := {s_k};
  while Q ≠ ∅ do
    u := DEQUEUE(Q); S = S − {u};
    C := S ∩ CHILDREN[u];
    if C ≠ ∅ then
      B := max{E(u → v) | v ∈ C}
      for each v ∈ C do
        ENQUEUE(Q, v); S := S − {v};
        E(u → v) := B; T*_k := T*_k ∪ {u → v};
```

If $P_k = (s_k, D_k)$ is a multicast i.e. $D_k \subset V - \{s_k\}$, we first obtain T''_k from T'_k by removing all nodes in $V - (D_k \cup \{s_k\})$ (and edges pointing at them) from which no nodes in D_k can be reached, and we then apply *Collapse* to T''_k to obtain $T^\star_k$.

3.2 Finding MinMax Energy T_ks with Minimum Heights

The energy reachability graph G may contain many different MinMax energy trees for P_k. It may be desirable to find a particular MinMax energy tree $T^\star_k$ by enforcing additional constraints. We focus on the problem of finding MinMax energy $T^\star_k$s with minimum height. This problem corresponds to finding MinMax energy trees with minimum communication delays. In this section we show how to find a MinMax energy trees with minimum height efficiently.

First, we find T'_k (for unicast and broadcast P_k) or T''_k (for multicast P_k) in $G = (V, E)$ using the modified Dijkstra algorithm presented in previous section, and also we compute the maximum $E_{max} = \max\{E(e) | e \in T'_k \text{ or } T''_k\}$ (depending on the type of P_k). Then, we obtain $G' = (V, E')$ from $G = (V, E)$ by removing all edges e with $E(e) > E_{max}$. We are now restricted to finding a minimum height MinMax energy tree in G'. For each edge $e \in E'$, we assign a unit cost $cost(e) = 1$. Then we apply

Dijkstra algorithm to find a shortest path tree T_k''' using *cost* as the length function. T_k''' is a minimum-height spanning tree of G', with respect to the length *cost*. If P_k is a unicast, then $T_k^\star$ is the path from s_k to d_k in T_k'''. If P_k is a broadcast, then $T_k^\star$ is T_k'''. If P_k is a multicast, then $T_k^\star$ is obtained from T_k''' by removing all nodes in $V - (D_k \cup \{s_k\})$ (and edges pointing at them) from which no nodes in D_k can be reached. We name this algorithm for computing $T_k^\star$ with minimum heights subject to MinMax energy as the *MinMax Energy-MinHop* algorithm.

3.3 Minimum Total Energy T_k Subject to MinMax Energy Constraint

One way of approximating $T_k^\star$ is to find a T_k with minimum total energy subject to the MinMax energy constraint. Such a T_k exists in G'. If P_k is a unicast, such a T_k can be computed in polynomial time by applying Dijkstra algorithm to G'. We call the problem of finding minimum total energy T_ks subject to MinMax energy constraint for multicast and broadcast P_ks the MinMax-MinTotal energy multicast (MMMTEM) problem and the MinMax-MinTotal energy broadcast (MMMTEB) problem, respectively. It turns out that MMMTEM and MMMTEB are computationally intractable. This can be shown by reducing the problem of finding minimum total energy broadcast tree (MTEB) to MMMTEB and MMMTEM.

4 Simulations

Simulation results show that the MinMax energy model is a better alternative for characterizing energy-efficient wireless networks. We considered two sets of networks (energy reachability graphs). The first contains general energy reachability graphs whose distances between nodes are random numbers in the range of $[0.01, 4.99]$. We call this set of networks *random general networks*. The second set contains Euclidean energy reachability graphs with nodes' coordinates randomly generated in the range $[0.01..4.99]$. We call this set of networks *random Euclidean networks*. We considered the number of nodes in the range of $[5, 200]$.

Similar to [1, 4], we ran 1000 simulations for each setup consisting of a network of a specified size and nodes' distances, parameter α, and a particular algorithm. We focus on the performance of the broadcast trees. For any given network instance G, we consider the broadcast from some random node v_k in G. We compare the broadcast trees produced by the algorithm *MinMax Energy* (which we call the MET trees), the algorithm *MinMax Energy-MinHop* (called the MMT trees), the algorithm BIP (Broadcast Incremental Power) [1] (called the BIP trees) and the algorithm IMBM (Iterative Maximum-Branch Minimization) [4] (called the IMBM trees), in terms of maximum node energy, height, total number of leaves, and total energy.

For each network instance, let E_{max}^{MET}, E_{max}^{MMT}, E_{max}^{BIP}, and E_{max}^{IMBM} be the maximum node energy consumption of MET, MMT, BIP, and IMBM tree, respectively. Note that the maximum node energy of MET tree and MMT tree are the same, i.e. $E_{max}^{MET} = E_{max}^{MMT} = E_{max}$.

In all the experiments, $E_{max}^{BIP} > E_{max}$ and $E_{max}^{IMBM} > E_{max}$. Furthermore, the network connected by MET or MMT trees will have the longest lifetime, followed by BIP, and IMBM (which can be 20 times shorter).

For random general networks, $\frac{E_{max}^{BIP}}{E_{max}}$ increases slowly from 1.16 to 1.21 and $\frac{E_{max}^{IMBM}}{E_{max}}$ increases sharply from 2.03 to 465.81 (see Fig. 3).

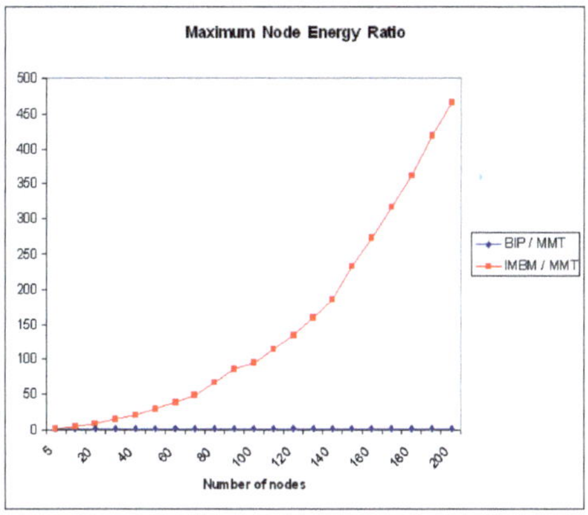

Fig. 3 Ratio between the maximum node energy of BIP and IMBM trees, and MMT trees for random general networks; $\alpha = 2$.

For random Euclidean networks, we observe that $\frac{E_{max}^{BIP}}{E_{max}}$ slowly increases from 1.1 to 1.31, and $\frac{E_{max}^{IMBM}}{E_{max}}$ sharply increases from 1.19 to 20.79 (see Fig. 4).

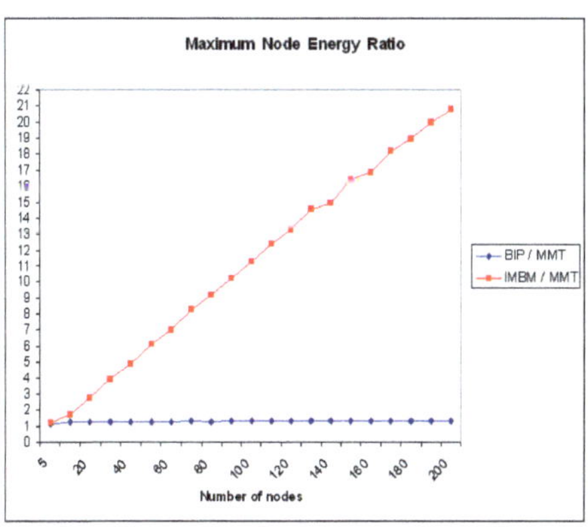

Fig. 4 Ratio between the maximum node energy of BIP and IMBM trees, and MMT trees for random Euclidean networks; $\alpha = 2$.

5 Conclusions

In this paper, we have addressed some issues associated with energy-efficient communications in all-wireless networks with the goal of achieving the longest network lifetime. We proposed a formal MinMax model, which can be used in network design, operation, dynamic energy consumption re-balancing, and incremental re-design. The key feature of this model is that it provides a unique way of unifying network design, use, management and modification. Our simulation results indicate that the MinMax model captures the essence of many aspects of optimal use of resources in all-wireless networks. Our theoretical study was based on an approach of combining locally optimal or near-optimal solutions into a globally near-optimal solutions. The complexities of several fundamental problems were analyzed. Further investigations are expected in the development of effective heuristic algorithms for the design and management of all-wireless networks with long lifetime based on the MinMax model.

References

1. Wieselthier, J.E., Nguyen, G.D., Ephremides, A.: On the construction of energy-efficient broadcast and multicast trees in wireless networks. In: INFOCOM. IEEE Computer Press, Los Alamitos (2000)
2. Cagali, M., Hubaux, J.P., Enz, C.: Minimum-energy broadcast in all-wireless networks: Np-completeness and distribution issues. In: MOBICOM, pp. 172–182 (2002)
3. Liang, W.: Constructing minimum-energy broadcast trees in wireless ad hoc networks. In: MOBICOM, pp. 112–122 (2002)
4. Li, F., Nikolaidis, I.: On minimum-energy broadcasting in all-wireless networks. In: LCN, pp. 193–202 (2001)
5. Wieselthier, J.E., Nguyen, G.D., Ephremides, A.: Algorithms for energy-efficient multicasting in static ad hoc wireless networks. Mobile Networks and Applications 6 (2001)
6. Stojmenovic, I., Seddigh, M., Zunic, J.: Internal nodes based broadcasting in wireless networks. In: HICSS (2001)
7. Stojmenovic, I., Seddigh, M., Zunic, J.: Dominating sets and neighbor elimination-based broadcasting algorithm in wireless networks. IEEE Trans. on Parallel and Distributed Systems 13, 14–25 (2002)
8. Cartigny, J., Simplot, D., Stojmenovic, I.: Localized minimum-energy broadcasting in ad-hoc networks. In: INFOCOM, vol. 3, pp. 2210–2217 (2003)
9. Thai, M.T., Li, Y., Du, D.Z., Ai, C.: On the construction of energy-efficient broadcast tree with hitch-hiking in wireless networks. In: IEEE International Performance, Computing, and Communications Conference, pp. 135–139 (2005)
10. Guo, S., Yang, O.: Localized operations for distributed minimum energy multicast algorithm in mobile ad hoc networks. IEEE Trans. on Parallel and Distributed Systems 18(2), 186–198 (2007)
11. Bein, D., Zheng, S.Q.: An effective algorithm for computing energy-efficient broadcasting trees in all-wireless networks. In: WWASN (2008)

12. Karp, R.M.: Reducibility among combinatorial problems. In: Miller, R.E., Thatcher, J.W. (eds.) Proc. of a Symp. on the Complexity of Computer Computations. The IBM Research Symposiam Series, pp. 85–103. Plenum Press, New York (1972)
13. Cormen, T.H., Leiserson, C.E., Rivest, R.L., Stein, C.: Introduction to algorithms, 2nd edn. MIT Press, Cambridge (2001)

Increasing the Network Capacity for Multi-modal Multi-hop WSNs through Unsupervised Data Rate Adjustment*

Matthew Jones, Doina Bein, Bharat B. Madan, and Shashi Phoha

Abstract. We propose to improve the quality of data for data fusion in a wireless sensor network deployed in an urban environment by dynamically controlling the transmission rate of the sensors. When nodes are grouped in multi-hop clusters, this mechanism will increase the number of messages being received at the cluster heads. We implement a previously proposed cross-layer data adjustment algorithm and integrate it into our multi-modal dynamic clustering algorithm MDSTC. Extensive simulations in NS2 using a simpler two-hop cluster show that using the data rate algorithm allows for better efficiency within the cluster.

1 Introduction

Data gathering is an important application of wireless sensor networks (WSNs), in which the data collected by individual sensors is sent to a so-called base station (BS). Clustering schemes such as LEACH [1], PEGASIS [2], and HEED [3] for homogeneous networks (where nodes have the same initial energy), respectively LEACH-B [4], SEP [5], M-LEACH [6], and EECS [7] for heterogeneous networks (where nodes start with various values for the initial energy) allow only the cluster heads, elected periodically, to aggregate the data from their cluster members

Matthew Jones
The University of Pennsylvania, School of Engineering and Applied Sciences
e-mail: majo@seas.upenn.edu

Doina Bein · Bharat B. Madan · Shashi Phoha
The Pennsylvania State University, Applied Research Laboratory
e-mail: {siona,bbm2,sxp26}@psu.edu

* This material is based upon work supported by, or in part by, the U. S. Army Research Laboratory and the U. S. Army Research Office under the eSensIF MURI Award No. W911NF-07-1-0376. Any opinions, findings, and conclusions or recommendations expressed in this publication are those of the authors and do not necessarily reflect the views of the sponsor.

F.M.T. Brazier et al. (Eds.): Intelligent Distributed Computing V, SCI 382, pp. 183–193.
springerlink.com

and send it to the BS. Another important application of wireless sensor networks is locating and tracking selective mobile targets crossing a sensor field [8, 9], while ignoring others. Instead of involving all the nodes in the network into the clustering process, in this paper we focus on allowing only a (changing) subset of nodes that are geographically near to the mobile target to self-organize into one or more clusters, thus creating time-critical collaborative inference in the vicinity of the target. The sensor node would be able to distinguish between various types of data patterns in its knowledge data base, which will allow it to selectively track a specific target among many mobile objects in an urban environment. A data pattern will be associated with a specific class of targets. The tracking algorithm would be able to predict the target location and velocity even when some nodes fail, and in some cases, move sensors along the predicted trajectory to obtain more accurate data [10].

Sensors are subject to a number of noise factors. Moreover, the sensed data is highly correlated in the vicinity of a stimulus. The sensors detect signal patterns in multiple, diverse, and spatially correlated sensor data streams that interact over the network. Based on these interactions, subsets of sensors may dynamically form *collaborative clusters* to estimate the position or the velocity of the target for tracking purposes. In [11] the authors present an information-driven dynamic sensor collaboration that is a Bayesian approach to dynamically fuse the data from sensors to track mobile targets; the approach relies on reliable estimates of densities for belief states. Tracking mobile targets in sensor networks using a clustering approach is also discussed in [12]. *Dynamic clustering* [13, 14, 15] generates flexible network topologies for sensors to cluster and collaborate on detecting targets moving through a sensor field. Motivated by challenging urban area applications that require adaptive sensor networks to dynamically cluster sensing, processing, and communication resources, a fusion-driven concept of the sensor network design is proposed in [16]. We have implemented this design in MDSTC [17] and develop a fusion-driven distributed dynamic network controller for a multi-modal [1] sensor network. While MDSTC deals with re-allocation of network resources to sparse regions, it does not address the issue of network congestion due to a burst of communication.

The authors of [18] proposed a cross-layer approach to the congestion control problem in wireless networks by requiring nodes to adjust their transmission rates in order to avoid buffer overflow at the receivers and ultimately at the cluster head. In the proposed control schemes (MWM, GMM, or MM), the network jointly optimizes both the users' data rates and the resource allocation at the underlying layers, which include modulation, coding, power assignment and link schedules. Various computational tradeoffs are further analyzed. Further transmission overhead could be obtained if 6LoWPAN [19] is implemented as an adaptation layer between the link and network layers of TCP/IP protocol to enable efficient transmission of IPv6 datagrams over 802.15.4 links. We address the issue of bursty communication by adding a node-centric congestion control mechanism to MDSTC by implementing and integrating the data rate adjustment algorithm. Extensive simulations in NS2 using a simpler two-hop cluster for mobile target tracking in an urban scenario show

[1] We denote by modality of the sensor the type of sensor, e.g. pressure, acoustic, video, infrared.

that using of the data rate algorithm allows for better efficiency within the cluster which leads to an improved estimated accuracy of target location.

Section 2 presents the communication mechanism within the cluster and the scenarios that motivates the need for an unsupervised data rate adjustment algorithm at each node. Section 3 shows how the data rate adjustment algorithm is integrated in MDSTC. Section 4 presents the experimental results. The paper is summarized and concluded in Section 5.

2 Communication Model

In this paper we make the following realistic assumptions: (i) the traffic generated by the sensors is bursty; when a traffic burst occurs, the data rate is much higher than the available link bandwidth, (ii) compressing the sensor data before transmitting it to the cluster head can partially alleviate this problem, and (iii) sensors in the vicinity of an event generate more traffic than the available bandwidth.

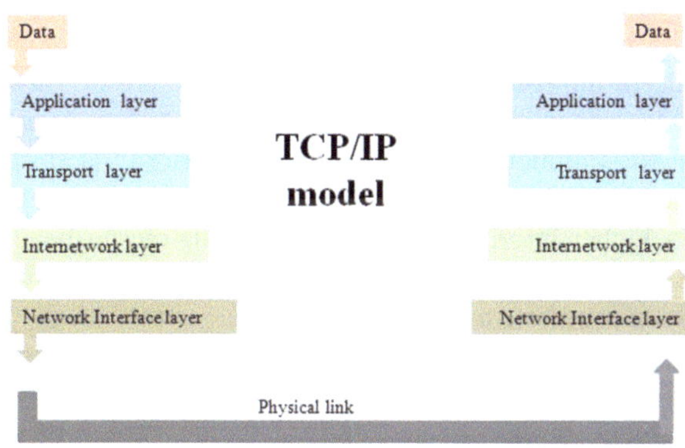

Fig. 1 Communication using TCP/IP model

In the TCP/IP model (see Figure 1) that is used for wireless sensor networks, each node has a receiving and a transmitting queue at each of the TCP/IP levels. The receiving queue at the MAC layer collects the received packets that are addressed to the node or need to be broadcast. Each node also has a timer implemented within its MAC layer. When the timer goes off, the specified node broadcasts information from its queue to the specified destination node. The amount of information sent from one node to another node is measured by a variable called *data rate*. Every node has its own date rate and the data rate specifies how much information a node should broadcast per timer event. Since a node has a finite number of processors (generally one CPU), it can only process a fixed amount of packets that it receives per time unit. Any packet received while the node is not serving any packet is placed in the nodes queue. Since the queue at the MAC layer is also of finite size, it can only hold a finite number of packets received per time unit; any extra packets are dropped. We used UDP at the transport layer, thus the data in the dropped packets is lost. Therefore we are interested in minimizing the packet drop rate while maintaining a high throughput for the communication links in the network.

Let us consider the relationship between the efficiency of a system and the defined queue structure at each node. When is the system most efficient? Let us assume that a node X is broadcasting packets to another node Y. The following two scenarios show the need to be able to adjust a nodes data rate in order to increase the system efficiency.

Case 1: Let us assume that Y's queue is full, since Y also gets packets from other nodes in the network. If X sends further packets, the extra packets will be dropped. In this case, not only the system efficiency decreases but also the accuracy of the system in fulfilling its purpose (e.g. data gathering, tracking). The accuracy of the system decreases due to lost packets: the fewer packets a node receives, the less information it has. For this case, we would like to have node X lower its data rate.

Case 2: Let us assume that Y's queue is nearly empty, so it can easily handle more packets from X. However, the data rate of X is still set to a low value and so X sends few packets per timer. Y can process all of the packets from X and possibly other nodes, but the system efficiency decreases because Y can handle more packets.

Let us consider a multi-hop cluster in which nodes are of various modalities (e.g. pressure, video, magnetic, acoustic). Additionally, each node in the cluster has a priority value; for simplicity we assume that the priority is fixed and depends only on the node's modality but more complex priority assignment schemes can be considered. A higher priority data should reach the cluster head with a higher probability than lower priority data. The priority of a data remains unchanged when it passes through intermediate nodes on its way to the cluster head.

We have used MDSTC [17], a fusion-driven distributed dynamic network controller for a multi-modal sensor network that incorporates distributed computation for in-situ assessment, prognosis, and optimal reorganization of constrained resources to achieve high quality multi-modal data fusion. In MDSTC, the design space of the data fusion algorithm is separated into two subspaces, Information Space and Network Design Space. The Network Design Space needs to adapt to the Information Space such that the statistical characteristics (predictability) of the ensemble of the original sensor data are preserved at the level of data fusion. Fig. 2 illustrates the concept of the closed-loop network control that manipulates the network topology τ, based on feedback information of evolving statistical patterns p_t derived from the data sequences $\{y_k\}$ from the sensors. The proposed architecture has the broader implication of providing insight and predictability in designing complex networks and evaluating their resilience to known or unknown operational perturbations. The adaptive parameters of the Network Design Space include the sensor positions, the resource assignments, and the network connectivity.

The MDSTC algorithm is efficient and reliable in clustering nodes that observe the same pattern of interest such that in any cluster only one node, called the *cluster head*, is elected distributively and will coordinate the communication among all

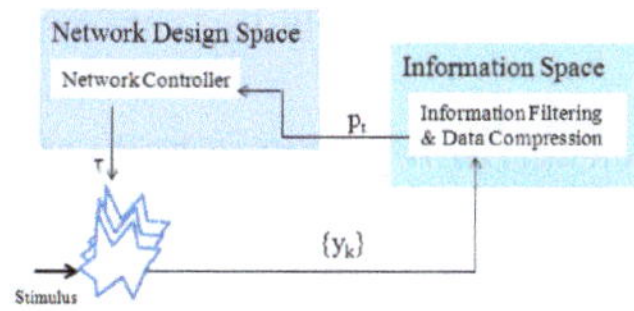

Fig. 2 Fusion-driven network control

other *cluster members*. MDSTC assumes that nodes have unique node IDs. Neighborhood information such as the list of the neighboring nodes and their physical distance to the current node is given by external protocols.

Once a cluster is being formed, the cluster head is responsible for managing the limited resources of the cluster such as the bandwidth while performing its function. We have designed MDSTC for target tracking, but other applications of the WSNs can be employed as well. Target tracking is realized at the cluster head in the following manner. Cluster members report to the cluster head their observed patterns of interest and also their geographical locations. The cluster head collects this information and using some triangulation scheme computes the estimation location and velocity for the target.

If not enough data is received at the cluster head (e.g. due to packet loss or corrupted), the estimation is done based on fewer packets received from fewer nodes, thus the error could be significant. In case two targets are present at the same time, the cluster head may also not be able to distinguish between the two and give erroneous information. This motivates our work in this paper, to allow the cluster head to manage the bandwidth within the cluster, to allow nodes to communicate efficiently to the cluster head and among themselves, and to allow higher priority nodes to have their packets received at the cost of packets coming from lower priority nodes.

3 Data Rate Algorithm

The cross-layer algorithm from [18] aims to optimize each nodes data rate value. Each node within the cluster determines its own data rate value. The algorithm works by considering a node's queue's remaining capacity to hold data. To quantify what it means for a system to be *more efficient* we use the following equation to define the utility of a node:

$$utility = priority \cdot \log(data\_rate \cdot packet\_received\_rate) \quad (1)$$

where $data\_rate$ is the data rate value of the node, $packet\_received\_rate$ is defined as the number of packets received for a node divided by the total number of packets sent to that node, and $priority$ is the priority of the data contained in the packet; if a node has a higher priority it means that its utility will be weighted more heavily within the cluster.

The cluster utility of the system is simply the summation of the utilities for each node within the cluster.

The value of the variable *data_rate* of some node affects not only the data rate in the utility, but also the value of the variable *packet_received_rate* of the parent [2] (or parents, if multiple communication paths are allowed in the multi-hop cluster). If too many packets are sent to a node's queue, they will be dropped. The packet received rate will be lowered because, even though these packets are sent to a node, the node is forced to drop them due to its queue being full.

The data rate algorithm works by adjusting the queue size of a node based on the queue sizes encountered by a packet along the path to the cluster head. The variable *avgq* is used to indicate the currently desired queue size and is periodically adjusted using its old value and the queue holding the most elements along the packet's path to the cluster head. To find the queue holding the most elements, the node will query the node (or nodes) along its path (or paths) to the cluster head, including itself, and compare their queue sizes; this will add a small overhead to the communication within the cluster but will not congest the network since clusters are formed only around the target and not throughout the network. Nodes located before the current node in the path (or paths) to the cluster head will have no influence. The algorithm will merge the old value for *avgq* with the one of the queue holding the most elements found, with 99% of the weight being placed on the old value:

$$avgq = .99 \times avgq + .01 \times max\_queue \tag{2}$$

The date rate is then calculated using the following equation:

$$data\_rate = max\_rate \times priority/avgq \tag{3}$$

where the variable *max_rate* indicates the maximum value the data rate can have. One has the ensure that the data rate remains below the threshold *max_rate*. This threshold has been set for our simulations to 1.0*e*6 (as in [18]), thus the data rate for our simulations is computed as

$$data\_rate = 1.0e6 \times priority/avgq.$$

We note that if the calculated queue size *avgq* is large, then the data rate *data_rate* will be small since we know that the queue towards which the node is sending information to cannot handle many packets, and we know this from the queue size we found. In this way, the cluster works more efficiently and the cluster head will offer a more accurate functionality.

[2] The parent node is either the cluster head or a neighboring node that is also part of the cluster and is physically closer to the cluster head.

4 Simulation Results

The scenario in Figure 3 programmed in NS2 simulates the positioning of the sensors in the lab as follows. Three types of sensors were deployed in city with roads arranged as a Manhattan-like grid: pressure sensors, video sensors (cameras), and magnetic sensors. A large brown block represents a building and a yellow dashed line is a road marking for a 20-feet wide two-way street.

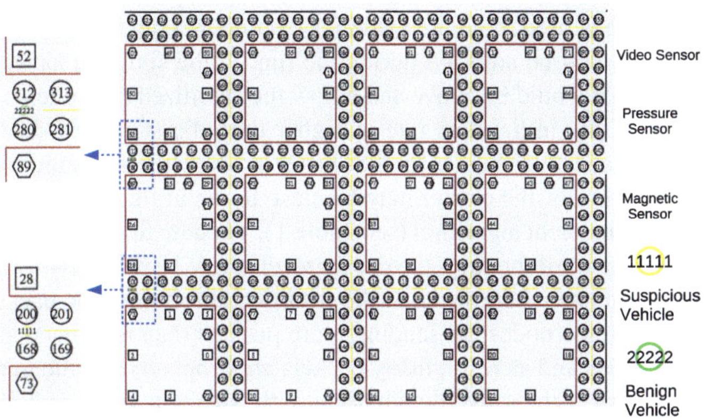

Fig. 3 Urban multi-modal sensor network

Nodes communicate wirelessly using the two-ray ground reflection radio model and the 802.11 channel access scheme. UDP protocol is used at the transport layer and Ad hoc On Demand Distance Vector (AODV) is the routing protocol.

During the operational phase we locate and track vehicles carrying large amounts of metallic material. Such a vehicle will trigger subpattern matches at the pressure, video, and magnetic sensors geographically closer to its moving location. Thus a composite pattern that involves patterns from these multi-modal sensors could be identified and a multi-modal cluster that contains the sensors that matched the subpattern be formed. Such a vehicle is deemed by our sensor network as suspicious and is marked with yellow color in Figure 3. Vehicles that do not carry large amounts of metal will not trigger a subpattern match at any magnetic sensor, thus a composite pattern that involves patterns from pressure, video and magnetic sensors cannot be identified. Such a vehicle is considered by our sensor network as benign and is marked with green in Figure 3.

To evaluate the performance of MDSTC with and without the data rate adjustment algorithm, we consider that the sensors that observe the same multi-modal pattern of interest form a two-hop cluster in which the cluster head (N_4) expects data from the cluster members (N_1, N_2, and N_3) (see Figure 4). N_1 and N_2 send packets to N_3 which simply forwards these packets to the cluster head N_4. N_3 also produces packets on its own. This simple two-hop cluster is the basic step for testing any multi-hop cluster for which the data rate adjustment algorithm is applied. Each node computes and later on adjusts its own data rate. Data rate is determined by its own queue and the nodes' queues along its path to the cluster head. Each cluster member adjusts dynamically its own transmission rate to increase the network utility function.

We ran extensive simulations in NS2 and then parsed the information in the event file in increments of 50 time units. The size of the packets sent by a node is 212 bytes. We set the maximum queue size at a node to be 50 and the initial data rate (defined as bytes per second) is set differently for each scenario. The stabilized data rate is the data rate obtained when using the data rate algorithm. If the adjustable data rate algorithm is used, then the value for the default data rate is not necessary since the data rate will be adjusted during the run. In the simulations we consider initial data rates that would be above and below the stabilized data rate.

As expected, if the initial data rate is higher than the stabilized date rate, the packet received rate is much higher for the cluster in which the adjustment algorithm is used and the utility of the cluster may be close to the utilities measured without using data rate adjustment algorithm (see Table 1). We note that, while the utility is almost the same for all of the runs, the packet receive rate is extremely low for all of the runs except the one where we used the data rate algorithm. The packet receive rate is low because the nodes are sending more packets than the queue at node N_3 can hold and process and as result many packets are dropped from the queue at N_3. The utility is relatively the same though because the data rate is set so high.

Table 1 Initial data rate higher than the stabilized data rate

Data Rate Adjusted?	Starting Data Rate	Stabilized Data Rate	Queue Size	Packet rcvd Rate	Total Utility
Yes	N/A	105k - 140k	50	99-100%	11.5 - 11.9
No	1,000,000	1,000,000	50	7-9%	11.2 - 11.4
No	500,000	500,000	50	15-16%	11.2 - 11.4
No	250,000	250,000	50	30-32%	11.2 - 11.4

If the set date rate is much lower than the stabilized data rate, the packet received rate is lower within the adjusted system and the utility for the data rate adjusted system is much higher than the other non-adjusted systems (see Table 2). We note

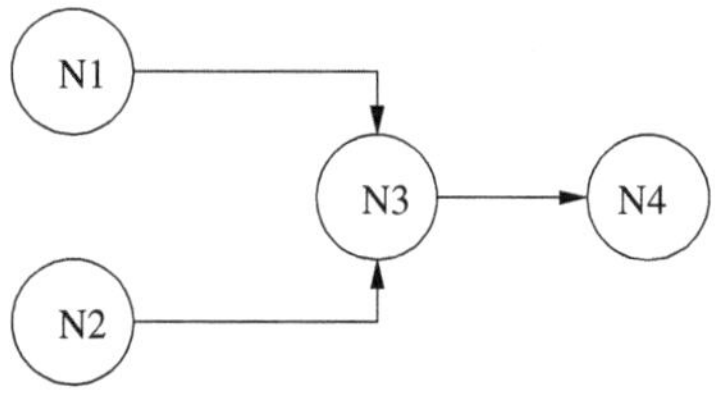

Fig. 4 Two-hop network with 4 nodes

that the packet received rate is relatively the same for all the runs, but the utility is significantly higher for the run in which the cluster uses the data rate algorithm. With a low data rate, the utility suffers, but the queue at node N_3 is able to process the packets without dropping any of them.

Table 2 Initial data rate lower than the stabilized data rate

Data Rate Adjusted?	Starting Data Rate	Stabilized Data Rate	Queue Size	Packet rcvd Rate	Total Utility
Yes	N/A	105k - 140k	50	99-100%	11.5 - 11.9
No	10,000	1,000,000	50	100%	9.2 - 9.3
No	25,000	500,000	50	100%	10.1 - 10.2
No	50,000	250,000	50	99-100%	10.8 - 10.9

From these results we conclude that the data rate algorithm allows the cluster to have a more efficient communication (minimize the number of packets lost or the number of retransmissions). This adjustment is done based on the queue sizes: if they are filled up, the data rate will drop allowing for the packets to be processed but if they are nearly empty, the algorithm will bump up the data rate. Overall, we observe that the number of packets dropped has greatly been reduced, the utility function has increased and functionality of the cluster that is target localization has greatly improved by reducing the position estimation error of the target.

5 Conclusion and Future Work

The inclusion of the data rate algorithm of [18, 20] into a multi-hop allows for better efficiency and accuracy within the cluster. The algorithm adjusts the data rate for each node by determining how many packets to be sent when the timer goes off. In this way, the algorithm can examine the queues of the nodes along the path that the packet will take to reach the cluster head and adjusts the data rate to ensure that an optimal amount of data is being sent each time. As of now, the data rate adjustment algorithm has only been implemented in NS2 but future plans include expansion to a real time experiment where clusters of a higher depth can be used.

Acknowledgements. We would like to thank Dr. Changhee Joo (Korea Institute of Technology) and Dr. Ness Shroff (Ohio State University) for their tremendous help in implementing their proposed data rate adjustment algorithm in NS-2.

References

1. Heinzelman, W.R., Chandrakasan, A.P., Balakrishnan, H.: Energy efficient communication protocol for wireless microsensor networks. In: Proceedings of the 33rd Hawaii International Conference on System Sciences, pp. 3005–3014 (January 2000)
2. Lindsey, S., Raghavenda, C.S.: Pegasis: power efficient gathering in sensor information systems. In: Proceedings of the IEEE Aerospace Conference, pp. 924–935 (March 2002)
3. Younis, O., Fahmy, S.: Heed: A hybrid, energy-efficient, distributed clustering approach for ad hoc sensor networks. IEEE Transactions on Mobile Computing 3(4), 660–669 (2004)
4. Depedri, A., Zanella, A., Verdone, R.: An energy efficient protocol for wireless sensor networks. In: Proceedings of Autonomous Intelligent Networks and Systems (AINS), pp. 1–6 (2003)
5. Smaragdakis, I.: Matta, and A. Bestavros. Sep: A stable election protocol for clustered heterogeneous wireless sensor networks. In: Proceedings of the 2nd International Workshop on Sensor and Actor Network Protocols and Applications (SANPA), pp. 1–11 (2004)
6. Mhatre, V., Rosenberg, C.: Design guideliness for wireless sensor networks: communications, clustering and aggregation. Ad Hoc Network Journal 2(1), 45–63 (2004)
7. Ye, M., Li, C., Chen, G., Wu, J.: Eecs: an energy efficient cluster scheme in wireless sensor networks. In: Proceedings of IEEE International Workshop on Strategies for Energy Efficiency in Ad Hoc and Sensor Networks (IWSEEASN), pp. 535–540 (2005)
8. Phoha, S., La Porta, T.F., Griffin, C.: Sensor Network Operations. John Wiley & Sons, Inc., Chichester (2006)
9. Biswas, P., Phoha, S.: Self-organizing sensor networks for integrated target surveillance. IEEE Transactions on Computers 55(8), 1033–1047 (2006)
10. Zou, Y., Chakrabarty, K.: Distributed mobility management for target tracking in mobile sensor networks. IEEE Transactions on Mobile Computing 6, 872–887 (2007)
11. Zhao, F., Shin, J., Reich, J.: Information-driven dynamic sensor collaboration for tracking applications. IEEE Signal Processing Magazine 19(2), 61–72 (2002)
12. Yang, H., Sikdar, B.: A protocol for tracking mobile targets using sensor networks. In: Proceedings of IEEE Workshop on Sensor Network Protocols and Applications, Anchorage, Alaska, USA (May 2003)
13. Friedlander, D., Griffin, C., Jacobson, N., Phoha, S., Brooks, R.: Dynamic agent classification and tracking using an ad hoc mobile acoustic sensor network. EURASIP Journal on Applied Signal Processing 4, 371–377 (2002)
14. Phoha, S., Jacobson, N., Friedlander, D., Brooks, R.: Sensor network based localization and target tracking through hybridization in the operational domains of beamforming and dynamic space-time clustering. In: IEEE Global Telecommunications Conference, vol. 5, pp. 2952–2956 (2003)
15. Phoha, S., Koch, J., Grele, E., Griffin, C., Madan, B.: Space-time coordinated distributed sensing algorithms for resource efficient narrowband target localization and tracking. International Journal of Distributed Sensor Networks 1, 81–99 (2005)
16. Phoha, S., Ray, A.: Dynamic information fusion driven design of urban sensor networks. In: IEEE International Conference on Networking, Sensing and Control, pp. 1–6 (2007)
17. Bein, D., Wen, Y., Phoha, S., Madan, B.B., Ray, A.: Distributed network control for mobile multi-modal wireless sensor networks. Journal of Parallel and Distributed Computing 71, 460–470 (2011)

18. Lin, X., Shroff, N.B.: The impact of imperfect scheduling on cross-layer rate control in multihop wireless networks. IEEE/ACM Transactions on Networking 14, 302–315 (2006)
19. Shelby, Z., Bormann, C.: 6LoWPAN: The Wireless Embedded Internet. Wiley Series on Communications Networking & Distributed Systems (2010)
20. Lin, X., Shroff, N.B., Srikant, R.: A tutorial on cross-layer optimization in wireless networks. IEEE Journal on Selected Areas in Communications 24(8), 1452–1463 (2006)

Alienable Services for Cloud Computing

Masoud Salehpour and Asadollah Shahbahrami

Abstract. Cloud computing enables delivery of services on-demand networks. Cloud servers with set of resources can response to different user requirements. However, allocating resources to various tasks is still a challenging problem. In this paper in order to provide dynamic resource allocation, we propose alienable services, which improves demanded resources allocation. Our experimental results show the performance of the proposed approach is a factor of 1.47 compared to the common service base models.

1 Introduction

Cloud computing is a set of network enabled services, and inexpensive computing infrastructures on demand, which could be accessed in a pervasive way [1]. There are six major subjects in the cloud computing namely, clients, applications, services, platform, storage, and infrastructure [2–5, 9]. The clients are consumers who receive services from the cloud environments. Services are set of software operations and components which are designed to support the clients requirements. In order to alleviate the burden and cost of software maintenance, customers can use different available applications software on the cloud environment. Platform in the cloud computing enhances development of applications without dealing with complexity of hardware and software management. In addition, the storage and infrastructure of the cloud computing can be used as a virtualization environment to provide services that can deliver storage or infrastructure requirements for customers [9, 14, 15].

However, designing an intelligent software model with an appropriate resource allocation and an adequate load balancing, is a bottleneck in the cloud computing [3, 6–8]. The existing models of resource allocation are typically provisioned to meet

Masoud Salehpour · Asadollah Shahbahrami
Department of Computer Engineering, Faculty of Engineering, University of Guilan, Rasht, Iran
e-mail: Salepour@live.com, Shahbahrami@guilan.ac.ir

F.M.T. Brazier et al. (Eds.): Intelligent Distributed Computing V, SCI 382, pp. 195–200.
springerlink.com

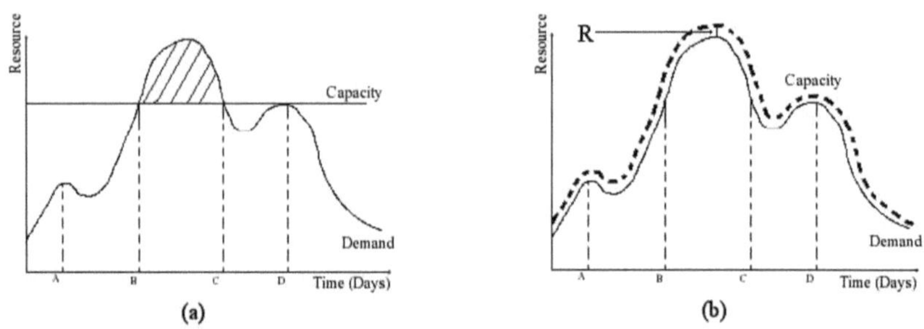

Fig. 1 (a) Static resources allocation in the cloud. (b) Final goal of the dynamic resource allocation.

the expected peak workloads and are not intelligent. These behaviors are depicted in Figure 1a. For example, requirements increase from A point to B point but the allocated resources are not adaptive. Therefore, resources are wasted or demanded resources are overloaded compared to provided resources capacity. This behavior can be seen from B point to C point. On the other hand, demanded resources are the equal to the provided resources capacity in the D point.

The final goal (ideal situation) of the dynamic resource allocation is illustrated in Figure 1b. The duration time for an allocated model to adapt with demanded loads is called "Responsiveness" time that is presented using *R* letter. Our contribution in this paper is proposing alienable services to provide services for clients' demands dynamically with adequate load balancing. The proposed model increases the performance of servicing and reduces response time. For example, our experimental results show the performance of the proposed approach is a factor of 1.47 compared to the common service base models.

The rest of this paper is organized as follows. In Section 2 we present the proposed approach and alienable services. We present our experimental results in Section 3. We discuss some related works in Section 4. Finally, the paper ends with some conclusions in Section 5.

2 The Proposed Approach

In this section we discuss our proposed approach, alienable service model. In order to show the position of the alienable services, we show a general architecture of a cloud computing in Figure 2. There are some components in this figure such as service layer (S1 and S2), and alienable services (small circular black points). An alienable service is a service that has all characteristics of a common service such as loose coupling, granularity, etc. In addition, the alienable service has four other features namely, message handling, specific virtual channel for data passing with storage's interface, predefined status, and pullulation and killing.

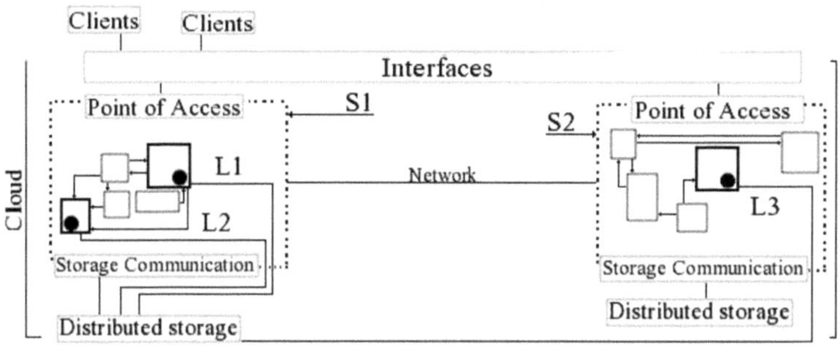

Fig. 2 Structural overview (architecture) of a cloud computing which is used in the cloud provider's server side with alienable services (circular black points).

Message handling ability adds communication protocols to facilitate the integration of all units in a software system for an alienable service. The message is a meaningful tag that the receiver can interpret this message and responses it. Our service's message is in eXtensible Markup Language (XML) format.

Specific channel for data passing with storage's interface, is to make our services able to transact with data storage system independently. This separate channel aids to ease of service composing for alienable services, as well. This is because an alienable service goes to a new service layer after pullulation, then without this feature, pullulation action makes an alienable service isolated from previous service layer storage system. The L1, L2, and L3 show these channels in Figure 2.

Predefined statuses of alienable services are namely, normal, peak, and idle. These statuses defined to ease of making decision about pullulation and killing. These behaviors are depicted in Figure 3.

In Figure 3, the λ shows the condition that an alienable service must pullulate. We define this situation by the name "Peak" as one of our predefine statuses. The λ parameter could be measured using units such as user-sessions per second or requests per second. The P parameter is the maximum load imposed by the demands. It relates to the maximum resource capacity required to successfully handle the clients' demands. In Figure 3, the T parameter is the duration for which the pullulation condition lasts. This is the length of the time period from the arrival of the demand till the load goes down to "Normal", as our another predefined status. The T parameter could indicate a basic shift in the service popularity levels. In Figure 3 the λ' shows the condition that a pullulated alienable service must kill and this status defined as "Idle" for an alienable service.

Alienable service model for allocating extra resources or deallocating unused resources, exploits pullulation and killing. Pullulation feature is for transferring to a new service layer, as resource allocation. Killing feature is to make free extra resources, as resource deallocation. Number of permitted pullulation for an alienable service must set depend on provider's environment.

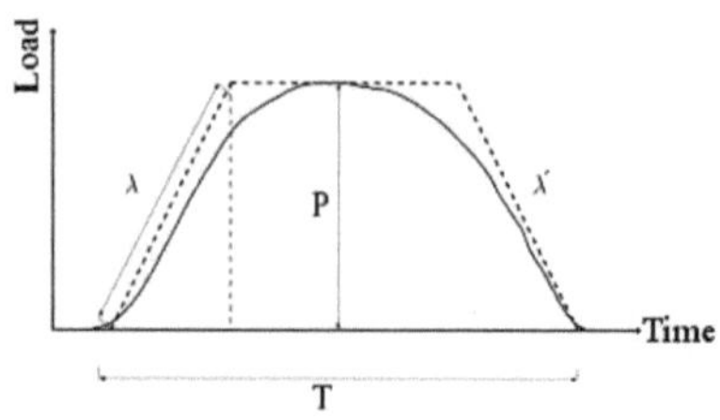

Fig. 3 Different parameters that affect statuses in an alienable service.

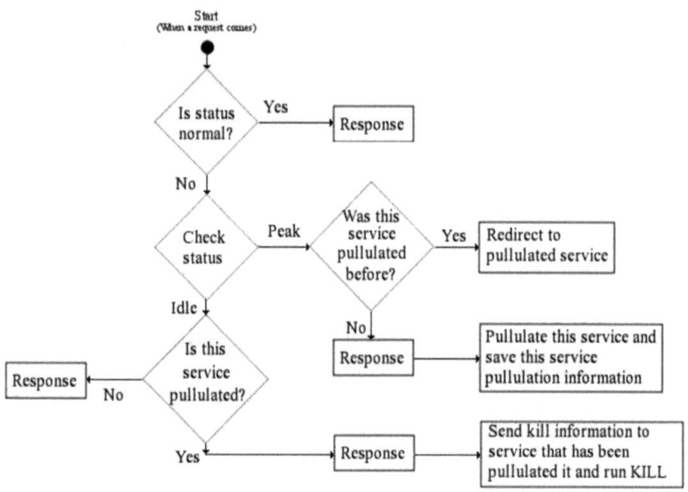

Fig. 4 Diagram of an alienable service's decision mechanism.

In addition, alienable service decision mechanism, makes it intelligent. When the workload of an alienable service increases more than its predefined status ("peak"), it pullulates itself to ameliorate resource allocation. This pullulation means transferring to an exist service layer, which this service layer is ready to accept an alienable service. After an alienable service pullulation, this service to transact with storage system uses "specific virtual channel for data passing with storage's interface" ability. To keep systems integrated after this pullulation, our model has message handling feature that enables an alienable service to interact with other service layers. When the status of an alienable service change to idle, then, this pullulated service kills itself. The diagram of this discussion is depicted in Figure 4.

3 Evaluation

In this section, we present first performance results of alienable service model and compare it to the non-alienable service base model.

Table 1 The result of most important parameters, λ (per minute), average of Responsiveness time (Second), overhead average (Second), and the percent of allocated resources.

Model	P	(λ)	**(R)**	**Overhead**	**Allocated resource (%)**
Non-alienable service	250	20	19.24	27.30	33.32
Alienable service	250	20	18.17	29.06	22.62
Non-alienable service	400	20	31.36	42.97	53.39
Alienable service	400	20	29.84	47.12	36.20
Non-alienable service	750	20	58.15	80.03	100.00
Alienable service	750	20	54.74	88.26	67.87

In our local cloud of servers, we had two groups. We used five nodes as our distributed servers per each group. The host operating system for each server node was Fedora Linux. Each node, comprises a 3 GB of memory, Intel 3.60 GHz CPU (2 CPU cores). All groups were connected through regular 1 GBit/s Ethernet links. We deployed alienable service model in one of these groups. The other group deployed by non-alienable service base model. We have chosen a service base model as our reference. This is because it is a software model and it is currently used in the distributed systems for cloud community. A simulator application has been created to model cloud users and this application sends requests to both groups, simultaneously.

In our experimental results, the most important parameters were total number of users request (P), the load growth rate (λ), responsiveness (R), and allocation overhead. Table 1 depicts the obtained results. According to Table 1, the performance of alienable service model in comparison to non-alienable service base model is a factor of **1.47** in a distributed server environment.

4 Related Work

The problem of resource allocation under hardware level constraints has been discussed in [12, 13]. The proposed solutions are not appropriate for resource allocation, due to concerns of performance limitation in large volume of service requests. An architecture of a resource consolidation system was proposed in [10], which is used as a common architecture in a majority of the later research. This propose suffers from a problem which is generally modelled as “Knapsack Problem”. [11] presents a resource management middleware layer for application placement in data-centers, but this middleware is complex in the process design. In addition, most of existing models can suffer from scalability and flexibility issues. These issues arise mainly due to centralized control which the existing models exploit for resource allocation. As the size of a data center increase the computation becomes more complex.

5 Conclusions

In the cloud computing, there are resources, which can be assigned to different applications. However, resource allocation is a challenging task. To facilitate dynamic resource allocation in the cloud computing, we proposed alienable services. The alienable services provide much more performance than common services. Based on alienable services model, we can conclude that it is possible to design a system to achieve dynamic resource allocation with a performance factor of 1.47 more than common service base models. We plan to improve alienable services ability to adapt to parallel data processing during the request execution automatically.

References

1. Wang, L., Laszewski, G., Kunze, M., Tao, J.: Cloud computing: A perspective study. In: Proc. Grid Computing Environments Workshop (2008)
2. Chang, M., He, J., Leon, E.C.: Service-oriented in the computing infrastructure. In: Proc. IEEE Int. Symp. on Service-Oriented System Engineering (2006)
3. Armbrust, M., Fox, A., Griffith, R., Joseph, A., Katz, R., Konwinski, A., Lee, G., Patterson, D., Rabkin, A., Stoica, I., Zaharia, M.: A view of cloud computing. Communications of the ACM 53(4), 50–58 (2010)
4. The NIST Definition of Cloud Computing, `http://www.csrc.nist.gov`
5. Deelman, E., Singh, G., Livny, M., Berriman, B., Good, J.: The cost of doing science on the cloud: the montage example. ACM/IEEE Conf. on Supercomputing, 1–12 (2008)
6. Randles, M., Lamb, D., Bendiab, A.T.: A comparative study into distributed load balancing algorithms for cloud computing. In: Proc. IEEE Int. Advanced Information Networking and Applications Workshop, pp. 551–556 (2010)
7. Zhang, B., Gao, J., Ai, J.: Cloud loading balance algorithm. In: IEEE Int. Conf. on Information Science and Engineering (2011)
8. Yau, S., An, H.: Adaptive resource allocation for service-based systems. Journal of Software Informatics 3(4), 483–499 (2009)
9. Riad, A., Hassan, Q.: Service-oriented architecture: A new alternative to traditional integration methods in b2b applications. Journal of Convergence Information Technology 3(1), 41 (2008)
10. Walsh, W.E., Tesauro, G., Kephart, J.O., Das, R.: Utility functions in autonomic systems. In: Proc. IEEE Int. Conf. on Autonomic Computing, pp. 70–77 (2004)
11. Adam, C., Stadler, R.: Service middleware for self-managing large-scale systems. IEEE Trans. on Network and Service Management 4(3), 50–64 (2008)
12. Yazir, Y.O., Matthews, C., Farahbod, R., Neville, S., Guitouni, A., Ganti, S., Coady, Y.: Dynamic resource allocation in computing clouds using distributed multiple criteria decision analysis. In: IEEE Int. Conf. on Cloud Computing, pp. 91–98 (2010)
13. Ranjan, R.: Coordinated resource provisioning in federated grids. Ph.D. thesis, The University of Melbourne, Australia (2009)
14. Crawford, C.H., Bate, G.P., Cherbakov, L., Holly, K.L., Tsocanos, C.: Toward an on demand service oriented architecture. IBM Systems Journal 44(1), 81–107 (2005)
15. Pasley, J.: How bpel and soa are changing web services development. In: IEEE Int. Conf. on Computing (2005)

Part VI
Trust and Security

Modeling Distributed Network Attacks with Constraints

Pedro Salgueiro and Salvador Abreu

Abstract. In this work we demonstrate how to model and perform the detection of Distributed Network attacks using NeMODe, a declarative system for Computer Network Intrusion Detection which provides a declarative Domain Specific Language for describing computer network intrusion signatures which span several network packets by stating constraints over network packets, thus, describing relations between several packets, in a declarative and expressive way.

Keywords: Constraint Programming, Intrusion Detection Systems, Domain Specific Languages.

1 Introduction

Maintaining the security of computer networks is a crucial task and plays an important role in keeping the users data safe and the network a safe place to work. Such task can be accomplished by Network Intrusion Detection System (IDS) e.g. Snort [1].

Distributed network attacks are very popular among *hacker* communities, since they are driven from several places and easily elude the detection mechanisms, since these attacks originate from several places at once, being difficult to identify the network traffic as an attack.

There are certain aspects that should be verified in order to maintain the security of the users data, as well as the quality and integrity of the services provided by a computer network. Being able to describe these aspects, together with a verification that they are met can be considered as an Network Intrusion Detection task.

Pedro Salgueiro · Salvador Abreu
CENTRIA and Dep. de Informática, Universidade de Évora, Portugal
e-mail: {pds, spa}@di.uevora.pt

F.M.T. Brazier et al. (Eds.): Intelligent Distributed Computing V, SCI 382, pp. 203–212.
springerlink.com

Describing those conditions, in terms of properties which must be verified in the network traffic, also describe the desired or unwanted state of the network, which can be induced by a system intrusion or another form of malicious access.

While using a declarative programming approach, such as Constraint Programming [2] or Constraint-Based Local Search Programming (CBLS) [3], we can describe those condition, in an easier, natural and expressive way.

NeMODe Intrusion Detection System is a declarative system that provides a Domain Specific Language enabling an easy and very descriptive way to describe the network intrusion signatures by following the constraint programming methodologies. Besides using constraint programming paradigm to describe the desired network situation, it also relies constraint programming to perform the detection of such intrusions, providing several back-end detection mechanisms based on Constraint Programming, such as Propagation Based solvers, using Gecode, and Constraint-Based Local Search, using Adaptive Search. NeMODe system is further described in [4, 5].

This paper is organized as follows. Section 1 introduces the work and makes a brief description of Network Intrusion Detection Systems and Constraint Programming. Section 2 describes the NeMODe. Section 3 presents some distributed network attacks and how to describe them in NeMODe. Section 4 presents the experimental results, Sect. 5 evaluates NeMODe while detecting distributed network attacks, and Sect. 6 presents the conclusions and the future work.

Throughout this paper, we mention some TCP/IP and UDP/IP technical terms, such as *packet flags ,URG, ACK, PSH, RST, SYN, FIN, acknowledgment, source port, destination port, source address, destination address, payload*, described in [6].

1.1 Intrusion Detection Systems

Network Intrusion Detection Systems are very important and one of the first lines of defense against network attacks or other type of malicious access, which constantly monitors the network traffic looking for anomalies or undesirable communications in order to keep the network a safe place.

There are several methods to perform network intrusion detection, but, among them two of them are more used [7]:

1. Based on the network intrusion signatures
2. Based on anomaly detection

On Network Intrusion Detection Systems based on signatures, the network attacks are described using their signatures, particular properties of network packets used to achieve the desired intrusion or attack, which are then looked in the network traffic. Intrusion Detection Systems based on anomaly detection, tries to understand the *normal* behavior of the systems by modeling its behavior using statistical methods and/or data mining approaches. The network behavior is then monitored, and if considered anomalous according the network model, the network is probably under some kind of attack. NeMODe uses use an approach based on signatures.

Snort [1] is a widely used Network Intrusion Detection System, primarily designed to detect signatures that can be identified in a single network packet, using efficient pattern-matching techniques to detect the desired intrusion signature.

Although Snort provides some basic mechanisms which allow the writing of rules that spread over several network packets, such as the `Stream4` or `Flow` preprocessors, they do so in a very limited and counter-intuitive way, not allowing the description of more complex relations between packets, such as the temporal distance between two packets.

Most of the work in the area of Intrusion Detection Systems consists in the development of faster detection methods [8], but there is also some work focused on how the network signatures are described and detected, such as in the work [9], where the authors present a declarative approach to specify intrusion signatures which are represented as a specialized graph, allowing the description of signatures that spread across several network packets.

1.2 Constraint Programming

Constraint Programming (CP) is a declarative programming paradigm consisting in the formulation of a solution to a problem specified as a *Constraint Satisfaction Problem* (CSP) [2], in which a number of variables are introduced, with well-specified domains and which describe the state of the system. A set of relations, called *constraints*, is then imposed on the variables which make up the problem. These constraints are understood to have to hold true for a particular set of bindings for the variables, resulting in a *solution* to the CSP.

There are several types of constraint solvers, in this work we use: (1) Propagation Based solvers; and (2) Constraint Based Local Search(CBLS).

1.3 Propagation-Based Solvers

Problems in Propagation-Based [2] solvers are described by stating constraints over each variable that composes the problem, which states what values are allowed to be assigned to each variable, then, the constraint solver will propagate all the constraints and reduce the domain of each network variables in order to satisfy all the constraints and instantiate the variables that compose the problem with valid results, thus reaching a solution to the initial problem.

Gecode [10] is a constraint solver library based on propagation, implemented in C++ and designed to be interfaced with other systems or programming languages.

1.4 Constraint Based Local Search

CBLS [3] is a fundamental approach to solve combinatorial problems such as Constraint Satisfaction Problems. CBLS is a method that can solve very large problems, although not a complete algorithm and unable to provide a complete or optimal solution. Usually, this approach initiates with an initial, candidate solution to the problem which is then iteratively improved though small modifications until some criteria is satisfied. The modifications to the candidate solution is usually driven by heuristics that guide the solver to a solution.

Adaptive Search (AS) [11] is a Constraint Based Local Search [3] algorithm, taking into account the structure of the problem and using variable-based information to design general heuristics which help solve the problem. The iterative repairs to the candidate solution in Adaptive Search are based on variable and constraint error information which seeks to reduce errors on the variables used to model the problem.

2 Intrusion Detection with Constraints

Detecting Network Intrusions with constraints consists on identifying a set of network packets in the network traffic, which identify and makes proof of the desired network signature attack, matching the desired network signature described through the use of constraints stated over a set of network packet variables, thus describing relations between several network packets.

In order to use the constraint programming mechanism to perform Network Intrusion Detection, there is the need to model the desired signature as a Constraint Satisfaction Problem (CSP). A CSP which models a network situation is composed by a set of variables, V, representing the network packets involved in the description of the network situation; the domain of the network packet variables, D; and a set of constraints, C, which relates the variables in order to describe the network situation. We call such a CSP a network CSP. On a network CSP, each network packet variable is a tuple of integer variables, 19 variables for TCP/IP[1] packets and 12 variables for UDP packets[2], representing the significant fields of a network packet necessary to model the intrusion signatures used in our experiments.

The domain of the network packet variables, D, are the values actually seen on the network traffic window, which is a set of tuples of 19 integer values (for the TCP variables) and 12 integer values (for the UDP variables), each tuple representing a network packet actually observed on the traffic window and each integer value represents each field relevant to intrusion detection. The packets payload is stored separately in an array containing the payload of all packets seen on the traffic window. The correspondence between the packet and its payload is achieved by matching the packet number, i, which is the first variable in the tuple representing the packets and the ith position of the array containing the payloads.

1 Here, we are only considering the "interesting" fields in TCP/IP packets, from an IDS point-of-view.

2 Here, we are only considering the "interesting" fields in UDP packets, from an IDS point-of-view.

Listing 1 shows a representation of such CSP, where P represents the set of network packet variables, where $P_{n,z}$, is each of the individual integer variables of the network packet variable, in a total of z fields for each network of the n variables, with $z = 19$ for TCP packets and $z = 12$ for UDP packets. D is the network traffic window, where $D_i = (V_{i,1}, \ldots, V_{i,z}) \in D$ is one of the real network packets on the network traffic window, which is part of the domain of the packet variables P. $Data$ is the payloads of the network packets present in the network window, where $Data_i$ is the payload of the packet $P_i = (V_{i,1}, \ldots, V_{i,z}) \in D$.

The associated domains of the network packet variables is represented by $\forall P_i \in P \Rightarrow P_i \in D$, forcing all variables belonging to P to obtain values from the set of packets in the network window D.

A solution to a network CSP, if it exists, is an assignment of network packet values, $D_i = (V_{i,1}, \ldots, V_{i,z}) \in D$, to each packet variable, $P_i = (P_{j,1}, \ldots, P_{j,z}) \in P$, that models the desired situation, thus identifying the network packets that identify the intrusion being detected.

Listing 1. Representation of a network CSP

$$P = \{(P_{1,1}, \ldots, P_{1,z}), \ldots, (P_{n,1}, \ldots, P_{n,z})\}$$
$$D = \{(V_{1,1}, \ldots, V_{1,z}), \ldots, (V_{x,1}, \ldots, V_{x,z})\}$$
$$Data = \{Data_1, \ldots, Data_x\}$$
$$\forall P_i \in P \Rightarrow P_i \in D$$

2.1 A DSL to Describe Network Signatures

The NeMODe Intrusion Detection System is a declarative system that provides a Domain Specific Language, following the constraint programming methodologies, enabling an easy and very descriptive way to describe the intrusion signatures that spread across several network packets by allowing to state of constraints over network entities and express relations across several network packets.

This Domain Specific Language (DSL) will then translate the *program* into constraints which are then solved by several constraint solving techniques, including Propagation-based systems, such as Gecode and Constraint-Based Local Search(CBLS), such as Adaptive Search. This DSL is further described in [4, 5].

2.2 Architecture

NeMODe is composed by a compiler, which reads a NeMODe program and parses it into a semantic model. Then, based on that semantic model, it is generated code for each of the available back-ends in system.

After all recognizers have been generated, each generated back-end receives as input the network traffic and produces a valid solution, if the intrusion described as a NeMODe Program exists on the network traffic that was given as input to each back-end detection mechanism.

All back-ends available in the system work in parallel, each one producing a solution to the problem. In a final step, the best solution produced is selected, which is simply the first solution to be produced.

Fig. 1 represents the architecture of the system and how the data flows between each component.

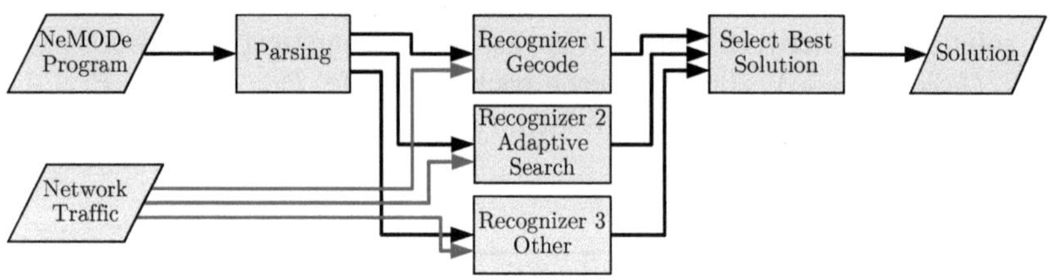

Fig. 1 NeMODe system architecture

3 Examples

So far, we have worked with some simple network intrusion signatures: (1) a distributed port-scan attack, and (2) a distributed SYN flood attack. All of these intrusion patterns can be described using NeMODe and the generated code was successful in finding the desired situations in the network traffic logs.

3.1 Distributed Port-Scan

A port-scan is an important step before a network attack, scanning for available services on a given computer, allowing to discover the potential vulnerabilities of the victim, helping the attackers to better the attack. A distributed port-scan is made from several computers, geographically distributed, all scanning the services of a single computer.

The port-scan attack can be described by a set of constraints between two network packets, the packet that initiates the TCP/IP connection and the packet that closes it, which closes the connection in a very short time after the TCP connection had been initiated. In order to detect if the network is under such attack we monitor if there are many of these sets of networks packet to appear on a network traffic in a short period of time.

Listing 2. A port-scan attack using NeMODe

```
portscan {
  C = { packet(A), tcp(A), syn(A), nak(A),
          packet(B), tcp(B), rst(B),
          time(A) - time(B) < usecs(100),
          related(A,B) },
    R := repeat(5,C),
    max_interval(R) < usecs(500)
  }
```

3.2 *Distributed SYN Flood Attack*

A Distributed SYN flood attack happens when several attackers, from several places, join forces to drive a distributed attack in order to initiate more TCP/IP connections than the server can handle and then ignoring the replies from the server, forcing the server to have a large number of half open connections in standby, which leads the service to stop when this number reach the limit of number of connections. This attack can be detected if a large number of connections is made from a single machine to other in a very short time interval. Listing 3 shows how a SYN flood attack can be described using NeMODe.

Listing 3. A SYN flood attack programmed with NeMODe

```
syn_flood {
  C = { tcp_packet(A), syn(A), nak(A) },
  R := repeat(30,C),
  max_interval(R) < usecs(500)
};
```

4 Experimental Results

Among others, we have tested the examples of Sect. 3, a Distributed SYN flood and a Distributed port-scan. All these intrusions were successfully described using NeMODe and valid Gecode and Adaptive Search code was produced and then executed in order to validate the code and ensure that it could indeed find the desired network intrusions.

The code was then run on a dedicated computer, an HP Proliant DL380 G4 with two Intel(R) Xeon(TM) CPU 3.40GHz and with 4 GB of memory, running Debian GNU/Linux 4.0 with Linux kernel version 2.6.18-5. The Adaptive Search was also run on a IBM BladeCenter H equipped with QS21 dual-Cell/BE blades, each with two 3.2 GHz processors, 2GB of RAM, running RHEL Server release 5.2, since it was recently ported to the Cell/BE architecture, presented in [12].

For the Distributed Port-scan attack, we created a log file composed of 400 TCP network packets while a computer was being under a Distributed Port-scan attack, being used as the network traffic. The used signature describes 52 TCP packet variables.

In the case of the Distributed SYN flood attack, we created a log file composed of 100 TCP network packets while a computer was being under a Distributed SYN flood attack, which was used as the network traffic. The signature used to model the problem was composed by 30 TCP network packet variables.

Table 1 presents the time(user time, in milliseconds) required to find the desired network situations for the attacks presented in is present work, using Gecode and Adaptive Search on a x86 architecture and also using Adaptive Search on the Cell/BE architecture. The number of network packet variables used to model the signatures and the number of network packets used to search the intrusions are also presented. The times presented are the average of 128 runs.

Table 1 Time(in milliseconds) necessary to detect the intrusions using Gecode and Adaptive Search, average of 128 runs

Intrusion to detect	Signature size	Window Size	Gecode x86 (ms)	A.S. x86 (ms)	A.S. Cell (ms)
Port-scan	52	400	127.3	674.9	62.5
SYN flood	30	100	56.6	6.1	22.5

5 Evaluation

The experimental results described in Sect. 4 shows that the performance varies in a great scale depending on the problem and the recognizer.

Table 1 shows that the Port-scan takes longer to be recognized than the Distributed SYN flood. This behavior is explained by the complexity of the problems, which is directly related to the way the problems are modeled, since the Port-scan is modeled by using more variables than the SYN flood attack, also, the constraints used to model the Port-scan are more complex than the ones used in the SYN flood, making the Port-scan much more complex than the SYN flood attack.

In the x86 architecture, the Port-scan performs better in Gecode than on Adaptive-Search, as for the SYN flood-attack, the opposite is verified. This shows that in the x86 architecture Gecode performs better on more complex, but when the complexity of problems decrease, Adaptive Search performs better.

While the performance of Adaptive Search Cell/Be version the x86 version can't be directly compared, it's possible to conclude that more complex problems takes advantage of the Cell/BE architecture and the performance doesn't degrades as much when the complexity increases.

As for the SYN flood in the Cell/BE A.S. version it presents a worst performance, which could be explained by signature used to model the problem, which is much simpler than the Port-scan attack. Also, the network packets that make part of the attack are much closer together in the network packet window, making the search of solution much simpler. Do to these two facts, the problem gets much less complex, not being able to take full advantage of the Cell/BE architecture.

The results obtained, are quite good, allowing us to start the detection of intrusions in real network traffic instead of log files.

6 Conclusions and Future Work

In this work we describe how to model two distributed network attacks in NeMODe, a system for Network Intrusion Detection based on Constraint Programming, allowing an intuitive an expressive way to describe such situation due to the possibility of the specification of relations between several network packets in an easy way.

This work shows that we can easily describe distributed network attacks or network situations, which spread across several network packets, using a declarative approach, and, from that single description, generate several Constraint Programming based network situation recognizers, using different Constraint Programming paradigms, which actually detect the desired intrusions, if they exist in the network traffic.

With this work, we also proved that we can easily describe and detect signatures which spread across several network packets, something which is hard to achieve in systems like Snort. Although the intrusions mentioned in this work can be detected with other intrusion detection systems, they are modeled/described with out relating several network packets, usually being described by specifying a set properties over a single network packet, which could lead to a large number of false positives.

The results obtained are very promising, providing a platform to start performing network intrusion detection on a live network traffic link in a near future, a very important future step.

We still need to model more network situations as a CSP to better evaluate the performance of the system. We also need to better evaluate the the work presented in this paper by comparing the obtained results with systems like Snort. Also, we have plans to implement new back-end detection mechanisms using different constraint programming paradigms.

Acknowledgments. Pedro Salgueiro acknowledges FCT –Fundação para a Ciência e a Tecnologia– for supporting him with scholarship SFRH/BD/35581/2007. The IBM QS21 dual-Cell/BE blades used in this work were donated by IBM Corporation, in the context of a SUR (Shared University Research) grant awarded to Universidade de Évora and CENTRIA.

References

1. Roesch, M.: Snort - lightweight intrusion detection for networks. In: LISA 1999: Proceedings of the 13th USENIX Conference on System Administration, pp. 229–238. USENIX Association, USA (1999)
2. Rossi, F., Van Beek, P., Walsh, T.: Handbook of constraint programming. Elsevier Science, Amsterdam (2006)
3. Van Hentenryck, P., Michel, L.: Constraint-based local search. MIT Press, Cambridge (2005)

4. Salgueiro, P., Diaz, D., Brito, I., Abreu, S.: Using Constraints for Intrusion Detection: the NeMODe System. In: Rocha, R., Launchbury, J. (eds.) PADL 2011. LNCS, vol. 6539, pp. 115–129. Springer, Heidelberg (2011)
5. Salgueiro, P.D., Abreu, S.P.: A dsl for intrusion detection based on constraint programming. In: Proceedings of The 3Rd International Conference on Security of Information and Networks, SIN 2010, pp. 224–332. ACM, USA (2010)
6. Comer, D.: Internetworking With TCP/IP Volume 1: Principles Protocols, and Architecture, 5th edn. Prentice Hall, Englewood Cliffs (2006)
7. Zhang, Y., Lee, W.: ntrusion detection in wireless ad-hoc networks. In: Proceedings of The 6Th Annual International Conference on Mobile Computing and Networking, p. 283. ACM, New York (2000)
8. Arun, K.S.P.: Flow-aware cross packet inspection using bloom filters for high speed data-path content matching. In: IEEE International Advance Computing Conference, IACC 2009, vol. 6-7, pp. 1230–1234 (2009)
9. Kumar, S., Spafford, E.H.: A software architecture to support misuse intrusion detection. In: Proceedings of The 18th National Information Security Conference, pp. 194–204 (1995)
10. Schulte, C., Stuckey, P.J.: Speeding up constraint propagation. In: Wallace, M. (ed.) CP 2004. LNCS, vol. 3258, pp. 619–633. Springer, Heidelberg (2004)
11. Codognet, P., Diaz, D.: Yet another local search method for constraint solving. In: Steinhöfel, K. (ed.) SAGA 2001. LNCS, vol. 2264, p. 73. Springer, Heidelberg (2001)
12. Diaz, D., Abreu, S., Codognet, P.: Parallel constraint-based local search on the cell/BE multicore architecture. In: Essaaidi, M., Malgeri, M., Badica, C. (eds.) Intelligent Distributed Computing IV. Studies in Computational Intelligence, vol. 315, pp. 265–274. Springer, Heidelberg (2010)

Gain the Best Reputation in Trust Networks

Vincenza Carchiolo, Alessandro Longheu, Michele Malgeri,
and Giuseppe Mangioni

Abstract. A generic user can find a great number of on-line systems to access a huge amount of information. Unfortunately, usually there is no control about the quality of information. To deal with this issue, several trust and reputation systems have been developed in the last years, adopting several metrics as the well-known EigenTrust. In this paper we investigate on the EigenTrust dynamics, showing that it exhibits some counter-intuitive behaviors especially if compared to our common sense notion of trust. and or trustworthy people.

1 Introduction

Today's virtual worlds allow people to perform activities in disparate application domains, as social networking, e-commerce, entertainment, e-learning and many others. The increasing complexity and amount of information (but also goods and money) exchange due to virtual interactions often demands for a robust trustworthiness framework that guarantees the reliability of often unknown, counterparts (people, peers, agents, websites). Well-known mechanisms to address this issue are trust and reputation [1–3], which have been successfully applied to large peer-to-peer and e-commerce systems. In particular, some issues about trust seem to be consolidated [4]:

- trust is based on the experience: each agent[1] can identify repeated experiences with others. On the other hand, *reputation* is used for non-direct interactions; agents exchange information about reputation through *recommendations*;
- trust is dynamic and non-monotonic, it changes over time - usually according with a new direct experience and/or some new recommendations. Trust values can also be affected by the joining and leaving of nodes in the trust network.

Vincenza Carchiolo · Alessandro Longheu · Michele Malgeri · Giuseppe Mangioni
Dip. Ingegneria Elettrica, Elettronica e Informatica, Facoltà di Ingegneria, Università degli Studi di Catania , Italy

[1] In this paper, we will use the term agent, peer, person or node indifferently.

F.M.T. Brazier et al. (Eds.): Intelligent Distributed Computing V, SCI 382, pp. 213–218.
springerlink.com

Since the early work by Marsh [5], where the first analytical formulation of trust was provided, many algorithms have been proposed [6–8]. One of the first and well-known is Eigentrust [6], a reputation-based metric which associates to each agent a global trust index using an algorithm inspired by PageRank [9].

The *dynamics* of trust network is an important question to deal with that has been only partially addressed by existing metrics. Actually the term has in this context (at least) two different meanings, the former takes into account the evolution of trust values over time, generally due to the change of agents behavior during their lifetime, whereas the latter, on which we focus, deals with the process of joining to (or leaving) the network.

In this paper we present a preliminary study of an agent's joining process to a trust network, in particular we consider two different types of network, i.e. a random and a scale-free network [10], performing exhaustive experiments with all possible ingoing links combinations between the joining node and all others, introducing the concept of *effort* as a measure of how many trust ingoing links the new agent has to get from others until he gains a certain value (possibly high) of trust, as it occurs in humans behavior. We adopt EigenTrust since it is one of the simplest metric, efficient in evaluating global trust; we also want to test whether this popular metric actually captures the common sense human gives to the term trust.

To our knowledge, our study has not been specifically considered in existing metrics [6–8]. Our approach is partially inspired by similar works proposed for the PageRank algorithm [11, 12].

The paper is organized as follows. In section 2 we cope with the problem of a node attaching to the network, while in section 3 we show our simulations and results, finally providing conclusion remarks and future works in section 4.

2 The Joining Process in a Trust Network

The question of how an existing trust network changes when a new agent (X in the following) joins and establishes trust links with others is not trivial, especially if we consider how many possible connections X can have in terms of both their number, which other agents are involved, and how much trust X gives to and receives from others.

Generally, if X is honest the goal of his joining process is to gain the highest trust. In particular, if X can choose which and how many agents trust him (incoming links) the trivial solution is that X becomes *fully* trusted by *all* nodes in the network. In real systems, though, gaining a good reputation conceals some costs, for instance in a P2P network a peer must always provide other peers with uncorrupted files to increase his reputation (hence, the trust). In general, the better the behavior according to some specific criteria, the higher the trust. This however requires the agent to somehow exert in achieving this goal, therefore we introduce the term *effort* to quantify the application with which X pursues his good behavior. More formally:

$$effort_X = \sum_j c_{jX} \tag{1}$$

where c_{jX} are the normalized local trust values X receives from his neighbors j according to the EigenTrust metric (see [6] for details).

According to this definition, a node X gains the best trust by receiving the highest direct trust from others (labels of X's ingoing links), though this means that X's effort will be the highest. Beyond this trivial solution, a better approach is to search for a trade off between the effort and the trust, hence the attaching strategy of the new node X can be reformulated as *how to get the best trust with the minimal effort*. This can be viewed as the multi–objective optimization problem of evaluating the direct trust c_{jX} a node X has to gain from his neighbors j to optimize the two following conflicting objectives:

$$\begin{cases} max\ t_X \\ min\ effort_X \end{cases} \tag{2}$$

In this preliminary work, rather than focusing on the multi-objective optimization problem we perform an exhaustive study on a small network. In particular, we consider two different types of network, i.e. random and scale-free networks [10], both with 20 nodes, then we perform experiments with all possible ingoing links combinations between the joining node X and all others, evaluating both the gained trust and the effort needed to get such a trust. The choice of a small network comes from the high number of possible links combination the node X can have (i.e., 2^N for N nodes).

A last question is whether the joining node is honest or malicious. If getting the best trust with minimal effort is a reasonable goal for honest agents, in the latter case we should study how to avoid that the node easily gains too much trustworthiness, since this could lead to subverting trust values in the network; at this stage, we assume that the agent is honest.

3 Simulations and Results

The goal of the simulations discussed in this section is to investigate what happens when the new agent X joins an existing trust network to gain the best trustworthiness with a minimal effort.

The initial scenario is as follows. Before X's arrival, in the network all existing arcs are labeled by 1.0 as trust value. We choose this setting in order to avoid that the distribution of trust values affects our simulation results. Note that the 1.0 trust value just refers to each *direct* trust, the *global* trust of each agent is instead evaluated with the EigenTrust metric and falls in the range $[0,1]$; also note that direct trust values may change as soon as X joins the network.

In order to reduce local effects and achieving significant results we built two group of networks: 10 random networks and 10 scale-free networks, each with 20 nodes. The experiment was to consider for each network all possible configurations where X is trusted by 1 up to 20 nodes in the network, in other words X's indegree ranges from 1 to 20. The label of any new arc is set to 1.0 (full trust), and we evaluated both X's gained (global) trust and his effort to get such a trust. Note that despite the little size of the networks, the number of configurations to be analyzed for each of them is $2^{20} = 1\ 048\ 576$.

Figures 1(a) and 1(b) reports X's global trust evaluated with the EigenTrust metric (y-axis = trust) as a function of the number of nodes that trust X (x-axis = effort). At a first glance, we can state that the more the effort the higher the trust, however we note that:

- given an effort value, a certain range of trust values can be achieved depending on which subset of nodes X has been trusted by. Of course, the number of subsets grows as effort increases, for instance if X is trusted by exactly 1 node we have 20 possible configurations, whilst being trusted by, say, 10 nodes results into 184 756 configurations. On the other hand, different configurations will lead to the same trust value. All this suggests that a good trust it is not only determined by the amount of effort X exerts but also by an appropriate choice of the subset of nodes X should be trusted by.
- Conversely, given a trust value, different efforts allow us to achieve it, for instance in fig. 1(a), the trust 0.1 can be acquired with an effort value of 10, 11 or 12. This suggests that X could achieve a desired trust value somehow limiting his effort (10 rather than 12 in the example).

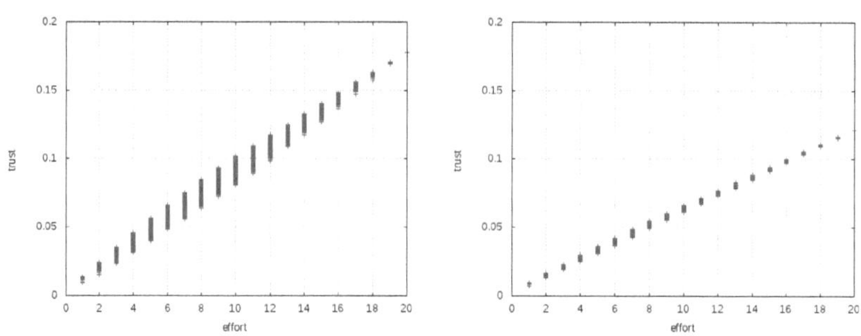

(a) Random Network (b) Scale Free Network

Fig. 1 Gained trust vs. effort

Both kinds of networks shown in figures 1(a) and 1(b) display similar results but, given an effort value, the range of trust values in the scale–free network is reduced than random one; this can be justified remembering that in scale–free networks a few nodes (called *hubs*) concentrates most connections, therefore when X connects to them his trustworthiness is not significantly affected by connections with other nodes. Finally, note that figures 1(a) and 1(b) represent two networks over the 10+10 examined in our simulations, however all others exhibited a similar behavior.

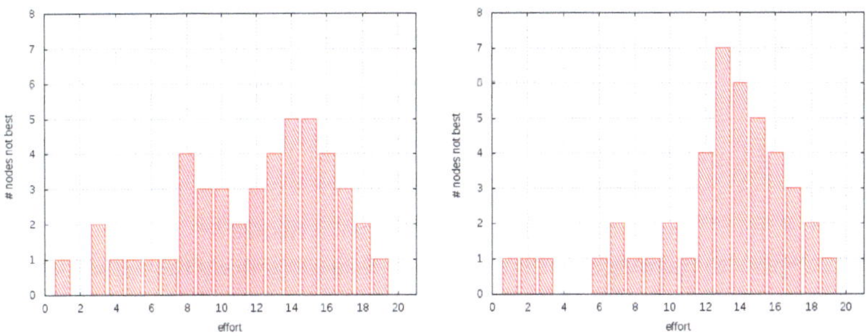

(a) Random Network (b) Scale Free Network

Fig. 2 Number of nodes selected but not best

In order to understand which subset of nodes (configuration) should be chosen to maximize trust achievable with a given effort, we may say that X should try to be trusted by the most trusted nodes; this idea agrees with humans behavior in social networks, where a newcomer usually tries to make friends with most authoritative persons in the network, aiming at obtaining the highest reputation (hence, trust) by others. Simulations though show a different, counter-intuitive behavior, i.e. the nodes to which X should attach to are not necessarily the most trusted (those with the highest trust values).

To quantify this, we first rank all nodes in the network according to their global trust value as provided by EigenTrust before the new node X joins the network (say B the ordered list of nodes). For each effort value e we consider the configuration that provides X with the highest trust (*best* configuration); we then define two lists, i.e. $\tilde{B}_e$ as the list of first e nodes in B (these are the *most trusted* nodes as stated above) and $\tilde{A}_e$ as those nodes X is trusted by, finally plotting in figure 2 how many nodes are in $\tilde{A}_e$ and not in $\tilde{B}_e$, (*#nodes not best* on y-axis). In such figure, we note that the in most cases X is trusted by nodes that are not necessarily the most trusted, which is a somehow counter-intuitive conclusion that occurs both for random and scale–free networks and in all networks we considered for simulations. In particular X is linked by nodes that are all the most trusted only in the 18.36% of possible cases, whereas a number of nodes *not best* from 1 to 3 occurs in the 58,16% of cases and the remaining 23.46% concerns with 4 to 7 nodes not best. This behavior is mainly due to the normalization of trust values, indeed adding new arcs make other's trust value decrease ($\sum_i t_{ij} = 1$).

4 Conclusions and Future Work

In this paper we presented a first study of a nodes joining process to an existing trust network based on the EigenTrust metric. According to human behavior, we introduced the concept of effort as a measure of how many trustworthiness links the new node has to get from others until he gains a certain trust; our basic idea is how to get the maximum trust with a minimal effort. Results of simulations on both random and scale-free networks shown that EigenTrust exhibits some counter-intuitive behavior, in particular searching for an ingoing link from the best trusted node does not imply a high trust for the new node.

Several questions that deserve future investigations arose from this work:

- apart from EigenTrust, which metric (GossipTrust, TrustWebRank...) fulfills with intuition? Do we need a new metric?
- which results could be expected in larger (real) networks?
- what happens to the outgoing links of the new node joining the network?
- what happens if a group of new nodes join the network simultaneously?
- how can be addressed the question of malicious nodes joining the network?

References

1. McKnight, D.H., Chervany, N.L.: The meanings of trust. Technical report, Minneapolis, USA (1996)
2. Misztal, B.: Trust in Modern Societies. Polity Press (1996)
3. Despotovic, Z., Aberer, K.: P2P reputation management: probabilistic estimation vs. social networks. Comput. Networks 50(4), 485–500 (2006)
4. Artz, D., Gil, Y.: A survey of trust in computer science and the semantic web. Web Semantics: Science, Services and Agents on the World Wide Web 5(2), 58–71 (2007)
5. Marsh, S.: Formalising trust as a computational concept. Technical report, University of Stirling PhD thesis (1994)
6. Kamvar, S.D., Schlosser, M.T., Garcia-Molina, H.: The eigentrust algorithm for reputation management in P2P networks. In: Proceedings of the Twelfth International World Wide Web Conference (2003)
7. Zhou, R., Hwang, K., Cai, M.: Gossiptrust for fast reputation aggregation in peer-to-peer networks. IEEE Trans. on Knowl. and Data Eng. 20(9), 1282–1295 (2008)
8. Walter, F.E., Battiston, S., Schweitzer, F.: Personalised and dynamic trust in social networks. In: Bergman, L.D., Tuzhilin, A., Burke, R.D., Felfernig, A., Schmidt-Thieme, L. (eds.) RecSys, pp. 197–204. ACM (2009)
9. Berkhin, P.: A survey on pagerank computing. Internet Mathematics 2(1), 73–120 (2005)
10. Newman, M.: The structure and function of complex networks. SIAM Review 45 (2003)
11. Csáji, B.C., Jungers, R.M., Blondel, V.D.: Pagerank optimization by edge selection. CoRR abs/0911.2280 (2009)
12. de Kerchove, C., Ninove, L., Dooren, P.V.: Maximizing pagerank via outlinks. CoRR abs/0711.2867 (2007)

A Security Approach for Credential-Management in Distributed Heterogeneous Systems

Marcus Hilbrich, Denis Hünich, and René Jäkel

Abstract. In distributed computing environments information is stored in different locations using different technical systems. Challenging is to allow easy access to heterogeneous resources without an uniform user management and to combine data from different resources for further processing. To delegate the user requests the according login data for different resources have to be present in a central location. The management of these highly sensible data comprises miscellaneous security aspects, e.g. to develop methods to store login data in a secure way. In this paper a security concept to deal with sensitive login data in a central component is presented. This includes a collaboration with the operators of the distributed resources and is discussed in the context of the knowledge infrastructure WisNetGrid.

Keywords: Security, Credentials, Cryptography, WisNetGrid, SSO, Grid, Knowledge Grid.

1 Introduction

One of the challenges of today's computer science is to enable various users to combine data from different resources from different fields of interest for further processing. In grid environments a specific technical approach is used to achieve particular project requirements. A collaborative work over different fields of interest and for diverse groups of users is often hard to achieve or even impossible due to technical and project-specific limitations.

To allow users to access data from different fields an access layer needs to be created to provide a common view to various resources. The underlying heterogeneous data layer can consist of different resource types at different locations. In this

Marcus Hilbrich · Denis Hünich · René Jäkel
Center for Information Services and High Performance Computing (ZIH),
Technische Universität Dresden

F.M.T. Brazier et al. (Eds.): Intelligent Distributed Computing V, SCI 382, pp. 219–224.
springerlink.com

context heterogeneous not only means different types of resources, it also implies different login mechanisms (e.g. username/password or the use of certificates) and providers.

For convenience, a central login location as entry point, realised in a web portal, is provided with an integrated service to manage the users login data. The placement of the users data at a central location (credential-store) is a security problem by design and most likely not accepted by users or forbidden by the resource operators.

In this paper we describe a way to avoid these security problems by handling credentials in a secure manner and discuss an architecture on top of distributed systems and a convenient mechanism to ensure accessibility to different kinds of resources with different access mechanisms. The approach presented here is part of the data access layer within the project WisNetGrid [1] (funded by the German Federal Ministry of Education and Research (BMBF)), which can use arbitrary grid and non-grid resources. Furthermore, a detailed description of the credentials and their usage is given and how these approach can be secured. Finally the presented security concept is discussed in the context of the German grid project WisNetGrid.

2 Related Work

To build a homogeneous access layer to connect various resources implies a delegation of user privileges from a central instance. A common approach for such architectures is the Single Sign on (SSO) concept. This concept allows to access all associated resources with the same login. Furthermore the authentication of a user is automated in case the login procedure was already processed for another resource. This way user rights delegation is implicitly usable to access further resources on request. Such a concept for grid-like environments is presented in [2]. Projects like MyProxy [3] or general concepts, such as role-based access control (RBAC) [4], can not be adopted easily, since a general access mechanism is not present for different middlewares used in grid environments, which still act independently from each other.

An other related project which uses distributed resources is the grid workflow system GWES [5], which allows to start grid jobs via a web-portal. Due to the use of one specific middleware it is not feasible to integrate different data management technologies.

3 Architecture of Target Systems

In the main scope of this work are architectures, where a central component is used as entry point for users to access different resources. As introduced in Sect. 1 this central component might be operated by a different authority as the under laying resource layer. As consequence access permissions can not be provided a priory by the central components, but the users can manage their own login data to be able to connect to the resource layer.

To allow users to access remote resources with their own user privileges the access management delegates existing login data, like certificates (used in D-Grid [6], which is a national German grid initiative), or username and password. Based on the authorisation, the user can make use of services provided by the central authority on accessible data. Specialised adaptors are needed to be implemented on each front-end in order to delegate the users request from the user-interface at the central component to the data sources. This way, many different resource technologies, such as dCache [7], iRODS [8], MySQL [9] or others, can be adapted. The interaction of the individual components is visualised in Fig. 1.

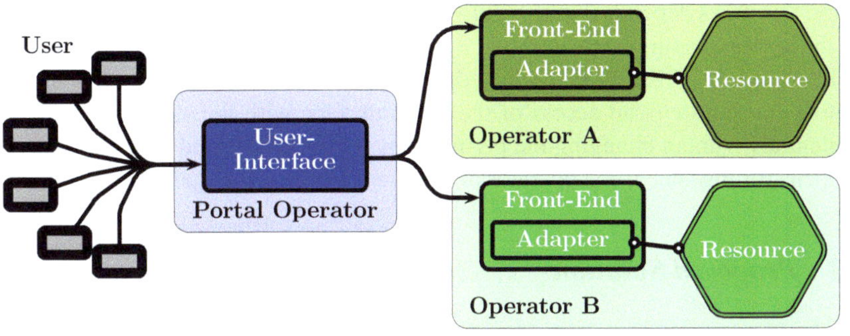

Fig. 1 Users can access heterogeneous resources by a central component (user-interface), which offers a homogeneous view to the data resources. The connection is realised by adapters to different resource types.

4 Managing Credentials

4.1 Delegation of Privileges

As consequence of the architecture described in Sect. 3 the user-interface has to act as the user, who triggered the request. The user has to commit the login data for the resource to the user-interface which creates a credential and stores it in the credential-store. The login data can be a grid proxy certificate, a private key, username/password or whatever is suitable to get access to connected resources. Then the user can start a request to a connected resource on the central server, which now holds the user login data to the resource persistent in its credential-store. The central server delegates the request together with the credential to the requested front-end and the matching adapter. There, the credential is read and used to accesses the resource.

It is not sufficient just to store the credentials in the credential store in a general readable format or as encrypted data next to the key (e.g. key stored on the same server). The resulting security problems are described in the following section.

4.2 Security Concerns

The access mechanism described above has a central component (user-interface), which contains the credential-store. If an attacker gets access to the credential-store in the user-interface the accounts of all connected resources are affected. The resources can be accessed by the attacker with the privileges and the identity of the users. As result all logins to the resources have to be disabled, which can not be done by the user-interface operator directly. This means the security level of the services of the resources depends on the security of the user-interface, which is not managed by the resource administrators.

Furthermore, if the user-interface operator is no longer reliable the resource operators can not deny an access of the user-interface with the stored compromised credentials without changing the permissions for every affected user account. This happens for instance if the operator of the user-interface changes.

4.3 Proposed Solution

A refined approach involves the front-end to secure the credentials to avoid the drawbacks described in the previous section. Moving the credential-store to the front-ends only relocate the problems. Therefore, the delegation process is divided into two steps. In the first step the encrypted credential is stored in the credential-store and in the second step the credential is decrypted on the front-end.

The encryption and decryption of the credentials is done with asymmetric cryptography [10]. The user-interface encrypts credentials with the public key of the chosen front-end and stores it in the credential-store. On a request the encrypted credential combined with the request data is transferred to the front-end. There the credential is decrypted with the private key of the front-end and passed to the matching adapter of the requested resource. Afterwards the login data and credential data are deleted on the front-end.

The separation of the delegation process demands a cooperation of the user-interface with the front-end, which is already present in form of the specialised adapter. No resource can be addressed without using implicitly having the private key of the assigned front-end and the resource operators can decide whether they trust the user-interface operator, or not. If they do, they provide an adapter at the local front-end. Moreover, the resource administrator has complete access control, which means by taking changes at the local front-end the user-interface operator can not further access the resource.

In contrast to the concept described in Sect. 4.1, Fig. 2 shows the concept using cryptography methods to protect the credentials.

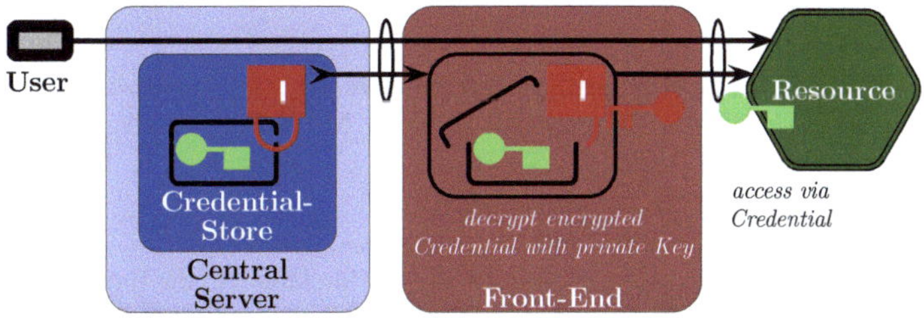

Fig. 2 The login data (green key) are encrypted with the public key of the front-end (red lock). This encrypted credential is stored on the central server. The front-end decrypts the encrypted credential with its private key (red key) to access the resource.

5 Conclusion

This paper discusses security aspects which arise for systems that allow various users to address resources with different access mechanisms. Since login data are in general very sensitive data, the central system stores only variants encrypted with the public key of the corresponding front-end. When the credential needs to be decrypted, e.g. in case of a request, it is only possible with the private key of the front-end and only at the front-end, where the key usually is located. The asymmetric key encryption ensures that the login data can not be misused directly, in case the central system is compromised. This concept can be easily adapted to other systems with a similar architecture.

The grid project WisNetGrid uses such an architecture. The WisNetGrid central server, where the user-interface is located, connects resources with different login mechanism and stores the encrypted credentials by using the public key of the corresponding WisNetGrid community server (front-ends). Here the login data are used for the login procedure and afterwards deleted, when the request have been processed. This way, high level services to perform searches over a variable amount of resources or the use of methods from knowledge generation and processing can be applied on user request.

References

1. WisNetGrid: Networks of Knowledge in the Grid, `http://wisnetgrid.org/`
2. Jensen, J., Spence, D., Viljoen, M.: Grid single sign-on in CCLRC. In: To appear in the Proceedings of the 2006 UK e-Science All Hands Meeting (2006)
3. Lee, A.J., Winslett, M., et al.: Traust: A Trust Negotiation-Based Authorization Service for Open Systems. In: SACMAT 2006: Proceedings Of The Eleventh ACM Symposium On Access, pp. 39–48. ACM Press, New York (2006)
4. Ferraiolo, D.F., Sandhu, R., Gavrila, S., Kuhn, D.R., Chandramouli, R.: Proposed NIST Standard for Role-Based Access Control (2001)

5. GWES: The Generic Workflow Execution Service, http://www.gridworkflow.org/kwfgrid/gwes-web/
6. Deutsche Grid-Initiative: The German grid initiative, http://www.dgrid.de/
7. de Riese, M., Fuhrmann, P., Mkrtchyan, T., Ernst, M., Kulyavtsev, A., Podstavkov, V., Radicke, M., Sharma, N., Litvintsev, D., Perelmutov, T., Hesselroth, T.: The dCacheBook for versions prior to 1.9.7-1. dCache.org (2010), http://www.dcache.org/manuals/Book-1.9.5/
8. Rajasekar, A., Moore, R., Hou, C.Y., Lee, C.A., Marciano, R., de Torcy, A., Wan, M., Schroeder, W., Chen, S.Y., Gilbert, L., Tooby, P., Zhu, B.: iRODS Primer: integrated Rule-Oriented Data System. Morgan & Claypool (2010); ISBN: 1608453332
9. MySQL: The world's most popular open source database, http://dev.mysql.com/
10. Menezes, A.J., Oorschot, P.C., Vanstone, S.A.: Handbook of Applied Cryptography (Discrete Mathematics and Its Applications). CRC Press (2001); ISBN: 0-8493-8523-7

Part VII
Business Processes

Human Drivers Knowledge Integration in a Logistics Decision Support Tool

María D. R-Moreno, David Camacho, David F. Barrero, and Bonifacio Castaño

Abstract. The logistics sector is a very dynamic and complex domain with large impact in our day-to-day life. It is also very sensitive to the market changes, so tools that can improve their processes are needed. This paper describes a decision support tool for a Spanish Logistics Company. The tool takes the knowlege from the human drivers, i.e. *the address used to deliver a package*, integrates and exploits it into the decision process. However, this is precisely the data that is often wrongly introduced by the drivers. The tool analyses and detects different mistakes (i.e. misspellings) in the addresses introduced, groups by zones the packages that have to be delivered, and proposes the routes that the drivers should follow. To achieve this, the tool combines three different techniques from Artificial Intelligence. First, Data Mining techniques are used to detect and correct addresses inconsistencies. Second, Case-Based Reasoning techniques that are employed to separate and learn the most frequent areas or zones that the experienced drivers do. And finally, we use Evolutionary Computation (EC) to plan optimal routes (using the human drivers past experiences) from the learned areas, and evaluate those plans against the original route executed by the human driver. Results show that the tool can automatically correct most of the misspelling in the data.

María D. R-Moreno · David F. Barrero
Computer Engineering Department. Universidad de Alcalá, Madrid, Spain
e-mail: (mdolores,david)@aut.uah.es

David Camacho
Computer Science Department. Universidad Autónoma de Madrid, Madrid, Spain
e-mail: david.camacho@uam.es

Bonifacio Castano
Mathematics Department. Universidad de Alcalá, Madrid, Spain
e-mail: bonifacio.castano@uah.es

F.M.T. Brazier et al. (Eds.): Intelligent Distributed Computing V, SCI 382, pp. 227–236.
springerlink.com

1 Introduction

The need for precise and trustable information in the logistics sector has driven the development of complex Geographic Information Systems (GIS), very useful for planning their routes. Those systems are combined with Global Positioning Systems (GPS) for positioning the different elements involved in the shipments. However, those systems although very useful, cannot make decisions based on, for example, shortcuts in the route that human operators (drivers) learn from the experience.

In our particular problem, the logistics company has a set of logistics distributors with General Packet Radio Service (GPRS) connexions and Personal Digital Assistants (PDAs) where the list of shipments is stored for each day. The list of tasks for each distributor is supplied each day from a central server. The order in which they should be carried out is decided by the driver depending on the situation: density of the traffic, state of the highway, hour of the day and place of the delivery. Although the final decision could depend by other topics such as the weather conditions or simply some kind of intuition, it will be finally taked by the driver based on his previous experiencies. However, this essential knowledge cannot be easily represented, handled and incorportated in a logistics tool.

Therefore, from the daily interaction between human drivers and the logistic tool, two main issues can be learned: the particular route that the drivers followed and the addresses that was delivered by the driver. In our previous work [10], it was shown how a particular representation for the drivers route can be used to be stored in a database of plans, and how they can be retrieved and later reused to plan new routes. However, from these initial results we detected that the main problem of reusing the drivers knowledge was related to the mistakes that appear in the addresses introduced. It avoids the system to correctly reuse the available knowledge.

Within this context, the company needs a decision support tool that, on the one hand, allows them to correct mistakes in the delivery addresses anotated by the users.And on the other hand, the company needs a tool to help them to identify which are the best drivers behaviors, learn from them and plan optimal routes.

In this paper we present a tool that combines techniques from three different AI areas to solve the above problems. First, Data Mining techniques to detect and correct inconsistencies in the addresses introduced by the drivers. Second, Case-Based Reasoning (CBR) techniques to separate and learn the most frequent areas or zones that the experienced logistic operators do. Third, we use Evolutionary Computation (EC) to plan optimal routes from the learning areas and evaluate them [10].

The rest of the paper is structured as follows. Section 2 presents an overview of the decision support tool. Section 3 describes the Record Linkage module developed to eliminate misspellings in the input addresses dataset. Section 4 presents the CBR and then section 5 describes its intregation into an Evolutionary Computation (EC) module used for route optimization. Section 6 shows an experimental evaluation of our application with the real data provided by the logistic company. Finally, some conclusions are outlined.

2 Intelligent Distributed Architecture

There are several available tools that address (partially) our described problem. For example, the ILOG SOLVER provides scheduling solutions in some domains such as manufacturing or transportation and the ILOG DISPATCHER provides extensions for vehicle routing [9]. There is a huge number of commercial and research software applications that provides different (usually *ad-hoc*) solutions to this problem, those are mainly based on the efficiency of the algorithms used [11], [12]. However, none of these tools, algorithms or solutions allow to automatically combine data processing to detect wrong adresses, the experience of the drivers learning from their past decisions, and planning optimal routes from the learnt knowledge as we propose.

Our application has been divided in five subsytems described as follows:

- The *Statistic Subsystem*: takes the data from the files provided by the logistic company in CSV format (Comma Separated Values). It analyses that the files are valid, loads the information into data structures and performs a first statistic analysis of each driver contained in the file. The analysis takes into account the number of working days, number of physical acts, average of physical acts performed, or average of the type of confirmation received by the client.
- The *Automated Data Processing Subsystem*: is based on Data Mining techniques. It takes as input a pair of addresses and it detects matching or differences. Statistics are calculated from the agreement on matching or differing addresses. A threshold is defined by the user, so values above that threshold are automatically replaced. Values below the threshold are shown to the user together to the possible solutions that the algorithm has found.
- The *Loading Data Subsystem*: is in charge of obtaining and loading the geographic information. It takes the latitude and longitude of each address and calculates the distances among right and corrected addresses. This information is provided by means of the Google Maps API. During the process of calculating the coordinates, we group the same addresses because a driver can delivery several packages in the same address. We name that *act*. So, in the database, the addresses will be loaded as *acts*, and an extra field will be added to represent the number of deliveries and pick ups for that act.
- The *Learning Working Areas (LWA) Subsystem*: creates zones or working areas for each driver using the data loaded and processed in the previous subsystems. To generate that subdivision we have used the zip code as the baseline. So, a zone or a working area might be representated using more than one zip code. It also allows us to visualize (using the Google Maps API) the different working areas.
- The *Learning Manager Task (LMT) Subsystem*: plans routes in a date selected by the user. We can specify the number of different plans we want as output, and compare them with the original plan performed by the driver. In the comparison and the generation of plans, we can use different parameters to measure the quality of the plans, such as the total distance, positive rate of LWA, time, etc. The planning algorithm can adapt its answer to the driver behavior and generate plans according to it, if that is what we want.

3 Automated Data Processing

Automatic detection of similar or duplicate entities is a critical problem in a wide range of applications, from Web learning to protein sequence identification [3, 8]. The identification of similar entities, or Record Linkage (RL), is necessary to merge databases with complementary information. Then, there is a need to algorithmically solve these problems given the costs of using human expertise.

The problem of RL was defined as the process of matching equivalent records that differ syntactically [8]. In [3], the entity matching issue is formulated as a classification problem, where the basic goal is to classify entity pairs as matching or non-matching, based on statistical unsupervised methods. Edit distances, such as Levenshtein [6] or Jaro-Winkler [13], have been primarily used to solve this problem in the past. A string edit distance is a metric that can be used to determine how closer are two strings. This value is obtained in terms of the number of character insertions, and deletions, needed to convert one into the other.

In our application, the percentage of mispelling in the addresses is around 45%. It is clearly the bottleneck of our application. After some preprocessing the percentage drops to 20%. But this is still very high when attempting to perform an automatic comparison of the routes.So we have implemented different RL methods and compare the results in our domain. We briefly describe them:

- The LEVENSHTEIN distance between two strings [6] is defined as the minimum number of edits needed to transform one string into another one. The allowable edit operations are insertion, deletion, or substitution of a single character.
- The JARO-WINKLER [5] distance is a measure of similarity between two strings. It is a variant of the Jaro distance metric and mainly used in the area of RL. The higher the Jaro-Winkler distance for two strings is, the more similar the strings are. The score is normalized such that 0 equates to no similarity and 1 is an exact match.
- The BLOCK DISTANCE, also known as Manhattan distance, represents distances between points in a city road grid. In the case of strings, it examines the absolute differences between caracters of a pair of strings.
- The COSINE distance is a measure of similarity between two vectors using the Cosine's angle between them. The result of the Cosine function is equal to 1 when the angle is 0, and it is less than 1 when the angle is of any other value. The value of the angles allows us to determine whether two vectors/words are pointing in roughly the same direction (same word).
- The DICE distance can be used for strings grouping two written letters (bigrams) of the 2 strings we want to compare. Once we have the set of bigrams for each string, then the DICE distance is the result of dividing the double of the number of character bigrams found in both strings by the number of bigrams in the first string plus the number of bigrams in the second string.
- The MONGE ELKAN distance [7] uses variable costs depending on the substring gaps between the words. This distance can be considered as an hybrid distance because it can accommodate multiple static string distances using several parameters.

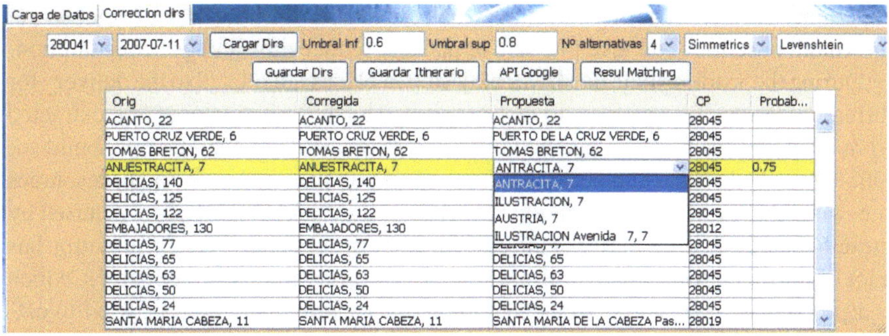

Fig. 1 Automated data processing interface.

Fig. 1 shows the API where the user can set some aspects of the algorithm such as the upper and lower threshold values he wants to apply. When the value returned by the algorithm is greater than the upper threshold value, the wrong street is automatically replaced by the string with the highest value found by the algorithm. The row that contains the change is marked as green. For values between the upper and lower threshold values, the application returns different alternatives and the user has to decide which one is the correct one. The row that contains this case is marked as yellow . If the value returned by the algorithm is smaller than the lower threshold value, the row is marked as red and no solutions are given. The user can change the lower threshold value or manually correct the street.

In Section 6 a set of experimental results will be shown to demonstrate how these techniques allow one to alleviate the problem of mismatching. The experiments have been carried out over a set of real data to stablish what is the best metric that should be used to our current problem.

4 Planning Routes from Past Experiences

In order to learn good behaviours from the drivers, we need to extract the knowledge from their experience and later reuse it for optimization and logistics issues. This knowledge is given by the files provided by the company that contain, among others, the following fields:

- *Delivery information.* Month, day, pick up and deliver time.
- *Address information.* Postal code, pick up and deliver address.
- *Comments or issues.* Any incidence or comment generated in the delivery process, for example, delays originated by the reception people, comments about the correct place or how to make de delivery in a particular place.

This information has to be loaded into the database. Then, the database contains the driver's knowledge and can be used to, for instance, estimate how much time is necessary to complete a particular ship; or it can be used to indirectly learn from experienced drivers. The drivers know which areas have thick traffic, and other useful information, so they can take alternative paths to reduce time and gas consumption.

During the working day, the driver may stop in places not related to the delivery for different reasons, for instance, to rest or to eat. This information could be extracted from the path variations in specific hours, and although it can theoretically build solutions from a distance perspective, it can be better in time and waste other resources (i.e. gas). However, this information has also some disadvantages, mainly caused by the deficiency and irregularity of the data given by the company. The information has a lot of noise: there are many mistakes in the zip codes and the street names, which is a problem difficult to correct. Even the delivery time, that could be used to calculate the time between some points in the path, is not reliable since sometimes the drivers annotate the hour after they have done some deliveries. For this reason this information has be cleaned.

Then, the following step is to group the deliveries in working areas. We define the concept of *working zone* or *working area* (WA) as a region in a city where one or several drivers work delivering objects (as mentioned before, we have called them *acts*). In other words, a WA is a region that concentrates a high number of acts. For this task we propose *Cased Based Reasoning* (CBR) techniques [1].

Usually, logistic companies define these WA and assign them to a set of drivers using a human expert. Minimizing the overlapping of these WAs among different drivers is essential to minimize the number of drivers and the deliver time. The automatic generation of these WAs is the target of our CBR algorithm. Then, these WAs are used later in the optimization process, i.e., the input of the optimization process is the output of the CBR. The WA that has been created by the CBR, in the context of this work, is named *learning working area*, or LWA.

The generation of the final processed and cleaned information, such as the working areas of the drivers, are carried out using statistical considerations (average of behaviours followed by all the drivers analysed) to minimize the noise.

Finally, each LWA learned is stored as a new case in the database, for each driver, the algorithm is applied and the LWA zones are generated. For more details about this algorithm the reader can refer to [10].

5 Route Optimization

One of the main features of the proposed system is its capacity to suggest efficient routes. The term efficient in this context should be interpreted as relative to a set of "good templates" extracted by the CBR module and provided to the system as input. This input is supposed to have been provided by a human expert. So, the goal of the described system is to improve a given set of human made routes rather than generating complete new routes. This characteristic is used by the optimization engine to guide its search.

The definition of the zones is provided by the CBR described above, so it does not have to handle this task. For this reason, the route optimization can be considered as a variation of the Travel Salesman Problem (TSP). The selection of the optimization algorithm is critical to the success of the system. In any case, the TSP is a well known problem and there are several tools that can be used. One of them is Evolutionary Computation (EC) [2], which provides a set of algorithms that have been successfully used in this scenario, in particular, *Genetic Algorithms* (GA).

GA is a stochastic search algorithm inspired by the biological evolution [4]. As other evolutionary algorithms, it uses a set of candidate solutions called population. In GA, each candidate solution is named chromosome or individual, and, as its name suggests, these chromosomes imitates biological chromosomes. They usually are a string of binary numbers that codify the solution. In order to evolve the population, GA introduces a selective pressure that imitates natural selection. Then, those fittest chromosomes are selected with some controlled randomness using a fitness function, that quantifies the quality of the chromosome. Finally, they are modified by introducing some random mutation and/or mixing two chromosomes in a process named crossover, or recombination.

The design of the GA we propose for route optimization is pretty simple. The chromosome represents the route, just a sequence of integers, each one representing an act. It is, essentially, a combinatory optimization problem. The fitness evaluation is done comparing the solution to the reference route. Of course, a route can outperform or not another route depending on the criteria that is applied. Our system uses four criteria or *qualifications*: Time, distance, zone and act. Using these qualifications we can evaluate different aspects related to the route, such as time or distance.

In order to provide a synthetic estimation of the quality of the individual, these qualifications are aggregated in a linear combination that conforms the fitness function. Each quantification is scalled by a factor, these factors are used to control the weight that the user wants to provide to each quantification, depending on his preferences.

The TSP is a problem where it is difficult, if not impossible, to know when a global maximum is achieved, because of the rapid grown of the search space with the number of acts. So, a convergence criteria is required to stop the execution of the GA. In this case, it is suppossed to have convergence if the average fitness in generation i is less than 1% better than the fitness in generation $i-1$.

In summary, the steps that we have followed in the GA are:

1. *Set the initial population*. A population is generated using the "good templates" extracted by the CBR module, in other words, the initial population is feed with solutions given by human experts. No repetitions of the same individual are allowed. We need to specify how many individuals will be part of the population (this value is set by the user).
2. *Run the algorithm*. Once the initial population is created, the following steps will be repeated.
3. *Selection process*. The fittest individuals (that is, the best fitness values) are selected. The number of individuals that will be saved in the next generation is given by an input parameter.

4. *Crossover*. To generate the rest of individuals we use random points in the actual individuals until we complete the initial population.
5. Results. After those iterations, the n first planning routes are presented to the user.

6 Experimental Results

One of the main problems we have encountered when testing, was the high number of irregularities found in the input data provided by the human experts. For this reason the input data had to be preprocessed before it was introduced in the CBR subsystem.

Without any preprocessing, the number of errors in the streets or zip codes is around 45%. It drops to 20% after some light preprocessing (i.e. removing special characters, acute words, etc.), such as using the address to update incorrect zip codes, or removing or adding some characters and searching again. This is still very high when attempting to perform an automatic comparison of the routes followed by the drivers and the ones generated by our tool. So, applying the RL algorithms described in section 3, when preprocessing has been applied, the number of streets that have to be manually corrected is 5%. Figure 2 summarizes these results.

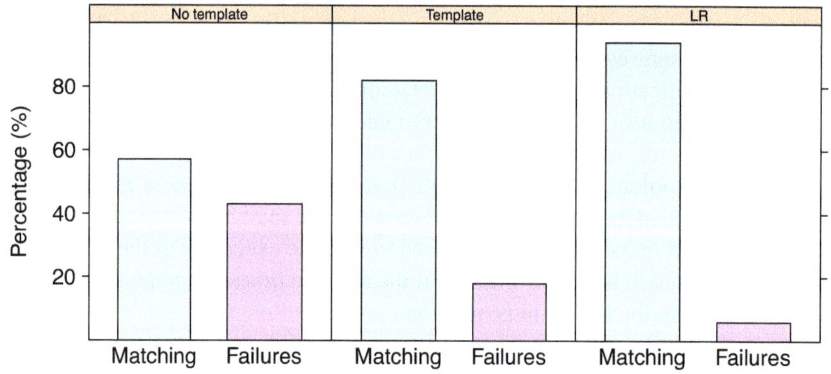

Fig. 2 Effects of three different data cleaning techniques. The percentage of correct and incorrect records in the data is shown.

It is interesting to empirically compare in more detail the RL algorithms described in section 3. Block Distance, Cosine and Dice algorithms get 52% of matches, which is not a notable result. MongeElkan algorithm is even worse, with only 40%. Therefore the only algorithms whose performance makes them suitable are JaroWinkler and Levensthein. In particular, their matches are 90% and 95% respectively. A summary of these results is depicted in Figure 3.

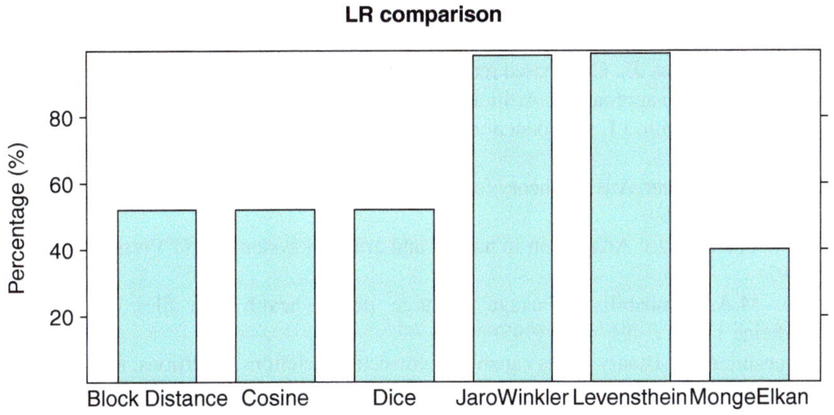

Fig. 3 Comparison of RL algorithms. The graph represents the percentage of matches.

7 Conclusions

In this paper we have presented an AI-based application developed for a Spanish logistics operator company. The main problems the company was interesting in solving are: reusing the experienced drivers knowledge about the route they follow every day and defining routes optimization criteria such as time or fuel.

From the daily interaction between human drivers and the PDAs they use to annotate the deliveries, two main issues can be learned: the particular route that the drivers followed and the addresses. But when dealing with these data, we have detected many mistakes in the addresses introduced so it avoids the system to correctly reuse the available knowledge.

To clean the mistakes on the data and learn the right drivers behaviours, our tool combines Record Linkage (RL), Case-Based Reasoning (CBR) and Evolutionary Computation (EC) techniques. Some of our initial results related to integration of CBR and EC-based plans were presented at [10]. This paper shows how our previous results can be improved using RL techniques. RL allows us to automatically correct the zip codes and the name of the streets that have been badly introduced. CBR is used to separate and learn the most frequent areas that the experienced drivers follow. These techniques allow one to separate the daily incidents that generate noise in the routes, from the decision made based on the knowledge of the route. The EC techniques plan optimal routes from the learning areas and evaluate those routes.

The tool has been tested using real data provided by the company. The results show that we can automate 95% of the data errors, leaving just 5% to the logistics users.

Acknowledgements. We want to thank Antonio Montoya and Antonio Zamora for their contributions in the tool developed. This work has been supported by the Espi & Le Barbier company and the public projects funded by the Spanish Ministry of Science and Innovation ABANT (TIN2010-19872) and by Castilla-La Mancha project PEII11-0079-8929.

References

1. Aamodt, A., Plaza, E.: Case-based reasoning: Foundational issues, methodological variations, and system approaches. Artificial Intelligence Communications 7(1), 39–52 (1994)
2. Eiben, A.E., Smith, J.E.: Introduction to Evolutionary Computing. Springer, Heidelberg (2009)
3. Fellegi, I.P., Sunter, A.B.: A theory for record linkage. Journal of the American Statistical Association (1969)
4. Holland, J.H. (ed.): Adaptation in natural and artificial systems. MIT Press, Cambridge (1992)
5. Jaro, M.A.: Probabilistic linkage of large public health data files. Statistics in Medicine 14(5-7), 491–498 (1995)
6. Levenshtein, V.: Binary codes capable of correcting deletions, insertions, and reversals. Soviet Physics Doklady (1966)
7. Monge, A., Elkan, C.: An efficient domain-independent algorithm for detecting approximately duplicate database records. In: Proceedings of the SIGMOD Workshop Data Mining and Knowledge Discovery, pp. 267–270. ACM, New York (1997)
8. Newcombe, H., Kennedy, J., Axford, S.J., James, A.P.: Automatic linkage of vital records. Science (1959)
9. Le Pape, C.: Implementation of resource constraints in ilog schedule: A library for the development of constraint-based scheduling systems. Intelligent Systems Engineering 3, 55–66 (1994)
10. R-Moreno, M.D., Camacho, D., Barrero, D.F., Gutierrez, M.: A decision support system for logistics operations. In: Corchado, E., Novais, P., Analide, C., Sedano, J. (eds.) SOCO 2010. AISC, vol. 73, pp. 103–110. Springer, Heidelberg (2010)
11. Sanders, P., Schultes, D.: Engineering fast route planning algorithms. In: Demetrescu, C. (ed.) WEA 2007. LNCS, vol. 4525, pp. 23–36. Springer, Heidelberg (2007)
12. Szczerba, R.J., Galkowski, P., Glickstein, I.S., Ternullo, N.: Robust algorithm for real-time route planning. IEEE Transactions on Aerospace and Electronics Systems 36, 869–878 (2000)
13. Winkler, W.: The state of record linkage and current research problems. Statistics of Income Division, Internal Revenue Service (1999)

A Method for Identifying and Evaluating Architectures of Intelligent Goods Services

Åse Jevinger, Paul Davidsson, and Jan A. Persson

Abstract. This paper presents a method for identifying possible architectural solutions for potential intelligent goods services. The solutions range from putting all intelligence at the goods level, to requiring no intelligence on the goods at all. The method is based on a general framework for describing intelligent goods systems, which involves several levels of intelligence related to both the goods and the local environments surrounding the goods. Furthermore, a number of quality attributes are identified, which may be used for evaluating and comparing the solutions. Based on these attributes, a quality evaluation of the architectural solutions related to a potential intelligent goods service is also provided, as an example.

1 Introduction

The concept *intelligent goods* generally refers to goods with some level of intelligence implemented on or close to the goods. It hereby represents a further development of bar codes or simplest forms of RFID tags. Intelligent goods enable new solutions and opportunities which sometimes are seen as important instruments for a more efficient future transport system. In order to find the best solution for different situations, services based on intelligent goods should be compared with other solutions in terms of costs, service quality etc. Furthermore, different implementations of intelligent goods also need to be compared. The purpose of this paper is to lay the foundation for such an analysis by proposing a method for how to identify the architectural solutions for potential intelligent goods services (Sect. 3). An investigation of how these solutions may be compared is also included (Sect. 4). The

Åse Jevinger
Blekinge Instisute of Technology, 374 24 Karlshamn, Sweden
e-mail: ase.jevinger@bth.se

Paul Davidsson · Jan A. Persson
Malmö University, 205 06 Malmö, Sweden
e-mail: {paul.davidsson, jan.a.persson}@mah.se

F.M.T. Brazier et al. (Eds.): Intelligent Distributed Computing V, SCI 382, pp. 237–242.
springerlink.com

method is based on a previously suggested intelligent goods framework [4] which comprises several levels of intelligence connected to the goods (Sect. 2). For our architecture analysis however, we also need a specification of the corresponding levels of intelligence of the local environments surrounding the goods. We will therefore use the framework for this purpose as well. The local objects surrounding the goods, (e.g. enclosing package, transport container, vehicle) will in this paper be denoted as housings. A housing may hold goods, other housings or both.

Different levels of intelligence on the goods and housings, open up for different implementations of a service. Furthermore, each implemented combination of intelligence levels may enable more than one physical architecture [2]. We denote each allocation of service functions and communications, in combination with the required levels of intelligence, as an architectural solution. Each architectural solution hereby includes a specification of the minimum levels of intelligence of both the goods and the housings, as well as the communication paths, the information to be transferred, service functions etc.

2 Intelligent Goods Framework

Previous studies of intelligent goods show different views on which capabilities are required for goods to be called intelligent, [3–5]. In [4], the capabilities of the goods are represented by a number of dimensions, each including different levels of intelligence. In this paper, we use these dimensions to describe the minimum capabilities of both the goods and the housings in the architectural solutions, see Table 1. Table 1 also indicates that the housings and the goods are required to have an ID. These IDs are static and are required since they identify the goods and the corresponding housings throughout the transport chain. On goods level, one ID may represent several goods units as long as they all belong to the same consignment and share the same origin and destination address. In our studies, we assume that the capabilities of the goods will follow the goods throughout the transport chain. These capabilities can be implemented either on the goods or close to the goods, e.g. on a pallet or transport container accompanying the goods. The order in which the capabilities are listed within each dimension in Table 1, corresponds to the level of intelligence, i.e. a lower number indicates a lower level within the same dimension. Please note that the sensor capability (G) in the form of position sensor is unnecessary when the housing is a terminal/warehouse. The corresponding capability would in this case be static information about the position of the terminal/warehouse.

The capability dimensions have dependencies, e.g. the ability to execute decision rules has to be combined with at least the ability to store decision rules. Based on these dependencies, presented in [4], we have identified the set of all valid combinations of capabilities. Apart from the combinations excluded due to the dependencies, a few redundant combinations have also been removed. This process has overall decreased the number of combinations from 4374 to 327. Fig. 1 shows the resulting combinations, i.e. which capability levels from the framework can be colligated.

Table 1 Capability dimensions for goods and housings

A. Memory storage:	**E. Processing:**
1. Ability to store ID	1. None
2. Ability to store goods data	2. Ability to execute decision rules
3. Ability to store algorithms/decision rules	3. Ability to execute algorithms
B. Memory dynamics:	**F. Autonomy:**
1. Static memory	1. None
2. Ability to change/add/delete data	2. Reactive capability (actions must be triggered by external entity)
3. Ability to change/add/delete alg./dec. rules	3. Proactive capability (no ext. trigger needed)
C. Communication out:	**G. Sensor:**
1. Data can be read by external units	1. None
2. Ability to send short-range messages	2. Sensor cap. and ability to read sensor data
3. Ability to send long-range messages	**H. Time:**
D. Communication in:	1. None
1. None	2. Ability to measure time intervals
2. Ability to receive short-range messages	3. Ability to determine actual time
3. Ability to receive long-range messages	

3 Architecture Analysis

A transport chain usually involves many different units and actors, which may receive and send service information. From our perspective, we are only interested in where these units/actors are situated, not which they are. On account of this, we have simplified the transport model by concentrating on which *levels* communicate, and identified three such levels; Goods level, Housing level and Enterprise resource planning (ERP) level. The architectural solution descriptions in this paper are all based on these three levels, and any communication within each level has been left out. By placing different capabilities on the housing and goods levels, it is possible to identify and compare different solutions for the same service. We will use the following service as an example to show how this can be done:

Condition and route deviations service. Reports are sent whenever the physical conditions exceed or fall below certain goods specific limits or when the route

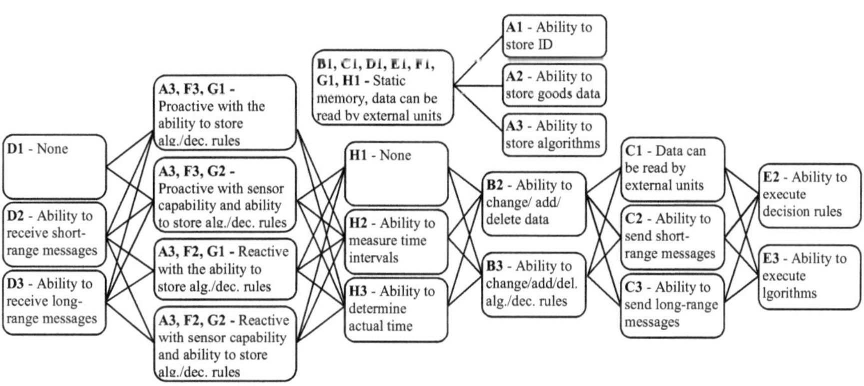

Fig. 1 All levels of intelligence included in the suggested framework

deviates from the one specified. Deviations are reported to ERP systems, housings, and to a questioning unit upon request.

Based on Table 1 and Fig. 1, we have started the analysis by listing all capability combinations on the goods and housing levels. From this set of solutions, the redundant or unrealistic ones have been removed, and the result shows all possible solutions for the service. The most relevant ones are presented in Fig. 2. The route information, used by the service include, for each transport link, the next stop position, the organization responsible (e.g. carrier or terminal operator) and the delivery time window. We assume that the route information is dynamic and that the ERP level is responsible for updating the routes. The capabilities of the ERP level have not been specified since these resources are considered always to be sufficient. Fig. 2 shows that services may be implemented in different ways and that there are many alternatives of how to partition the capabilities between the Goods and Housing levels. The identified solutions can be used to investigate the advantages and disadvantages with different implementations of the same service.

If a service is dependent on the capabilities of a housing, reloading during transport might cause these capabilities to disappear. Since our architecture analysis involves different solutions of the same service, using different levels of intelligence on Goods level, such scenarios can be evaluated based on these. We also believe that our analysis can be equally applied in cases where several levels of housings are used (packet, pallet, container etc.).

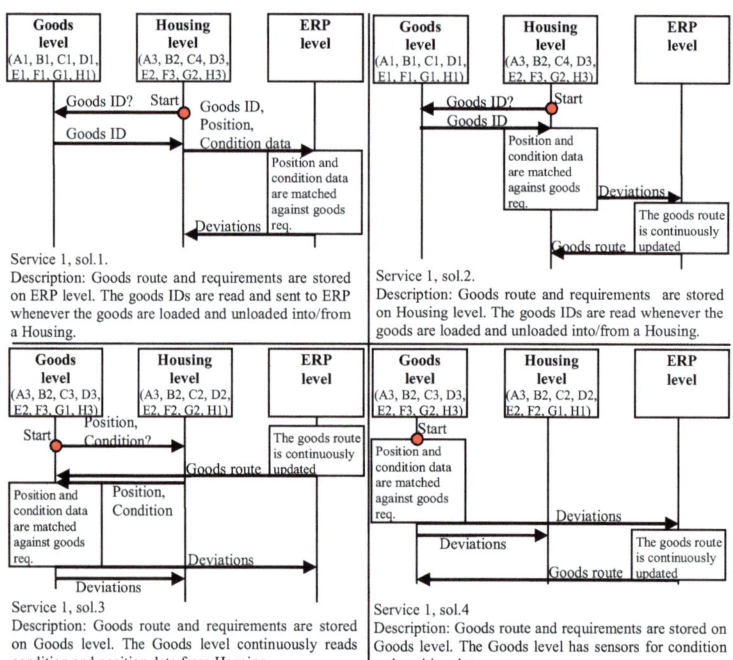

Fig. 2 Architectural solutions

Table 2 Quality attributes

Quality attribute	Description
Time to response (TtR)	Time to get service response, assuming all required capabilities are supported. The value may depend on wireless coverage, location of information entities, updating procedures etc.
Reliability of response (RoR)	Service response depending on changing capabilities, e.g. information may be sent in real time for some links but only at terminals for others, depending on the transport vehicle capabilities.
Quality of response (QoR)	Quality of service response, assuming all required capabilities are supported, e.g. lost goods are more easily tracked if the response includes the current position of the goods.
Integrity (I)	Information integrity concerning stored information entities and messages transmitted. The value may depend on the number of intermediaries, location of information entities etc.
Short-range wireless com. (SW)	Amount of short-range wireless communication needed.
Long-range wireless com. (LW)	Amount of long-range wireless communication needed.
Goods processing power (GP)	Amount of processing power needed at Goods level.
Vehicle processing power (VP)	Amount of processing power needed on vehicles.

4 Quality Evaluation

We have specified a number of quality attributes for evaluating and comparing the performance of different architectural solutions, see Table 2. We will use these to evaluate the identified architectural solutions using the Analytic Hierarchy Process (AHP), similar to the way it is done in [1]. AHP provides an approach to select the most suitable alternative from a number of alternatives evaluated with respect to several criteria. Our analysis will show a way to select the most suitable architectural solution for the particular service under specific circumstances. However, we do not have any measured data to base the analysis on and we will therefore use estimated, relative values instead. The purpose of this section is to show how a quality evaluation can be performed and real implementations represent future work. In accordance with AHP, we have started by assigning priorities to the quality attributes, see Table 3, first part. The first set of priorities (P1) reflects a situation where the long-range wireless communication resources are scarce, and a relatively short time to response and high integrity are required. The second (P2) reflects a situation where the processing power on the goods is low.

Table 3 The first two rows shows the priorities and the last four rows presents the estimated, normalized values of the quality attributes for each solution of service no 2, as well as the results

	TtR	RoR	QoR	I	SW	LW	GP	VP	Result P1	Result P2
P1	0.2	0.1	0.1	0.2	0	0.4	0	0		
P2	0.1	0.1	0.1	0.1	0	0.1	0.4	0.1		
Sol. 1	0.1	0.2	0.2	0.3	0.25	0	0.45	0.2	0.12	0.28
Sol. 2	0.3	0.2	0.2	0	0.25	0.4	0.45	0.1	0.26	0.30
Sol. 3	0.3	0.2	0.2	0.3	0.1	0.3	0.1	0.3	0.28	0.2
Sol. 4	0.3	0.4	0.4	0.4	0.4	0.3	0	0.4	0.34	0.22

The next step of AHP involves collecting metrics for each of the quality attributes, for each architectural solution. Since we will perform our analysis without metrics, we will skip this step and instead use estimated, normalized values of the quality attributes. These values will, in combination with the priorities above, be used for calculating an overall performance result for each architectural solution. Using the AHP process and considering that it is desirable with low values for TtR, SW, LW, GP and VP and high values for RoR, QoR and I, the normalized values are given in Table 3, last part.

In the final step, we calculate the performance result for each architectural solution. This is done by multiplying the priorities with the corresponding normalized values, for each architectural solution. These are thereafter summed and presented in Table 3 (result columns). As expected, when long-range communication resources are scarce, solution no 4 is the most suitable and when processing power on the goods are low, solutions no 1 and 2 are superior.

5 Conclusions

This paper suggests a new method for identifying possible architectural solutions for potential intelligent goods services. The method is based on a previously published intelligent goods framework and is illustrated using a specific potential intelligent goods service. It has been shown that a number of alternative implementations may exist for a service, and that different situations may require different implementations. In order to decide which solution is the best, the solutions must be evaluated and compared with respect to the different situations. To lay the foundation for such an analysis, the paper also provides a set of quality attributes which are used in a quality evaluation of a potential intelligent goods service.

References

1. Davidsson, P., Johansson, S., Svahnberg, M.: Using the analytic hierarchy process for evaluating multi-agent system architecture candidates. In: Müller, J.P., Zambonelli, F. (eds.) AOSE 2005. LNCS, vol. 3950, pp. 205–217. Springer, Heidelberg (2006)
2. Hong, L., Hickman, M., Weissenberger, S.: A structured approach for ITS architecture representation and evaluation. In: IEEE 6th International Conference on Vehicular Navigation and Information Systems, pp. 442–449. IEEE, Los Alamitos (1995)
3. Huschebeck, M., Piers, R., Mans, D., Schygulla, M., Wild, D.: ICSS - Impact assessment study on the introduction of intelligent cargo systems in transport logistics industry, `http://www.intelligentcargo.eu/` (cited April 1, 2009)
4. Jevinger, Å., Davidsson, P., Persson, J.A.: Agent based intelligent goods. In: 6th Workshop on Agents in Traffic and Transportation@AAMAS (2010)
5. McFarlane, D., Sarma, S., Chirn, J.L., Wong, C.Y., Ashton, K.: Auto ID systems and intelligent manufacturing control. Journal of Engineering Applications of Artificial Intelligence 16(4), 365–376 (2003)

Proposal of Representing BPMN Diagrams with XTT2-Based Business Rules*

Krzysztof Kluza, Tomasz Maślanka, Grzegorz J. Nalepa, and Antoni Ligęza

Abstract. Business Processes describe operations of an organization. However, they do not take into account such aspects as detailed logic specification or process implementation. Business Rules are generally better suited for these purposes. Although there is ongoing research on integration of Business Processes with Business Rules, there is no standardized methodology for this integration. We propose a new method of providing visual rule modeling for specification of Business Process logic. In the paper, an outline of the integration of BPMN Business Processes with XTT2 Business Rules is sketched. The method allows for translating selected BPMN models to the XTT2 Business Rules, and executing them using the HeaRT rule engine.

1 Introduction

Business Process Model and Notation (BPMN) [5] has emerged as a de facto standard for Business Process (BP) modeling. It captures processes describing activities of organizations and provides a visual notation emphasizing control flow.

In the Business Rules (BR) approach, the precise semantics of BP model elements and the logic of particular tasks in the process can be specified using rules. As BPMN is not suitable for BR modeling, Rule-Based Systems (RBS) [1], which constitute a mature and well established technology, can be used for this purpose. Despite the ongoing research on integration of BP with BR, there is no standardized and coherent methodology for this integration. The idea is simple – the BPMN model is to define the behavior, and the particular rules are to describe the process logic. However, the integration details are not well specified [3].

Krzysztof Kluza · Tomasz Maślanka · Grzegorz J. Nalepa · Antoni Ligęza
AGH University of Science and Technology,
al. A. Mickiewicza 30, 30-059 Krakow, Poland
e-mail: {kluza,gjn,ligęza}@agh.edu.pl

* The paper is supported by the BIMLOQ Project funded from 2010–2012 resources for science as a research project.

F.M.T. Brazier et al. (Eds.): Intelligent Distributed Computing V, SCI 382, pp. 243–248.
springerlink.com

In this paper, we propose a new method of integration providing visual rule modeling for BP logic specification. The approach uses XTT2 Business Rules and the HeKatE rule engine (HeaRT) [3]. The primary goal of the research is to run HeaRT inference for selected models created using BPMN. This paper is intended as a summary publication from the Technical Report [2].

The paper is organized as follows. Section 2 presents the motivation and main concepts of the proposed method. In Section 3 the integration of BPMN with the XTT2 method is proposed. Section 4 contains an evaluation of the proposal with the related works. The paper is summarized in Section 5.

2 Motivation

As software systems become increasingly complex, in order to deal with this complexity the design process has become more declarative. Such a declarative specification of an application can be decomposed to specification of processes and rules. However, this decomposition faces two problems: the design-implementation gap and lack of dedicated tools. Although the process can be well modeled, it does not assure that the system will be well implemented. This issue is often called the semantic gap between the design and the implementation. The precise semantics of BP model elements and the logic of particular tasks in the process can be specified using rules, called Business Rules (BR). However, there are very few tools which provides specification of BR for BP.

In the Hybrid Knowledge Engineering (HeKatE) project [4] the EXtended Tabular Trees ver. 2 (XTT2) method and the HeKatE Run Time (HeaRT) environment have been developed. The goal of the research presented in this paper is to develop a method, which combines the XTT2 method with BPMN and allows for translating the BPMN models to the XTT2 rules. Then, it will be possible to run selected BPMN models using the HeaRT inference engine. In such an approach, BP integrated BR can be used as an executable specification. Moreover, both processes and rules can be modeled visually using BPMN and XTT2.

XTT2 is a knowledge representation that incorporates the attributive table format [4]. In XTT2, similar rules are grouped within separated tables, and the system is split into a network of such tables representing the inference flow. This visual table-based representation can be automatically transformed into HeKatE Meta Representation (HMR), which is suitable for direct execution by the HeKatE Run-Time (HeaRT) [3], a dedicated inference engine. HeaRT also provides a verification module – HeKatE Verification and Analysis (HalVA) [2]. The module implements simple debugging mechanism that allows tracking system trajectory, and logical verification of models.The network of decision tables grouping rules can be modeled using a visual editor. Moreover, the XTT2 structure has a formalized description in the ALSV(FD) logic [4].

3 BPMN and XTT2 Integration Proposal

Our solution uses the XTT2 Business Rules that determine flow of Business Processes, specified using the BPMN notation. The proposed BPMN and XTT2 integration approach addresses two problems noted in the previous section:

- *The semantic gap.* Because all BP elements are mapped to rules, especially tasks are defined using rule tables or networks, there is no semantic gap between the high level specification of processes and the low level specification of business logic. In our approach, rules can precisely define both the semantics of BPMN elements and the logic of process tasks.
- *Lack of dedicated tools.* The approach uses the BPMNtoXTT2 translator, developed to provide a translation from the XML format of the BPMN model to the XTT2-based Business Rule representation.

In our approach, a BPMN model has to be designed first. This can be done in a BPMN editor that allows for exporting diagrams to an XML-based BPMN serialization. Then, such a model is transformed into the XTT2 knowledge representation. Particular tasks can be specified manually using the XTT2 tables or networks. Such an XTT2 model can be executed in the HeaRT rule engine.

3.1 BPMN to XTT2 Translation

Selected BPMN elements can be translated to the XTT2 rule representation using the BPMNtoXTT2 translator. Each element and its sequence flows can be described by the proper XTT2 table filled with rules, which are defined by the logic function appropriate for the particular BPMN element. The detailed specification of the translation and the logic functions for elements can be found in the Technical Report [2]. Below, an example of such logic function is presented:

Exclusive split gateway

$$in(g) = \{f_0\},\ \ out(g) = \{f_1, f_2, \dots, f_n\}$$
$$Cond(f_i) = c_i \qquad i \in [1, n-1]$$

`precondition:` f_0

`postcondition:` $f_1 \veebar f_2 \veebar \dots \veebar f_n$

`logic function:`

$$f_1 \leftarrow f_0 \wedge c_1$$
$$f_2 \leftarrow f_0 \wedge \neg f_1 \wedge c_2$$
$$\vdots$$
$$f_n \leftarrow f_0 \wedge \neg f_1 \wedge \dots \wedge \neg f_{n-1}$$

`notes:` f_n is default path;
$\veebar$ symbol is used to denote *xor* logic operation.

In the example precondition denotes if the logic of an entity may be executed, while postcondition describes the state after execution of this entity logic. The logic function defines the entity logic respecting the precondition and postcondition.

Since BPMN does not specify the logic of particular tasks, currently it has to be implemented manually. In the proposed approach, it can be specified either using Business Rules in the form of the XTT2 table or network or as a HeaRT callback. As each BPMN element is represented by the proper XTT2 table, an exemplary XTT2 table for exclusive split gateway is presented in Table 1.

Table 1 BPMN to XTT2 translation for the exclusive split gateway

<table>
<tr><th>BPMN Element</th><th>XTT2 Table</th></tr>
<tr><td colspan="2">Exclusive split gateway</td></tr>
<tr><td>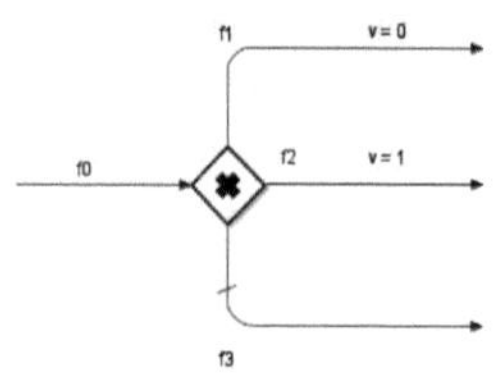
</td><td>
<table>
<tr><th>f_0</th><th>v</th><th>f_1</th><th>f_2</th><th>f_3</th></tr>
<tr><td>$= 1$</td><td>$= 0$</td><td>$:= 1$</td><td>$:= 0$</td><td>$:= 0$</td></tr>
<tr><td>$= 1$</td><td>$= 1$</td><td>$:= 0$</td><td>$:= 1$</td><td>$:= 0$</td></tr>
<tr><td>$= 1$</td><td>$\notin [0, 1]$</td><td>$:= 0$</td><td>$:= 0$</td><td>$:= 1$</td></tr>
</table>
</td></tr>
</table>

The BPMNtoXTT2 translator is implemented in the Prolog programming language. The acceptable input is an XML-based BPMN diagram serialization exported from the BPMN editor, e.g. Eclipse STP BPMN modeler. Currently, the translation works only for a subset of the BPMN elements and supports only instances of the Partially Ordered BP diagram class [8] with sequence flow conditions specified according to ALSV(FD).

Let us show an exemplary BPMN diagram (see: Fig. 1) consist of *start* and *end* events; an exclusive split gateway g_1 that has one conditional path ($condAttr = 1$) and default path; two activities a_1 and a_2 and an exclusive join gateway. Elements are connected by $f_1, f_2, f_3, f_4, f_5, f_6$ sequence flows.

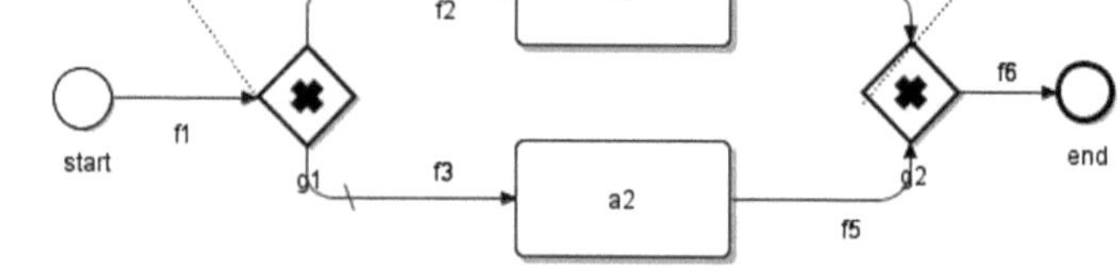

Fig. 1 An example of the BPMN diagram for BPMNtoXTT2 translation

Attributes that represent diagram entities have the same names as diagram entities: $f_1, f_2, \ldots,$ end. They are all of the boolean type representing *true* or *false* values.

Table 2 presents the rules obtained from the exemplary BPMN diagram.

Table 2 An example of the XTT2 rules obtained from the BPMN diagram

Table	Rules	Action
xtt_start	$start = 1 \rightarrow f_1 := 1$	
xtt_g$_1$	$f_1 = 1,\ condAttr = 1 \rightarrow f_2 := 1$	
	$f_1 = 1,\ f_2 = null \rightarrow f_3 := 1$	
xtt_a$_1$	$f_2 = 1,\ condAttr = 1 \rightarrow a_1 := 1,\ f_4 := 1$	*on_a*$_1$
xtt_a$_2$	$f_3 = 1 \rightarrow f_5 := 1, a_2 := 1$	*on_a*$_2$
xtt_g$_2$	$f_4 = 1 \rightarrow f_6 := 1$	
	$f_5 = 1 \rightarrow f_6 := 1$	
xtt_end	$f_6 = 1 \rightarrow end := 1$	

4 Evaluation

Current implementation of the translator supports only a small subset of BPMN diagrams that are Partially Ordered BP diagrams. This limitation is caused by loops, which are not allowed in the HeaRT rule engine. In fact, this limitation of the solution exists also in several simulation tools, which are not prepared for loops in BP models as well as in the Business Process Execution Language (BPEL) tools [8]. In the future, this problem can be resolved by introducing either a translation process of loops to another form of the XTT2 rules without loops, or a new inference mode allowing loops. Alternative for this could be rule engine, which can process rules reacting on events (e.g. on *event* if *conditions* then *conclusions*). This would allow for much more natural expressing of the BPMN events and loops in particular.

There are some developed solutions concerning BPMN models and rules. However, there is no standardized and coherent methodology for their integration. In Corel iGrafx Process[1] rules can be modeled using BPMN gateways, which controls the flow according to the value of attribute. However, it is not possible to group such defined rules in decision tables, or reuse them in another gateways, and this does not allow for using rules in BPMN tasks. IBM WebSphere Business Modeler Advanced[2] supports BR tasks in BPMN models. However, this solution does not provide a visual specification of rules. Business Process Visual Architect[3] allows a user to depict single rules and rule grids within a BPMN diagram. But, it does not use such rules in the simulation or execution process. Although Drools[4] constitute one of the running example of rules and processes integration, it does not support visual rule modeling, especially it does not provide a decision table editor.

---

[1] See: `http://www.igrafx.com/`

[2] See: `http://www-01.ibm.com/software/integration/wbimodeler`

[3] See: `http://www.visual-paradigm.com/product/bpva/`

[4] See: `http://www.jboss.org/drools`

The solution described in this paper, in contrast to the above mentioned tools, uses rules for specifying the semantics of BPMN elements as well as for defining the tasks in the process. It provides a visual environment for designing rules. Furthermore, these rules can be directly used in the execution process.

5 Conclusions and Future Work

In this paper, preliminary results of the research concerning integration of Business Rules with Business Processes are presented. We outline a proposal of a new method, which combines BPMN Business Processes with XTT2-based Business Rules. The method uses the HeaRT rule engine, for inference in selected BPMN models. The aim is to provide the executable environment for Business Processes.

Future work includes extending one of the existing BPMN tools, and its integration with the HeKatE Qt Editor (HQEd) for XTT2. Moreover, the unified methodology of BPMN modeling supported by XTT2-based Business Rules will be developed. This will allow for verification and quality assurance of Business Processes.

References

1. Ligęza, A.: Logical Foundations for Rule-Based Systems. Springer, Heidelberg (2006)
2. Maślanka, T., Ligęza, A., Kluza, K., Nalepa, G.J.: BPMN to XTT2 translation proposal. Tech. Rep. CSLTR 1/2011, AGH University of Science and Technology (to be published, 2011)
3. Nalepa, G.J.: Architecture of the HeaRT hybrid rule engine. In: Rutkowski, L., Scherer, R., Tadeusiewicz, R., Zadeh, L.A., Zurada, J.M. (eds.) ICAISC 2010. LNCS (LNAI), vol. 6114, pp. 598–605. Springer, Heidelberg (2010)
4. Nalepa, G.J., Bobek, S., Ligęza, A., Kaczor, K.: HalVA – rule analysis framework for XTT2 rules. In: Bassiliades, N., Governatori, G., Pasckhe, A. (eds.) RuleML2011 - International Symposium on Rules. LNCS. Springer, Heidelberg (2011)
5. Nalepa, G.J., Kluza, K., Ernst, S.: Modeling and analysis of business processes with business rules. In: Beckmann, J. (ed.) Business Process Modeling: Software Engineering, Analysis and Applications, Business Issues, Competition and Entrepreneurship. Nova Science Publishers (to be published, 2011)
6. Nalepa, G.J., Ligęza, A.: HeKatE methodology, hybrid engineering of intelligent systems. International Journal of Applied Mathematics and Computer Science 20(1), 35–53 (2010)
7. OMG: Business Process Model and Notation (BPMN): Ftf beta 1 for version 2.0 specification. Tech. Rep. dtc/2009-08-14, Object Management Group (2009)
8. Ouyang, C., Wil, M.P., van der Aalst, M.D., ter Hofstede, A.H.: Translating BPMN to BPEL. Tech. rep., Faculty of Information Technology, Queensland University of Technology (2006)

Proposal of Formal Verification of Selected BPMN Models with Alvis Modeling Language*

Marcin Szpyrka, Grzegorz J. Nalepa, Antoni Ligęza, and Krzysztof Kluza

Abstract. BPMN is a leading visual notation for modeling business processes. Although there is many tools that allows for modeling using BPMN, they mostly do not support formal verification of models. The Alvis language was developed for modeling and verification of embedded systems. However, it is suitable for the modeling of any information systems with parallel subsystems. The goal of this paper is to describe the concept of using Alvis for a formal verification of selected BPMN models. In the paper a translation from BPMN to Alvis model is proposed. The translation is discussed and evaluated using a simple yet illustrative example.

1 Introduction

Modeling of business process can be considered on two levels. The first one concerns the visual specification with a modeling notation e.g. BPMN. The second one is related to the *correctness* of the process model. This general feature is of great importance in case of safety critical systems. The analysis of the formal aspects of the model opens up possibility of optimization of the process model.

The goal of the paper is to describe the possibility of using Alvis for a formal verification of BPMN models. Alvis [6] has been originally defined the modeling and verification of embedded systems, but it is also suitable for the modeling of any information systems with subsystems working in parallel. One of the main advantages of this approach is the similarity between BPMN and Alvis models. The Alvis model resembles the original BPMN one from the graph structure point of view.

Marcin Szpyrka · Grzegorz J. Nalepa · Antoni Ligęza · Krzysztof Kluza
AGH University of Science and Technology,
al. A. Mickiewicza 30, 30-059 Krakow, Poland
e-mail: {mszpyrka, gjn, ligeza, kluza}@agh.edu.pl

* The paper is supported by the BIMLOQ Project funded from 2010–2012 resources for science as a research project.

F.M.T. Brazier et al. (Eds.): Intelligent Distributed Computing V, SCI 382, pp. 249–255.
springerlink.com

Thus, after a verification of the Alvis model, it is easy to link the model properties to the properties of original BPMN model.

The paper is organized as follows. Section 2 provides a short presentation of BPMN and Alvis modeling languages. It presents a BPMN model of the student's project evaluation process. Selected Alvis features that are essential from the considered problem point of view are described as well. The transformation method from BPMN to Alvis is described in Section 3. We have limited the presentation to describing a simple case study. However, it is possible to use this approach with more complex models. A short summary is given in the final section.

2 BPMN and Alvis Modeling Languages

Business process [7] can be defined as a collection of related, structured tasks that produce a specific service or product (serve a particular goal) for a particular customer. Business Process Model and Notation (BPMN) [5], is a visual notation for modeling business processes. The notation uses a set of predefined graphical elements to depict a business process and how it is performed. For the purpose of this research, only a subset of BPMN elements is considered i.e. elements used to model orchestration business processes, such as flow objects (*events*, *activities*, and *gateways*) and connecting objects (*sequence flows*). The model defines the ways in which individual tasks are carried out. Gateways are to determine forking and merging of the sequence flow between tasks depending on some conditions. Events denotes something that happens in the process. The icon within the event circle denotes the event type, e.g. envelope for *message event*, clock for *time event*.

Let us analyze a BPMN use case describing a student project evaluation process. The diagram shown in Fig. 1 depicts the evaluation process of a student's project for the Internet technologies course. The process is applied to the website evaluation. At the beginning, the syntax is automatically checked. Every website code in XHTML needs to be a *well-formed* XML and *valid* w.r.t. XHTML DTD.

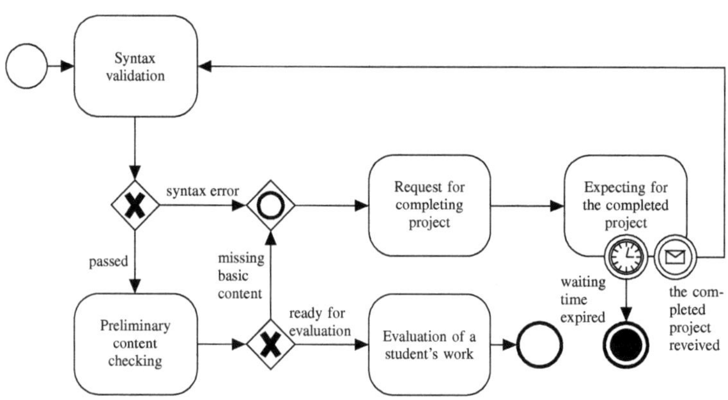

Fig. 1 An example of the student's project evaluation process

If the syntax of the project file is correct, preliminary content checking is performed. Then, if the project contains expected elementary tags (e.g. at least several headings, an image and a table), it can be evaluated and a grade can be given according to some specified evaluation rules [3]. On the other hand, if the project contains any syntax error or lacks some basic required content, it is requested to be completed. After receiving the completed project, the whole process starts from the syntax checking again. However, if the completed project is not received on time, the process is terminated (thus, the author of the project does not get a credit).

The key concept of Alvis [6] is an *agent* that denotes any distinguished part of the system under consideration with a defined identity persisting in time. An Alvis model is a system of agents that usually run concurrently, communicate one with another, compete for shared resources etc. To describe all dependences among agents Alvis uses three model layers: graphical, code and system one. The *code layer* is used to define the behavior of individual agents. Each agent is describe with a piece of source code implemented in Alvis Code Language (AlvisCL) [6]. From the code layer point of view, agents are divided into *active* and *passive* ones. *Active agents* perform some activities and each of them can be treated as a thread of control in a concurrent or distributed system. *Passive agents* do not perform any individual activity, but provide a mechanism for the mutual exclusion and data synchronization. The *graphical layer* (communication diagram) is used to define interconnections (communication channels) among agents. A communication diagram is a hierarchical graph whose nodes may represent both kinds of agents (*active* or *passive*) and parts of the model from the lower level. From users point of view, the *system layer* is predefined and only graphical and code layers must be designed. Alvis provides a few different system layers. The most universal one is denote by α^0 and makes Alvis similar to other formal languages. The layer is based on the following assumptions: each active agent has access to its own processor and performs its statements as soon as possible; the scheduler function is called after each statement automatically; in case of conflicts, agents priorities are taken under consideration.

3 BPMN to Alvis Transformation

The transformation procedure starts with preparing the *initial set of agents*. Initially, each activity is treated as a potential active agent in the corresponding Alvis model. Because the names of agents in Alvis are programming identifiers that must start with an upper-case letter, the initial set of agents is as follows: {*SV*, *PCC*, *RCP*, *ESW*, *ECP*}, where *SV* stands for *Syntax validation*, etc. In the second stage, the initial set of agents is optimized. Consider the *Request for completing project* activity: From the data flow point of view, the corresponding Alvis agent works like a buffer that collects a signal/value and then sends it to the next agent.

The agent realizes three steps (entering the loop, *in* and *out* statements) in every cycle of its activity. These steps do not provide any essential information from the verification point of view, but influence the state space size. Without losing any possibility of the Alvis model verification, we can join the *RCP* and *ECP* agents into one (*ECP* name has been chosen for the agent that represents these two activities).

For each agent identified in the previous stage we have to define its interface i.e. the set of ports. To do this, we consider the set of *surrounding edges* for a given BPMN activity. *Surrounding edges* are these ones that go to or from the activity, but if an edge goes from the activity to a gateway, instead of the edge we consider edge going from the gateway. Each surrounding edge is transformed into a port of the corresponding agent. Moreover, we have to add names (identifiers) for each port. The result of this step for the *Syntax validation* activity is shown in Fig. 2. The edge drawn with dashed line is not a surrounding edge for the considered activity.

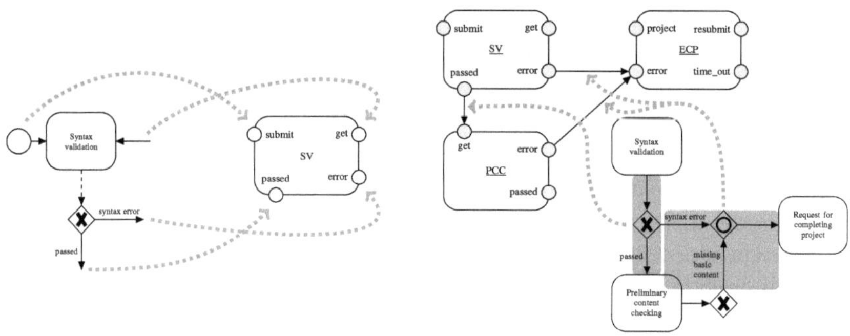

Fig. 2 Transformation of the *Syntax validation* activity to the *SV* agent (left), Generation of communication channels (right)

To complete the *SV* agent definition it is necessary to define its behavior:

```
data Project =  None | Defective | CorrectSyntax
             | CorrectContent String String deriving(Eq,Show);
data Grade = Grade Char deriving(Show);}
-- ...
agent SV {
  pr :: Project = None;
  in submit pr;
  loop {
    if(pr == Defective) { out error; }
    else { out passed pr; }
    pr = None;
    in get pr; }
```

It should be underlined that the BPMN XOR gateway is represented here by the *if else* statement. The *SV* agent: waits for a project to evaluate; collects it via the *submit* port; uses the *pr* parameter to store the project; if the project is *defective*, sends a signal via the *error* port, otherwise sends the project via the *passed* port; waits for revised project (if necessary).

Other agents in the Alvis model are defined in very similar way. Let us focus only on the most interesting features of the code layer. In the model, a student is treated as a part of the system environment. Thus, a project (submission or resubmission), a grade and a timeout are sent via border ports. The ports are specified as follows:

```
in submit   Project [0]   durable;          out time_out [] [];
in resubmit Project [1..] signal durable; out grade Grade [];
```

It means that a value of the *Project* data type can be sent once (at the beginning) via the *submit* port and may be sent (it's not necessary) via the *resubmit* port every one day. In both cases, if a *Project* value is provided by the environment, it waits for serving. A signal (without any particular value) can be sent any time via the *time_out* port and a value of the *Grade* data type may be sent via the *grade* port.

The last stage of the transformation procedure is to define communication channels in the Alvis model graphical layer. In most cases it is easy to point out pairs of ports that should be connected. In the considered example, the most interesting is the transformation of the OR gateway (see Fig. 2). It is necessary to connect two ports (*SV.error* and *PCC.error*) with the *ECP.error* port.

The transformation of a BPMN model into an Alvis one is only a half-way to the formal verification of the BPMN model. Next, the Alvis model is transformed into a *Labelled Transition System* (LTS) that is used for a formal verification. An LTS graph is an ordered graph with nodes representing states of the considered system and edges representing transitions among states. A state of a model is represented as a sequence of agents states. A state of an agent is four-tuple: agent mode (e.g. running, waiting), its program counter (point out the current step/statement), context information list (additional information)and a tuple with parameters values (see [6]). The initial state for the considered system is:

```
SV:  (running,1,[],-)
PCC: (running,1,[],-)
ECP: (running,1,[],-)
ESW: (running,1,[],(-,-))
*:   submit/0, resubmit/1
```

It means that all agents are *running* and are about to execute their first steps. The last line means that a value will provided from the environment to the *submit* port immediately, and in 1 day a value via the *resubmit* port may be provided.

An LTS graph is verified with the CADP toolbox [1]. CADP offers a wide set of functionalities, ranging from step-by-step simulation to massively parallel model-checking. Let us focus on the considered model safeness properties. The LTS graph contains 1438 nodes. Many of them are results of the Alvis language specificity. For example, Alvis distinguish the states before and after entering a loop that do not correspond to different *states* of the BPMN model. From the BPMN model point of view, an analysis of the corresponding Alvis model deadlocks is very important. Deadlocks are state without any transition going from them. Moreover, states that represent a situation when the environment generates signals but the Alvis model does not respond to them are also treated as deadlocks. The LTS graph for the model contains only one state without a transition going from it:

```
SV:  (W,7,[in(get)],-)
PCC: (W,2,[in(get)],-)
ECP: (F,0,[],-)
ESW: (W,1,[in(get)],(-,-))
```

It represents the situation after the *ECP* agent time out. Moreover, the LTS graph contains 4 other deadlocks that represent situations after grading a project. From the LTS graph analysis, we can also conclude that our system always provide a grade for a project with correct content, and that for any received project, the system provide a grade or a request for a revised project.

4 Conclusion and Future Work

Practical analysis of business process models is needed in case of many systems, e.g. control or embedded systems. The analysis of the formal aspects of the model allows for optimization of the model. To perform it, translations of the process model to some formal specification can be considered. Alvis language was developed for modeling and verification of embedded systems. It is suitable for the modeling of any information systems with parallel subsystems. The goal of this paper is to describe the concept of using Alvis for a formal verification of selected BPMN models.

In the paper, a proposal of the translation from BPMN to Alvis model is discussed and evaluated using an example. The transformation of a BPMN model into an Alvis model allows for formal verification. The Alvis model is transformed into a Labelled Transition System that is used for a formal verification. There are two possible approaches to a formal verification of an LTS graph. It can be also encoded using the Binary Coded Graphs format and verified with the CADP toolbox.

As future work, a heterogeneous verification approach is considered. In this case the XTT2 rule language [4] is proposed to model selected activities. XTT2 rules (and tables) can be formally analyzed using the so-called verification HalVA framework [2]. In this approach table-level verification would be performed with HalVA and the global verification would be provided translation to Alvis model.

References

1. Garavel, H., Mateescu, R., Lang, F., Serwe, W.: CADP 2006: A toolbox for the construction and analysis of distributed processes. In: Damm, W., Hermanns, H. (eds.) CAV 2007. LNCS, vol. 4590, pp. 158–163. Springer, Heidelberg (2007)
2. Nalepa, G.J., Bobek, S., Ligęza, A., Kaczor, K.: HalVA – rule analysis framework for XTT2 rules. In: Bassiliades, N., Governatori, G., Pasckhe, A. (eds.) RuleML2011 - International Symposium on Rules, Lecture Notes in Computer Science, Springer, Heidelberg (accepted for publication 2011)
3. Nalepa, G.J., Kluza, K., Ernst, S.: Modeling and analysis of business processes with business rules. In: Beckmann, J. (ed.) Business Process Modeling: Software Engineering, Analysis and Applications, Business Issues, Competition and Entrepreneurship. Nova Science Publishers (to be published, 2011)
4. Nalepa, G.J., Ligęza, A.: HeKatE methodology, hybrid engineering of intelligent systems. International Journal of Applied Mathematics and Computer Science 20(1), 35–53 (2010)

5. OMG: Business Process Model and Notation (BPMN): Ftf beta 1 for version 2.0 specification. Tech. Rep. dtc/2009-08-14, Object Management Group (2009)
6. Szpyrka, M., Matyasik, P., Mrówka, R.: Alvis – modelling language for concurrent systems. In: Bouvry, P., González-Vélez, H., Kołodziej, J. (eds.) Intelligent Decision Systems in Large-Scale Distributed Environments. SCI, vol. 362, pp. 315–341. Springer, Heidelberg (2011)
7. White, S.A., Miers, D.: BPMN Modeling and Reference Guide: Understanding and Using BPMN. Future Strategies Inc., Lighthouse Point (2008)

Part VIII
Applications

Robust 3D Hand Detection for Gestures Recognition

Tudor Ioan Cerlinca and Stefan Gheorghe Pentiuc

Abstract. The aim of this paper is to present a fast, robust and adaptive method for real-time 3D hand detection. Unlike traditional 2D approaches which are based on skin or features detection, the proposed method relies on depth information obtained from a stereoscopic video system. As a result, it provides the 3D position of both hands. It also deals with changes of lighting and it is capable of detecting the hands from long distances. The experimental results showed that our method can be successfully integrated in various and complex vision-based systems requiring real-time recognition of hands gestures in 3D environments.

1 Introduction

Needless to say, speech is the most natural method of interaction between people. Even so, speech will always be accompanied by hand gestures which strengthen the speaker's ideas. Moreover, people sometimes express their feelings or communicate with each other only through hand gestures. Nowadays, complex vision-based systems perform gestures recognition, allowing people to interact with machines in a very natural way. Gestures recognition can be performed both in 2D [2–4] and 3D [5–7] and it is used in different areas such as: remote control, sign language recognition, game interaction etc. Hand detection is one of the most important stages in the process of gestures recognition. Regardless of the complexity of gestures, hands should always be quickly and accurately detected. 3D gestures recognition systems use different strategies for hand detection: some of them detect the hands directly within the depth image and give 3D coordinates while others use the 2D image for detection and the depth image or blob matching techniques for distance estimation [1]. Some other systems use 3D models of the hand and different tracking algorithms (e.g.: Shape Flow) [5].

Tudor Ioan Cerlinca · Stefan Gheorghe Pentiuc
Stefan cel Mare University of Suceava, 13 University Street, RO-720229 Suceava, Romania
e-mail: tudor_c,pentiuc@eed.usv.ro

F.M.T. Brazier et al. (Eds.): Intelligent Distributed Computing V, SCI 382, pp. 259–264.
springerlink.com

2 Method for 3D Hand Detection

The proposed method for 3D hand detection is based on a classical 2D image segmentation algorithm, namely region growing. Unlike the classical algorithm which analyzes the color component of the 2D images, our method analyzes the depth information taken from a stereoscopic video system. The 3D approach has a major advantage: an object can be correctly identified, even when it shows an uneven distribution of color components. Moreover, two partially overlapping objects (with a similar distribution of color components), that can be found at different distances from the stereoscopic video system, are correctly identified and not labeled as one. The 3D hand detection is a 7 stages process as follows: image acquisition, head detection and tracking, hands position estimation, depth map filtering, 3D region growing, hand components connecting, hand isolation and validation.

2.1 Head Detection and Tracking

The 2D images are not used for the actual detection of hands, but only to estimate their position based on head position and size. Head detection and tracking was achieved with an improved version of the algorithm provided by the OpenCV library [9]. Because the face detection runs quite slow when working with big images, we decided to implement a procedure that will automatically narrow the searching area. First time, the algorithm searches for the human face across the entire image. After that, based on the previous head position and size, the new searching area will be automatically determined so as to cover all places in which the head could move in the time elapsed between the acquisition of the previous and the current image [8]. This results in a significant increase of the detection speed.

2.2 Hands Position Estimation

The main purpose of this stage is to estimate the 2D circular area in which the hands can be located. The estimation is done through a probabilistic procedure which is entirely based on a series of statistical reports that exist between different parts of the human body [10]. On the base of these statistical reports and head's location and size, we determine the circular area in which to search for the hands. Except the cases when the human operator is too close to the stereoscopic video system, this stage increases the detection speed by reducing the searching area.

2.3 2D and 3D Filtering

The proposed hand detection method is designed for 3D gestures recognition. We assume that the hands will always be located closer to the camera than the head. The

3D filter operates on the depth map and automatically removes all objects located in the back of the human head. The filter also removes those objects which are located in the proximity of the stereoscopic video system. The 2D filtering purpose is to remove those objects that are outside the areas where the hands can move. A point P in the reference image may belong to the human hands only if it can be located inside the circular area determined in the section 2.2 and its depth satisfies the following condition: $Th_{depth} < D_P < D_{head}$, where: D_P is the depth of the P point, D_{head} is the depth of the head and Th_{depth} is the minimum allowed depth (usually 50cm). Figure 1 presents an example of a depth map and the filtering results.

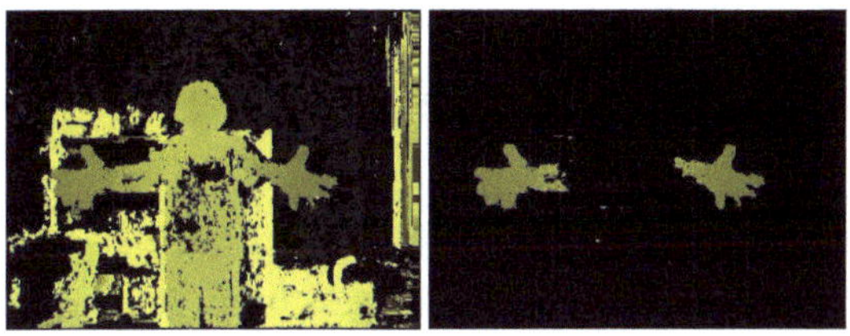

Fig. 1 Depth map filtering

2.4 3D Region Growing

The main purpose of this stage is to group together all the points which belong to the same object. Thus, we developed an algorithm which is based on a classical 2D image segmentation algorithm, namely region growing. The 3D segmentation algorithm consists of four stages, as follows:

1. *choose a seed P_{seed}(a representative point for a particular object) and initialize a new region R;*
2. *add to the region R, all neighboring points, whose distance from the stereoscopic video system is approximately equal to that of point P_{seed} ($D(P,P_{seed}) < T$);*
3. *for each point that was added to the region, repeat step no. 2, until no more points can be added to the region;*
4. *repeat step no. 1 until all points have been assigned to a region.*

The results of the algorithm highly depend on the threshold value T. For example, if the threshold T is too small and a particular object contains points that don't meet the condition $D(P,P_{seed}) < T$ then these points will not be added to the corresponding region of that object. As a result, a single object may be split into smaller adjacent objects. To remove this drawback, the 3D region growing will be followed by a region merging stage. Figure 2 presents the results of this stage (different objects are colored with different colors).

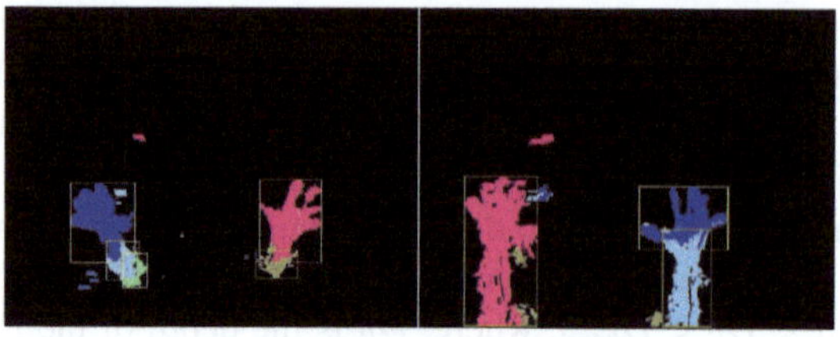

Fig. 2 3D Region growing

2.5 *3D Region Merging*

Ideally, the 3D region growing algorithm will detect the hand as a single object. The experimental results showed that sometimes, the algorithm will not detect a single object, but a set of smaller objects. The main goal of this stage is to connect different objects that belong to the hands, lower-arms and upper-arms. First, a specialized filter removes all objects which have the area less than a threshold value $Th_{MinArea}$. The threshold can be automatically determined, at each step, on the base of the estimated size of the hands. Next, the remaining objects will be connected when possible. Two objects O_1 and O_2 may be connected only when they are close to each other $(Dist(O_1,O_2) < Th_{MaxDist})$ and their average depth is approximately equal $(|Depth(O_1) - Depth(O_2)| < Th_{MaxDepth})$. The result of this stage is a set of objects that belong to the hands, lower-arms and upper-arms and can be connected together.

2.6 *Hand Isolation and Validation*

As we already mentioned, usually, the 3D region growing and merging algorithm does not detect only the hand, but a bigger region that may correspond to the hand, lower-arm and upper-arm. In this stage, we determine the exact location of the hand, on the base of the followings: previous locations, hand estimated size, lower-arm, upper-arm and hand current location and orientation. Validation is an optional stage which operates on the 2D reference images and uses the results presented in the section 2.1. Basically, the validation is done by comparing the color distribution in the HSV (Hue Saturation Value) color space of two regions: head and what we detected as hand. In this stage we remove those objects which were mistakenly detected as hand, lower-arm or upper-arm.

3 Experimental Results

Figure 3 presents few experimental results and a sample trajectory which was drawn on the base of a series of 3D hand coordinates obtained with our detector. The performances of our detector were tested with images grabbed in real-time from the stereoscopic video system. We conducted different tests in different conditions (changes of lighting, different distances from stereoscopic video system etc.) and determined the positive and negative detection rates.

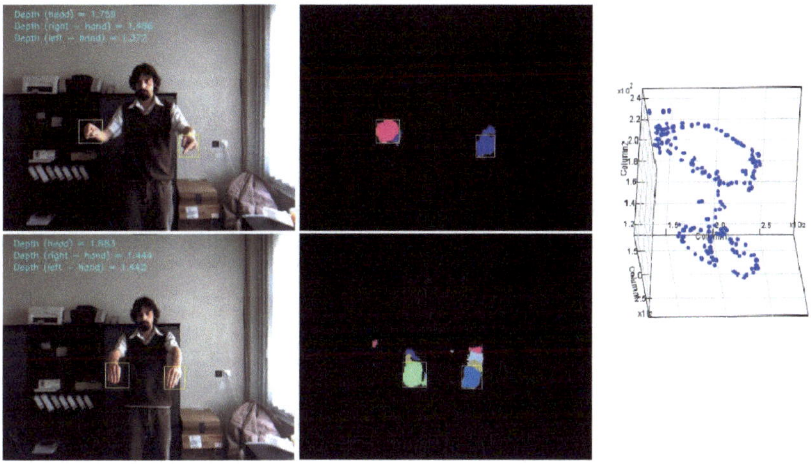

Fig. 3 Experiments with the proposed method for 3D hand detection. A sample trajectory

4 Conclusions and Future Work

This paper presents a fast, accurate and robust method for real-time 3D hand detection. The proposed method deals with changes of lighting and it is capable of detecting the hands from long distances. It is fully adaptive and requires no training. The experiments we have conducted have shown that the hands are properly detected in almost all conditions and therefore, our method can be successfully integrated in various and complex vision-based systems that require real-time recognition of hands postures and gestures in 3D environments. Compared to existing approaches, our method presents several advantages, as follows:

- the detection is done using the depth image and does not rely on shape matching or complicated 3D models of the hand which may limit the number of postures that can be recognized;
- hands are accurately detected regardless of their positions/postures;
- the method doesn't rely on skin detection which may be susceptible to errors when the scene contains objects with a color distribution close to that of the hands.

Future work will be focused on the integration of our method into a vision-based system that will enhance the Human-Computer Interaction through complex hand gestures.

Acknowledgements. This paper was supported by the project "Progress and development through post-doctoral research and innovation in engineering and applied sciences- PRiDE - Contract no. POSDRU/89/1.5/S/57083", project co-funded from European Social Fund through Sectorial Operational Program Human Resources 2007-2013.

References

1. Argyros, A.A., Lourakis, M.I.A.: Binocular Hand Tracking and Reconstruction Based on 2D Shape Matching. In: 18th International Conference on Pattern Recognition, pp. 207–210 (2006), doi:10.1109/ICPR.2006.327
2. Qing, C., Georganas, N.D., Petriu, E.M.: Hand Gesture Recognition Using Haar-Like Features and a Stochastic Context-Free Grammar. IEEE Transactions on Instrumentation and Measurement, 1562–1571 (2008), doi: 10.1109/TIM.2008.922070
3. Lei, S., Yangsheng, W., Jituo, L.: A real time vision-based hand gestures recognition system. In: Cai, Z., Hu, C., Kang, Z., Liu, Y. (eds.) ISICA 2010. LNCS, vol. 6382, pp. 349–358. Springer, Heidelberg (2010), doi:10.1007/978-3-642-16493-4_36
4. Ankit, C., Raheja, J., Karen, D., Sonia, R.: Intelligent Approaches to interact with Machines using Hand Gesture Recognition in Natural way:A Survey. International Journal of Computer Science & Engineering Survey (IJCSES) 2(1) (February 2011)
5. Hahn, M., Kruger, L., Wohler, C., Kummert, F.: 3D Action Recognition in an Industrial Environment. In: Proc. of 3rd Int. Workshop on Human-Centered Robotic Systems (HCRS 2009), Bielefeld, Germany (2009)
6. Agnes, J., Sebastien, M., Olivier, B., Jean-Emmanuel, V.: HMM and IOHMM for the Recognition of Mono- and Bi-Manual 3D Hand Gestures. In: FG Net Workshop on Visual Observation of Deictic Gestures (2004)
7. Van den Bergh, M., Van, G.L.: Combining RGB and ToF cameras for real-time 3D hand gesture interaction. In: 2011 IEEE Workshop on Applications of Computer Vision (WACV), pp. 66–72 (2011), doi: 10.1109/WACV.2011.5711485
8. Cerlinca, T., Pentiuc, G.: Hand Posture Recognition under Unconstrained Scenes for Human Robot Interaction. In: International Conference on Development and Application Systems- Suceava, Romania, May 27-29 (2010)
9. Paul, V., Michael, J.: Robust Real-time Object Detection. In: Second International Workshop on Statistical and Computational Theories of Vision Modeling, Learning, Computing, and Sampling, Vancouver, Canada, July 13 (2001)
10. Art II Impressionism Unit Proportions of the Human Figure, `http://www.immaculateheartacademy.org/Outside2/Art/Proscia/Art%20II%20Impressionism%20Unit%20Proportions%20of%20the%20Human%20Figure.htm`

A Swarm Simulation Platform for Agent-Based Social Simulations

Raul Cajias, Antonio González-Pardo, and David Camacho

Abstract. The social sciences have long employed simulations and statistical models in their research methodology as testing and validation tools. Recent advances in complex networks have produced a set of new graph-based tools with which to study and characterize larges-scale, complex social data. We propose a swarm simulation platform that allows quick prototyping of agent-based social models, using an iterative simulation development cycle. This platform is designed to provide meaningful graph-based metrics of the social ties formed. A description of the system is given, as well as experimental results obtained from sample models.

Keywords: Agent-Based Social Simulation, Swarm computing, Graph Theory.

1 Introduction

The current boom in social-oriented online technologies has generated a wealth of social data. As a consequence, social computing now attracts increasing attention from a number of different fields ranging from the social sciences and social simulation, to statistical physics and computer science. Specialty journals like *Social Networks* and several conferences (e.g the International Sunbelt Social Network Conference) have helped bring attention to the field [9].

One major computational approach to social simulation is the field of agent-based social computing (ABSS). Complex systems like social interactions may be too difficult to understand using regular statistical and mathematical tools. ABSS borrows

Raul Cajias · Antonio González-Pardo · David Camacho
Departamento de Ingeniería Informática, Escuela Politécnica Superior,
Universidad Autónoma de Madrid,
C/Francisco Tomás y Valiente 11, 28049 Madrid, Spain
e-mail: {raul.cajias,antonio.gonzalez,david.camacho}@uam.es

F.M.T. Brazier et al. (Eds.): Intelligent Distributed Computing V, SCI 382, pp. 265–270.
springerlink.com

from agent-based computing, computer simulation and social sciences to provide a framework of social simulation based on agent-based modeling and artificial intelligence. The result is a methodological approach to the testing, refinement and extension of theories [19]. Agent-based models are characterized by their flexibility, allowing to incorporate a wide variety of actor-driven micro-mechanisms influencing the formation of ties , as well as permitting social influences to be controlled to assess its effect on the network as a whole [17, 7]. While social sciences have a long history of employing computer simulations, agents-based simulations have only been partially developed [4]. On the other hand, great strides have been made in the field of complex networks—again, spurred by the growing popularity of online social networks. These works tend to focus on the use of graph topology to extract patterns that may be used in the mining of knowledge. In the past, the topological characteristics of a social graph have been used to generate dynamic models [1, 6, 2]. Other node-centric models have been developed to explore how assortivity influences graph evolution [11, 12].

In social graphs, some nodes tend to group together in highly-connected clusters, known as communities. In social graphs, it has been observed that the stability, age and size of communities found in social and non-social relationship graphs over time, had a particular signature that could help in differentiating them [20]. Building on the swarm agent platform MASON [10], we have developed a swarm ABSS platform designed to explore different social models using graph-mining tools. As a first contribution, we present the platform, and provide a baseline test to determine whether a given model present the same high-order characteristics.

This article is structured as follows: the next section will give a description of the simulation platform . Section 3 gives an overview of the simulation design process, while section 4 provides a detailed description of three sample social models implemented with the platform. Finally chapter 5 draws out our conclusions and ongoing efforts.

2 Description of the Simulation Platform

The platform is comprised of two main components—a simulation engine and a graph-mining toolkit. Together, they allow social models to be quickly implemented, meassured and compared. A simulation consists of a static number of swarm agents navigating a terrain, where they interact with each other and choose to establish or sever social ties. Arbitrary proximity constraints can be placed in the interaction of agents as it has special significance to social interactions [3, 7, 16].

To run a simulation, agents are given movement, and interaction patterns. Movement dictates how agents navigate their environment—random, follow-the-leader, etc—while interaction determines the tie formation protocol each agent abides by. Likewise, environment size, simulation duration, and other external attributes are defined prior to the simulation. During the simulation, snapshots of the social ties formed are taken in the form of an undirected graph, where nodes represent agents, and edges the existence of social ties between them. Once the simulation ends, a list of such snapshots in chronological order is returned. At this point, the simulation may be analyzed from a dynamic networks perspective, using the graph-based toolbox provided by the platform.

A main concern in the network analysis step is to determined whether the model implemented presents basic social networks properties. Indeed, in social networks, the age of a community seems to be positively correlated to its size—bigger communities tend to last longer—while its member stability is negatively correlated to its lifespan—communities that last longer have a very active membership. Non-social networks, like router networks or the world wide web, on the other hand present negative size/age correlations and positive member stability/lifespan correlation [20].

In order to study the evolution of simulated social models, a number of graph metrics are provided. After each simulation cycle, the graph cluster index, average path length and node degree distribution are calculated from the social graph. These metrics allow the model to be weighted against known small-world and free-scale graphs topologies [1], both of which are known to be characteristic of real social networks.

The work by [18] suggests that these metrics may not be enough to fully characterize real-world social networks, and instead advocates the use of *high-order* structures, so two community detection algorithms are used: Girvan-Newman algorithm and the Cluster Percolation Method (CPM). Non-overlapping communities can be detected using the Girvan-Newman algorithm[5], which has a long track in the social sciences field, while overlapping communities can be detected with CPM [14]. Tracking community evolution can be done by either monitoring community overlaps across time-steps with *CPM*—see [8], or with the *CommTec* algorithm proposed by [20], which identifies core community nodes and follows their community membership through time.

As proposed in [20], tracing of communities over time can yield two important metrics: the correlation between the size and age of communities, or *growth* and the correlation between member stability and lifespan of communities, or *metabolism*. Size of a community at a given time $C^{(t)}$ refers to the number of members it has, while age refers to how long the community has existed. The lifespan of a community is the number of future communities that can trace their origin to it. Member stability (MS) is an index referring to how much a community changes over time, calculated with the following formula:

$$MS(C^{(t)}) = \frac{C^{(t)} \cap (C_1^{(t+1)} \cup C_2^{(t+1)} \ldots \cup C_n^{(t+1)})}{C^{(t)} \cup (C_1^{(t+1)} \cup C_2^{(t+1)} \ldots \cup C_n^{(t+1)})}$$

Where $C_1^{(t+1)} \cup C_2^{(t+1)} \ldots \cup C_n^{(t+1)}$ is the union of all communities in the next time-step that can trace their origin to $C^{(t)}$

3 Experimental Results

3.1 Random Model

To establish a baseline for models, we first implement a swarm model to produce an Erdös-Rényi random graph[15]. Whereas the former presents each agent with an uniform connection probability p over $n-1$ other agents, this model works on the subset of n agents within a predefined range.The model is implemented as follows: a number of swarm agents are created and scattered in a finite plane, which they navigate in random trajectory. Agents come in contact with one another by getting within a predefined range. At each simulation time-step, agents scan their surrounding and create or drops ties with agents found, with a probability p.

Figures 1(a), 1(b) displays the growth and metabolism graphs from a sample run. In this case, both graphs present a positive correlation, a behavior not observed in social or non-social graphs. The model was implemented using 100 agents which ran for 300 steps. From trial runs, the majority of results did not conform to social or non-social behavior—76% of the simulations rans

3.2 Christakis-Fowler Model

The *Christakis-Fowler* model (CF) is a parametrized social network model that takes into account individual personality traits, as well as peer influence when forming social ties. A thorough study of the effect each parameter has on the model, from a statistical perspective, can be found in [13].

The model was implemented with parameters from [13] for validation purposes, and simulations were run using 1000 agents moving randomly on a finite plane for 300 steps. Figure 1(c) displays the age/size graph taken from a sample run, while figure 1(d), the stability/trace span of the same run. As observed in social graphs, the growth of this model positive, while metabolism negative. This behavior was observed in 90% of the simulation ran.

We have observed how higher-order topological characteristics from a graph can be used to characterize social ties. Furthermore these experiments have helped us validate our social model implementation, as well provided a common ground on which to compare different models.

4 Conclusions and Further Work

Social network analysis is currently a booming field of research attracting talent from areas like the social sciences, computer science, statistical physics and many others. Recent advances in the field of complex networks have helped understand the mechanics of social networks, offering new ways of measuring and modeling them. To take advantage of these tools, we have developed a new agent-based social simulation platform that incorporates graph-based metrics to the evaluation of their models.

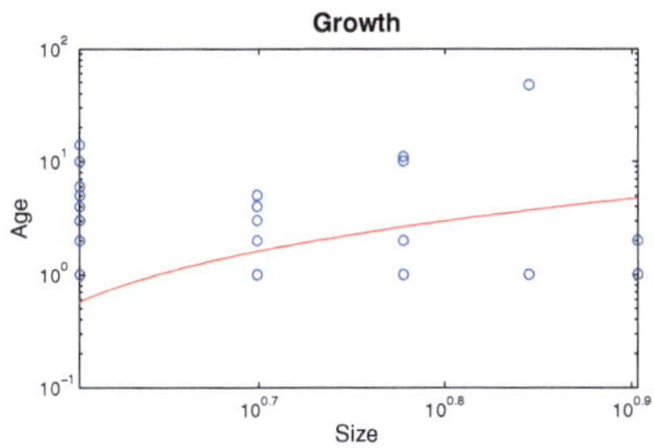

(a) Random Model. Size and age

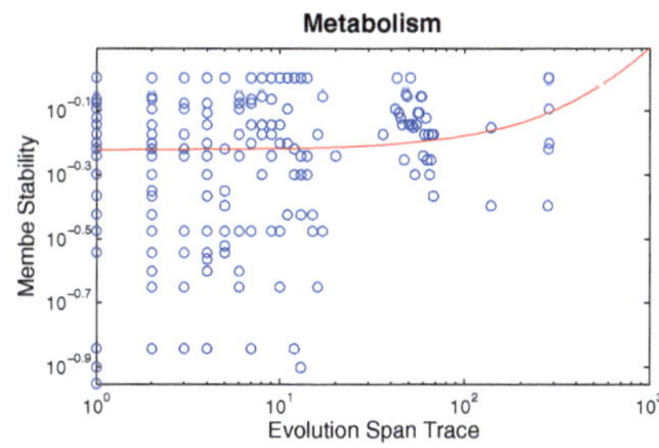

(b) Random Model. Stability and evolution trace span

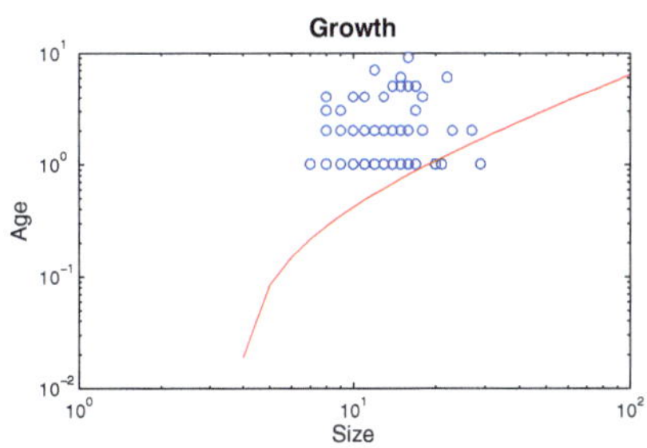

(c) CF Model. Size and age.

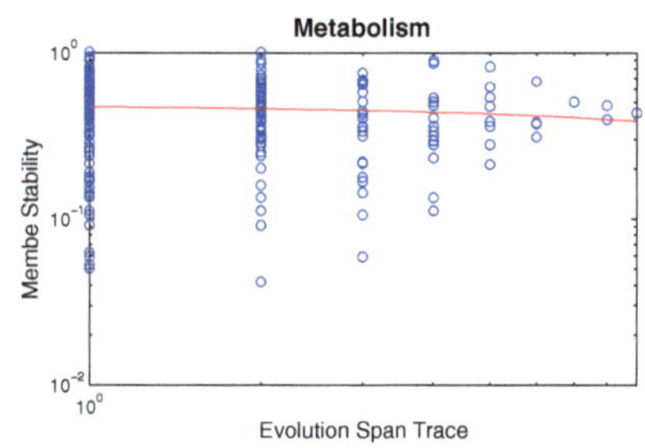

(d) CF Model. Stability and evolution trace span

Fig. 1 Growth and metabolism of the CF and random models.

Using this new platform, we have simulated two swarm models. In these models, communities and their evolution have been tracked over time to test for graph topologies typically found in social networks. We found that these metrics where able to correctly differentiate a social model from a non-social. future work, we plan to extend the metrics used to provide a more robust baseline social test, as well as the ability to compare models by low- and high-order characteristics. We believe this new platform will become very valuable in understanding complex dynamic models.

Acknowledgements. This work has been partially supported by the Spanish Ministry of Science and Innovation under ABANT (TIN2010-19872) and by Jobssy.com (FUAM-076912).

References

1. Albert, R., Barabási, A.: Statistical mechanics of complex networks. Reviews of Modern Physics 74(1), 47–97 (2002)
2. Barabasi, A., Jeong, H., Neda, Z., Ravasz, E., Schubert, A., Vicsek, T.: Evolution of the social network of scientific collaborations. Physica A: Statistical Mechanics and its Applications 311(3-4), 590–614 (2002)
3. Brown, D.J., McFarland, D.D.: Social distance as a metric: A systematic introduction to smallest space analytics. In: Bonds of Pluralism: The From and Substance of Urban Social Networks. Wiley, Chichester (1973)
4. Davidsson, P.: Agent based social simulation: a computer science view. The Journal of Artificial Societies and Social Simulation 5(1) (January 2002)
5. Girvan, M., Newman, M.E.J.: Community structure in social and biological networks. Proceedings of the National Academy of Sciences of the United States of America 99(12), 7821–7826 (2002)
6. Jeong, H., Néda, Z., Barabási, A.L.: Measuring preferential attachment in evolving networks. EPL (Europhysics Letters), 567 (2003)
7. Latane, B., Liu, J., Nowak, A., Bonevento, M., Zheng, L.: Distance matters: Physical space and social impact. Pers. Soc. Psychol. Bull. 21(8), 795–805 (1995)
8. Lazer, D., Pentland, A., Adamic, L., Aral, S., Barabasi, A.-L., Brewer, D., Christakis, N., Contractor, N., Fowler, J., Gutmann, M., Jebara, T., King, G., Macy, M., Roy, D., Van Alstyne, M.: Computational social science. Science 323(5915), 721–723 (2009)
9. Lewis, K., Kaufman, J., Gonzalez, M., Wimmer, A., Christakis, N.: Tastes, ties, and time: A new social network dataset using facebook.com. Social Networks 30(4), 330–342 (2008)
10. Luke, S., Cioffi-Revilla, C., Panait, L., Sullivan, K., Balan, G.: Mason: A multiagent simulation environment. Simulation 81(7), 517–527 (2005)
11. Newman, M.E.J.: Assortative mixing in networks. Physical Review E (2002)
12. Newman, M.E.J.: Mixing patterns in networks. Physical Review E 67(2), 26126 (2003)
13. Noel, H., Nyhan, B.: The unfriending problem: The consequences of homophily in friendship retention for causal estimates of social influence. In: Sunbelt Social Networks Conference (2010)
14. Palla, G., Barabasi, A.-L., Vicsek, T.: Quantifying social group evolution. Nature 446(7136), 660–667 (2007)
15. Renyi, A., Erdos, P.: On random graphs i. Publicationes Mathematicae, 290–297 (1956)
16. Preciado, P., Snijders, T., Burk, W., Stattin, H., Kerr, M.: Does proximity matter? distance dependence of adolescent friendships. Social Networks (2011)
17. Snijders, T., van de Bunt, G., Steglich, C.: Introduction to stochastic actor-based models for network dynamics. Social Networks 32(1), 44–60 (2010)
18. Sridharan, A., Gao, Y., Wu, K., Nastos, J.: Statistical behavior of embeddedness and communities of overlapping cliques in online social networks. arXiv 1–9 (September 2010)
19. Li, X., Mao, W., Zeng, D., Wang, F.-Y.: Agent-Based Social Simulation and Modeling in Social Computing. In: Yang, C.C., Chen, H., Chau, M., Chang, K., Lang, S.-D., Chen, P.S., Hsieh, R., Zeng, D., Wang, F.-Y., Carley, K.M., Mao, W., Zhan, J. (eds.) ISI Workshops 2008. LNCS, vol. 5075, pp. 401–412. Springer, Heidelberg (2008)
20. Yi, W., Bin, W., Shengqi, Y.: Commtracker: A core-based algorithm of tracking community evolution. Journal of Frontiers of Computer Science and Technology 3(3), 282–292 (2009)

A Decentralized Technique for Robust Probabilistic Mixture Modelling of a Distributed Data Set

Ali El Attar, Antoine Pigeau, and Marc Gelgon

Abstract. This paper deals with a machine learning task, namely probability density estimation, in the case data is composed of subsets hosted on nodes of a distributed system. Focusing on mixture models and assuming a set of local probability distribution estimates, we demonstrate how it is possible to combining local estimates in a dynamic, robust and decentralized fashion, through gossiping a global probabilistic model over the data set. Experiments are reported to illustrate the proposal.

1 Introduction

In the field of machine learning, computing a multivariate probability distribution estimate from a given training set is a fundamental task. It has numerous applications in the context of web-based content sharing services, as social networks or recommender systems.

This paper considers the case where the data set to be modeled is distributed over a set of nodes. This distribution is uncontrolled, as local data sets are generated locally, independently, and may be viewed as independent contributions to a global, yet physically distributed, data set (or set of data sources). We assume probabilistic mixture models can be estimated locally on nodes, from local data ; this is a well-known task in statistical pattern recognition, for which many good solutions exist [1]. From there, our goal is to aggregate these local models into a global one, reflecting the distribution of the global data set.

The contribution of the paper is a novel scheme to conduct global model estimation from local model estimates, characterized as follows:

- we conduct local-to-global aggregation in a decentralized fashion based on a gossip protocol to propagate models randomly into the neighborhood [2].

Ali El Attar · Antoine Pigeau · Marc Gelgon
LINA (UMR CNRS 6241)
Université de Nantes
Nantes, France
e-mail: {firstname.name}@univ-nantes.fr

F.M.T. Brazier et al. (Eds.): Intelligent Distributed Computing V, SCI 382, pp. 271–276.
springerlink.com

- the proposed aggregation scheme is statistically robust, i.e. it can cope with a minority of *outlier models* (such outlier models may result from local outlier data).The proposed gossiping process both conducts this iterative outlier identification/rejection task and computes mixture aggregates.

The remainder of the paper is organized as follows. Section 2 presents related work. Section 3 discloses the technical details of our solution. Section 4 reports experimental result, while section 5 concludes.

2 Related Work

The properties of decentralized clustering algorithm [3, 4] concern the following points: the underlying distributed environment, the message exchanged between nodes and the number of rounds of communication required. Their mechanisms are however similar, consisting in iterating between two steps: first an aggregation function is applied independently on each node, and second, a communication protocol is applied to communicate the local clustering parameters on the network.

Several works [4, 3] focus on the estimation of a model mixture with an EM algorithm. Their solution consists to dispatch only the parameter on the network while each node carried out a local E or M step directly on the data. In our context, these data are not available since we work only with local Gaussian models.

Detection of outliers in the decentralized data sets is an important property for our objective. Numerous proposals [5, 6] have been proposed in the litterature, based on existing approaches already used in the scope of centralized clustering. An interesting method [6], presenting the advantage to avoid critical arbitrary parameters, is based on a sampling approach. In this work, we propose to use this approach to discard the outlier models.

Once the local aggregation obtained, the last step of a decentralized algorithm is the communication protocol to dispatch the global estimation in the network. For that, we use a specific implementation of a gossip protocol, the peer sampling service (PSS) [7], which is a probabilistic way to choose a member pairs to communicate and exchange an information.

3 A Distributed and Robust Estimation Algorithm

Let us assume a peer-to-peer network composed of n nodes, where each node i is defined with the following properties: a local data set D_i, a mixture model M_i and a list of c neighbors L_i.

The models M_i are a precondition for our algorithm: we assume data set D_i on each node is modelled as a probabilistic mixture, using for instance an EM algorithm or a Bayesian form thereof, supplying model parameters M_i.

Overall, the scheme iterates over the 5 following steps. For node i:

1. **retrieve all the models of its neighbors L_i:** each node $j \in L_i$ sends its local model M_j to node i. Let $M_{neighbor}$ be the set of L_i's models $\cup$ model M_i;
2. **detect outliers locally:** filtering outliers out of $M_{neighbor}$ is carried out with a model clustering process. The chosen algorithm is adapted from a sampling method [6]. Note that node i itself can be considered as an outlier. Let M_{robust} be the set of models built from $M_{neighbor}$ where outlier models have been removed;
3. **update its model M_i:** we aggregate the set M_{robust} to obtain a new local model. To this aim, we resort to a technique exploiting approximate KL divergences between mixtures [8], that enables to aggregate mixture models. It operates similarly to a k-means algorithm;
4. **apply a gossip protocol:** in order to guarantee a correct propagation of the new model M_i, the node i updates its list of neighbors. The proposed method is founded on the peer sampling service [7]. In a nutshell, the principle consists in exchanging a part of the neighbors list of M_i with one of its neighbors. The point is to increase the convergence speed of the clustering process and the quality of the global model estimate by changing the topology of the network. Each new aggregation of models for model M_i is then carried out with different neighbors, avoiding to fall in a local configuration;
5. **compute the convergence criterion:** before executing a new iteration of the estimation process, we check whether models in the neighborhood of M_i are all similar enough. Roughly, our approach consists in verifying that the whole part of L_i presents a distance with the node i lower than a fixed threshold. Practically, the distance used is an approximation of the Kullback-Leibler divergence, the *KL matched bound* [9] (see below).

In the following, we explain in details the main steps of our algorithm. The similarity measure between mixture models is first presented, followed by the detection of outliers and the estimation algorithm of Gaussian models.

Similarity of Gaussian Components

The aggregation or filtering of mixture models raises the need for a metric between Gaussian components. Based on the experiments proposed in [9], which assessed several approaches to compare Gaussian mixture models, we chose an approximation of the Kullback-Leibler divergence, the *KL matched bound* criterion. For further details on this criterion, see [9].

Detection of outliers

Our algorithm is an adaptation of the sample algorithm [6] to deal with Gaussian models. Its objective is to build a sample with no outlier in order to obtain a baseline before to apply our aggregation algorithm. The principle is to draw randomly several samples of a set of Gaussian model, to improve them with an iterative algorithm and to select the one presenting the best similarity with the initial data set. Such a method presents the interesting advantages to not depend on the space dimension and on strong arbitrary parameters. Indeed, for this latter, it is based on a simple hypothesis, the percentage of outliers supposed on the network.

More formally, the initial step of the algorithm is to draw T samples of size s_{size} from a set of Gaussian model $M_{neighbor} = \{M_1, \ldots, M_L\}$. Then for each sample $S_t = \{M_1, \ldots, M_r, \ldots, M_{s_{size}}\}, t \in [1,T], M_r \in M_{neighbor}$, the following steps are applied until convergence:

1. compute a model M_c obtained with the merge of all $M_r \in S_t$;
2. compute the KL_{match} divergence between M_c and each model $M_l \in M_{neighbor}$. An ascending sorted list of models is then computed;
3. update the sample S_t with the first s_{size} of the sorted list obtained previously.

The convergence is achieved when each sample S_t remains stable (the set of models S_t is not changed). Finally, the sample minimizing its KL_{match} divergence with all the $M_l \in M_{neighbor}$ is selected. This set of models is supposed to be cleared of outliers and is then used as a baseline for the final aggregation of the set of models $M_{neighbor}$.

Mixture Model Estimation

Once a node retrieves the models of its neighbors and filters the outliers, the next step is the computation of a new local mixture model . This problem can be formulated as transforming a mixture model f into another mixture g with less components, while minimizing a Kullback-Leibler divergence involved by the simplification process. Our solution is based on an aggregation algorithm [8], presenting the advantage that only model parameters are accessed to group components, i.e. neither access to data nor sampling are required. Our adaptation concerns the use of a different KL-approximation, the KL_{match}.

The search for optimal g is composed in two alternating steps. The first one consists in determining the best mapping m between the components of f and g such that criterion (1) is minimized:

$$d(f,g) = \arg\min_{m} \quad d(f,g,m) = \arg\min_{m} \quad KL_{match}(f||g) \tag{1}$$

The second step updates the model parameters of g, again from the sole model parameters of f :

$$\mu_{g_j} = \frac{1}{\pi_{g_j}} \sum_{i \in m^{-1}(j)} \pi_{f_i} \mu_{f_i} \tag{2}$$

$$\Sigma_{g_j} = \frac{1}{\pi_{g_j}} \sum_{i \in m^{-1}(j)} \pi_{f_i} \left((\Sigma_{f_i} + (\mu_{f_i} - \mu_{g_j})(\mu_{f_i} - \mu_{g_j})^T) \right) \tag{3}$$

where μ_{g_j}, Σ_{g_j} and π_{g_j} are respectively the mean, the covariance and the mixing proportion of components g_j, and $\pi_{g_j} = \sum_{i \in m^{-1}(j)} \pi_{f_i}$.

These two steps are iterated until the convergence of the criterion defined in equation 1. The initial parameters of the reduction model g are set with the kmeans++ [10].

4 Experiments

To evaluate our proposed algorithm, we report experiments on a synthetic data set with known ground truth. We ran our algorithm on a peer to peer network composed of 200 nodes, characterized as follows:

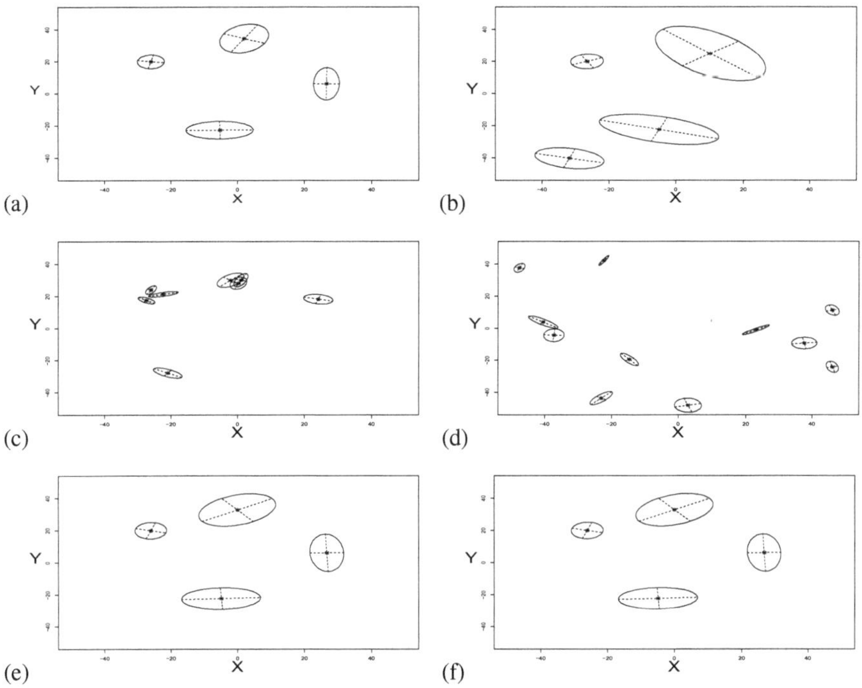

Fig. 1 Figure (a) shows the global view obtained without the outliers models while figure (b) presents the global view obtained with the outliers models. Figures (c) and (d) show examples of initial models on two nodes (figure (d) represents an outlier). The models obtained on these nodes, after 12 iterations of our algorithm, are depicted in figures (e) and (f).

- each node has $c = 20$ neighbors and contains a Gaussian mixture model with a number of components varying between 3 and 12;
- 20% of models are considered as outliers. The components of the outlier nodes are all generated randomly;
- the number of nodes exchanged with the gossip protocol is set to $c/2$, the half of the neighbors list.

In order to evaluate the results of our distributed algorithm, we compute our global model without the outliers in a centralized way, presented in figure 1(a). This model, composed of 4 distinct components, represents the real global data distribution on the network. To demonstrate that our sampling algorithm succeed to discard the outliers, we also compute the global view with the outliers models, depicted on figure 1(b). The result shows the impact of the outliers on the aggregation process: the models obtained present strong differences.

Figures 1(c) and (d) present two examples of initial models on the peer to peer network. These models are composed of different number of components (Figure 1(c) and (d) with 8 and 11 components respectively). An example of outlier models is showed in figure 1(d).The results obtained after 12 iterations of our algorithm is showed on figures 1(e) and (f) for the initial models showed respectively in figures 1(c) and (d). Both models seem identical to the initial centralized global model without outliers (figure 1(a)). This result confirms that our sampling algorithm is effective to discard the outliers from the aggregation process. Finally, to verify the convergence of our algorithm, we compute the evolution of an average KL divergence between each local model and the real global model: this criterion tends quickly toward zero all along the iterations.

5 Conclusion

This paper presents a new method to estimate a probabilistic mixture model from a distributed data set. The novelty of the task resides in the building of local aggregation models with the sole use of their parameters. Our proposal is of much interest in decentralized collaborative systems, for data clustering or recommendation and provide interesting properties with regard to data privacy and network load.

References

1. Bishop, C.M.: Pattern Recognition and Machine Learning. Springer, New York (2006)
2. Kermarrec, A.-M., van Steen, M.: Gossiping in distributed systems. Operating Systems Review 41(5), 2–7 (2007)
3. Gu, D.: Distributed EM algorithm for Gaussian Mixtures in Sensor Networks. IEEE Transactions on Neural Networks 19(7), 1154–1166 (2008)
4. Kowalczyk, W., Vlassis, N.A.: Newscast EM. In: NIPS, pp. 713–720. MIT Press, Cambridge (2004)
5. Chandola, V., Banerjee, A., Kumar, V.: Anomaly detection: A survey. ACM Comput. Surv. 41(3), 1–58 (2009)
6. Rousseeuw, P.J., Driessen, K.V.: A fast algorithm for the minimum covariance determinant estimator. Technometrics 41(3), 212–223 (1999)
7. Jelasity, M., Voulgaris, S., Guerraoui, R., Kermarrec, A.-M., van Steen, M.: Gossip-based peer sampling. ACM Trans. Comput. Syst. 25(3), 8 (2007)
8. Goldberger, J., Roweis, S.T.: Hierarchical clustering of a mixture model. In: NIPS, pp. 505–512. MIT Press, Cambridge (2004)
9. Hershey, J.R., Olsen, P.A.: Approximating the Kullback Leibler divergence between Gaussian mixture models. In: IEEE International Conference on Acoustics, Speech and Signal Processing, ICASSP 2007, vol. 4, pp. IV-317–IV-320 (2007)
10. Arthur, D., Vassilvitskii, S.: k-means++: the advantages of careful seeding. In: SODA 2007: Proceedings of the Eighteenth Annual ACM-SIAM Symposium on Discrete Algorithms, pp. 1027–1035. Society for Industrial and Applied Mathematics, PA (2007)

Part IX
MASTS'2011

Reuse by Inheritance in Agent Programming Languages

H.R. Jordan, S.E. Russell, G.M.P. O'Hare, and R.W. Collier

Abstract. The need to modularise and thereby reuse complex agent programs has long been recognised in agent programming language research. Current approaches to agent modularity fall into two main categories: compositional; and environment-based. Motivated by a problem which requires two variants of the same agent to be built, this paper proposes a set of language extensions which add a complementary modularity style - reuse by inheritance - to agent programming languages. The extensions are designed so that they can be implemented in a preprocessor and added easily to an existing language, without affecting its type system, and with little or no change to its underlying interpreter.

1 Introduction

Agent programming languages (APLs) have so far been successful in the laboratory. However, industry adoption of APLs has been disappointing, especially when compared with their object-oriented (OO) counterparts.

Experience of commercial-scale software has led researchers to realise that "most software systems are ... variants of systems that have already been built" [8], and families of related software systems - also known as software 'product lines' - have become a hot topic of research. The ultimate purpose of systematic reuse is to improve quality [10], by investing in the development of reusable software assets [8]. However, this investment can rarely be justified in academic environments, and APL designers have therefore placed comparatively little emphasis on the language features needed to support variability.

H.R. Jordan · S.E. Russell · G.M.P. O'Hare · R.W. Collier
University College Dublin

F.M.T. Brazier et al. (Eds.): Intelligent Distributed Computing V, SCI 382, pp. 279–289.
springerlink.com

As will be discussed in section 2, sophisticated modularity constructs are available in many APLs. This research has been driven by the need to decompose complex agents into manageable and reusable pieces. However, the provided modularity constructs are mostly compositional, and support for variability through specialisation [18] is absent from several major languages.

Even in stand-alone applications, where variability is not required, modularity by specialization or inheritance can provide several benefits. Empirical evidence suggests that appropriate use of inheritance can make maintenance significantly easier [6]. It has further been argued that inheritance and composition are just two of many modularity dimensions, and that programming languages should support as many of these as possible, in order to help software engineers achieve better separation of concerns [19].

The structure of this paper is as follows. Section 2 discusses related work in the agent-oriented and OO paradigms, and section 3 describes the proposed language extensions. Section 4 presents a case study in which the extensions are applied to two contrasting agent languages, and used to implement a small application. Section 5 concludes and discusses possible further work in this area.

2 Related Work

The primary approach to reuse in APLs is modularity. Central to this approach is the concept of 'capabilities': reusable clusters of beliefs, plans, and events, which form the building blocks of agents [4] [3]. Capabilities provide both runtime encapsulation, and highly flexible compositional reuse; multiple instances of the same capability may be instantiated within a single agent, and nested capabilities allow reuse at a wide range of granularities. This flexibility comes at a cost to the programmer, as the external interface of each capability must be specified manually. Capabilities are implemented in the JACK and Jadex languages, which are based on Java and therefore also support object oriented reuse mechanisms such as composition, inheritance, and component-based development.

In response to a practical need to decompose complex agents into manageable pieces, a compositional modularity system for Jason has been developed [9]. This system is based loosely on capabilities, with the added restriction that nested capabilities are disallowed in order to simplify the event routing within each agent. Like several APLs, Jason employs a subset of Prolog as its belief base language; however this can cause an agent to be sensitive to the order in which its beliefs are adopted, which presents an additional (and as yet unsolved) barrier to modularisation.

3APL has a 'goal-oriented' compositional modularity system [15], in which a goal may be 'dispatched' to a module, and single-mindedly pursued until it is either achieved or all applicable plans contained within the module have failed. 3APL modules may not encapsulate beliefs, thus any belief queries or updates required by a 3APL module must operate on the agent's global belief base.

In 2APL, agents are composed exclusively of modules, which are either 'created' or 'included' together to form an agent program [7]. Created modules contain separate belief bases, which may be declared private or public, while included modules are simply merged with the belief base of the including agent. Control passes to a created module instance by means of an 'execute' command, and is returned when a specified halting condition is met; an agent may execute several module instances simultaneously in parallel. Lone instances of 'singleton' modules can be exchanged at runtime between agents, without loss of state.

An alternative approach to intra-agent reuse is to modularise the agent interpreter itself. At runtime, an agent can be viewed as a collection of beliefs, goals, plans, and environment interaction modules; each of these could easily be described using an existing interpreted language [12]. For example, beliefs are highly suited to knowledge representation languages, while procedural languages might be most appropriate for plans. However, this kind of modularity makes it difficult to encapsulate related data and operations together, and requires coordinated development across multiple programming languages.

Shared environments can also be considered as reuse mechanisms, which allow heterogeneous agents to share common action and perception routines. Standardised agent-environment interaction mediators [1] [14] enable environment code to be easily reused across different systems, agent platforms, and APLs. Extending this reuse model, it has recently been proposed to encode artefact 'usage protocols' or manuals in a specialised plan language [14], and store them in the environment alongside the artefacts on which they act.

An alternative to the modular approaches that are employed by most specialised AOP languages is the use of inheritance. Inheritance, including abstract (or 'virtual') methods, was first popularised as a means to support variation in simulation languages [5]. In modern OO programming languages, inheritance is often implemented in conjunction with static typing; for example in Java, use of the 'extends' keyword yields a new subtype which is at least syntactically compatible with the supertype. However, inheritance need not imply subtyping, and it has been argued that the two concepts *should* be separated: type should be determined by behaviour, and should thus be based on specification, not implementation [17].

3 Overview of Language Extensions

In this section, language extensions which add single inheritance to existing APLs, and can be implemented entirely in a preprocessor, are described. The choice of single inheritance restricts the complexity of the inheritance hierarchy, and in our view yields programs which are easily understood. The abstract syntax of these extensions is given as a set of transformations on a generalised agent programming language X, yielding an extended language X'; and a preprocessor is specified, which translates any valid X' program to its equivalent in X.

A practical implementation of these extensions should employ a suitable concrete syntax to match the style and conventions of the chosen target language.

3.1 Abstract Syntax of a Generalised APL

For the proposed extensions to have the widest possible applicability, minimal assumptions are made on the syntax of the base language X. Let an agent a written in X consist of 'constructs' c, which may be named or anonymous, and categorised as imperative (e.g. plans), or declarative (e.g. goals). The abstract syntax of X can be given in a BNF-like notation as:

```
                   a ::= <c>*                                              (1)
                 <c> ::= <named-c> | <anonymous-c>
           <named-c> ::= <imperative-c> | <declarative-c>
      <imperative-c> ::= <imperative-c-name>
                         <imperative-c-body>                               (2)
 <imperative-c-body> ::= <statement>*                                      (3)
```

The proposed extensions can now be described in terms of this syntax.

3.2 Motivation and Intent

3.2.1 Agent Identification

All agents participating in inheritance relationships must be uniquely named for identification purposes. If language X does not already include an `<agent-name>` declarative construct, an `agent` keyword, which introduces the agent's name, is added to definition 1.

3.2.2 Direct Invocation

To take full advantage of well-known reuse patterns such as Template Method, the target language should support direct invocation of imperative constructs. However, languages such as AgentSpeak(L) do not support direct invocation. In these cases, definition 2 is extended with the `partial` keyword, which marks the imperative construct that follows as directly invokable; and 3 is extended to allow the body of one imperative construct to invoke another, via `<imperative-c-name>`.

3.2.3 Establishing Inheritance

In OO languages, a single inheritance hierarchy is typically defined by allowing each class to declare one superclass. Following this convention, an X' agent may have at most one superagent. Definition 1 is therefore (further) extended with an optional `extends` keyword, which introduces the name of this superagent.

3.2.4 Deferring Implementation to Subagents

Template Method, and other reuse patterns, require an agent to defer some or all of its functionality to a subagent. In support of this, the optional `abstract` keyword is added to definitions 1 and 2. When applied to the name of an imperative construct, `abstract` declares that its corresponding body will be implemented in a subagent. Any agent which includes an abstract construct, or which inherits an abstract construct from a superagent without implementing an associated body, must itself be marked `abstract`, which signifies that the agent is incomplete and thus cannot be instantiated. An agent which inherits an abstract construct from a superagent, and implements this deferred functionality, must declare this implementation using the `implements` keyword, which is further added to definition 2.

3.2.5 Re-implementing Inherited Functionality

The author of a superagent may wish to provide default functionality, which can optionally be overwritten by a subagent. In support of this, definition 2 is further extended with the `override` keyword. This keyword indicates that the following imperative construct redefines and replaces a construct inherited from a superagent.

3.3 *Extension Syntax*

The abstract syntax of an agent a', written in an extended language X' which supports the inheritance features introduced above, can now be described:

```
                 a' ::= [[abstract] agent <agent-name>
                        [extends <superagent>]] <c>*                    (4)
                <c> ::= <named-c> | <anonymous-c>
          <named-c> ::= <imperative-c> | <declarative-c>
     <imperative-c> ::= [abstract | implements | override] [partial]
                        <imperative-c-name> [<imperative-c-body>]      (5)
<imperative-c-body> ::= (<statement> | <imperative-c-name>)*          (6)
```

Note that, depending on the features of X, a concrete implementation of this syntax may omit the `agent <agent-name>`, `[partial]`, and `| <imperative-c-name>` portions of this grammar, in definitions 4, 5, and 6 respectively.

3.4 *Preprocessor Algorithm*

Preprocessing begins with a set A' of agents $a'_1 ... a'_n$, each written in the extended language X'. The preprocessor then maps A' to a new set A, where every agent in A is described in the base language X, and $|A| \leq |A'|$.

1. Check that the keyword `abstract` has been correctly applied in each element a'_i of A'.

   - An error is flagged if a'_i declares an abstract imperative construct with a non-empty body.
   - An error is flagged if a'_i declares an abstract imperative construct, but is not itself declared as abstract.

2. Determine the superagent and subagent(s) of every element of A'.

   - An agent with no superagent is 'root' and is in the set R.
   - An agent with no subagent is 'leaf' and is in the set L.

3. Check that the inheritance relationships between the elements of A' form one or more acyclic graphs.

   - Any detected cycles (for example a'_1 extends a'_2, a'_2 extends a'_3, a'_3 extends a'_1) are flagged as errors.

4. For each non-abstract element l_i in the set L:

   a. Find the shortest path on the inheritance graph from l_i to an element r_i in R.
   b. Create a new, empty agent a_i in the set A.
   c. For each agent a'_j on this path from r_i to l_i, copy the contents of a'_j into a_i.
      - If a construct c' in a'_j has the same name as a construct c in a_i:

	c' modifier			
	`abstract`	`implements`	`overrides`	none
c `abstract`	error	c' replaces c	error	error
c unmodified	error	error	c' replaces c	error

      - An error is flagged if a named construct of a'_j is modified with `implements` or `overrides`, but no construct of the same name exists in a_i.

   d. Remove the modifiers `implements` and `overrides` from all constructs in a_i.
   e. Any abstract imperative constructs remaining in a_i are flagged as errors.

5. If X' includes the `partial` extension, resolve partial imperative constructs. For each agent a_i in A:

   a. An error is flagged if a_i defines two or more partial imperative constructs of the same name.
   b. For each partial imperative construct p in a_i:
      i. All occurrences of the name of p outside of p itself are replaced with a copy of p's body[1].
      ii. The definition of p is then erased completely.

6. For all agents in A, erase the X' extensions, such that A is a valid X program.

[1] This mechanism does not allow recursive partial plans to be defined, however we do not consider recursion to be an essential feature of a high-level APL.

4 Case Study

The benefits of the proposed extensions are illustrated with an example problem, based on a grid environment of 20 × 12 squares, and inspired by the 'vacuum world' of Russell and Norvig. The grid initially contains 64 pieces of randomly-regenerating 'dust', 32 fixed obstacles, and eight cleaning robots. Each robot can move north, south, east, and west; sense its immediate surroundings with a small field of vision; and clean its current square. Within this environment, the robots must clean the grid of dust, as quickly as possible.

Since the dust randomly regenerates as it is cleaned, the large grid must be repeatedly explored to discover new dust. The robots can only clean slowly, which suggests dividing the robots into 'cleaner' and 'explorer' groups with a central non-situated coordinator. The cleaner and explorer agents have shared navigation and communication functionality, which motivates the use of the proposed extensions. To demonstrate the extensions' applicability, outline solutions to this problem will now be presented in two APLs.

4.1 AF-AgentSpeak Solution

AF-AgentSpeak is a variant of AgentSpeak(L) [13], similar to Jason [2], developed to test the capabilities of the Agent Factory Common Language Framework (CLF) [16], and extended as described in section 3. Due to space constraints, we assume a basic understanding of AgentSpeak(L).

To illustrate our extended language, figure 1 shows part of an AF-AgentSpeak solution to the cleaning problem. In this solution, cleaner and explorer agents share a common, extensible navigation model that combines: receipt of target coordinates; moving to the target; performing the required task; and requesting a new task from the coordinator.

The complete design consists of *Cleaner*, *Explorer*, and *Coordinator* concrete agents. *Cleaner* and *Explorer* are subagents of *AbstractVacBot*. In conformance with the rules of section 3, *AbstractVacBot* is declared `abstract` because it defers implementation of the abstract partial plan `@endOfStep` to its subagents.

The `@endOfStep` partial plan is implemented differently in the *Cleaner* and *Explorer* agents. Informally, when the agent reaches its target, `@endOfStep` is invoked by a plan inherited from *AbstractVacBot*. In the case of the cleaner, `@endOfStep` cleans the dirty target square; while for the explorer, reaching the target equates to completing the assigned exploration task (not shown).

In comparison with a Jason implementation of a similar design, the solution presented here reuses, rather than duplicates, almost the entire functionality of the *AbstractVacBot*. This should lead to long-term maintainability benefits [6]. In comparison with a Jason implementation using compositional modularity [9], our solution allows the specified relationships between the superagent and its subagents to be strictly enforced by the preprocessor.

```
#abstract #agent AbstractVacBot

// ... code for updating map & moving removed ...

#abstract #partial @endOfStep;
#abstract #partial @customPercept(?item, ?xi, ?yi);

#partial @retask <-
    query(manager(?n, ?a) & type(?type)),
    .send(request, agentID(?n, ?a), newTask(?type));

#agent Cleaner #extends vacworld.AbstractVacBot

#implements #partial @endOfStep <-
    if (square(here, dust)) {
        durative(eis.perform(clean))
    };

#implements #partial @customPercept(?item, ?xi, ?yi) <-
    .nil;
```

Fig. 1 AF-AgentSpeak Program

4.2 AF-TeleoReactive Solution

AF-TeleoReactive (AF-TR) is an implementation of Nils Nilsson's teleo-reactive formalism [11] that has also been built using the CLF [16]. The teleo-reactive formalism was designed to provide a highly reactive goal-oriented language ideally suited to dynamic environments, and AF-TR enables this programming style on a modern multi-agent platform.

At its heart, AF-TR combines a logical belief base that consists of grounded predicate formulae, with a procedural language of teleo-reactive functions, which represent an agent's reactions to the state of its environment. Functions consist of an ordered set of n production rules. Each rule consists of a context K_i and an action or function call A_i.

The interpreter cycle of an AF-TR agent consists of three steps: sensing, deliberation, and action. In the first step, percepts are gathered, new beliefs are derived from them using the inference rules, and truth maintenance rules are applied to the entire belief base to ensure the accuracy of any persistent beliefs. In the second step, the rules of the `main` function are scanned in order; if K_i evaluates to true, A_i is selected. If A_i is a function, the interpreter binds the required variables, and scans the rules of A_i as before. In the third step, the resolved action is executed and the cycle is complete.

Figure 2 shows part of an AF-TR solution to the cleaning problem. As with the AF-AgentSpeak solution in section 4.1, the AF-TR design consists of an *AbstractVacBot* agent, which acts as a superagent to the *Cleaner* and *Explorer* agents. The

```
#abstract #agent AbstractVacBotVac #extends
        com.agentfactory.eis.AbstractEisAgent
  module eis -> com.agentfactory.eis.EISControlModule;
  module nav -> vacworld.NavModule;

  #override function main{ ... }

  function progress{ ... }

  #abstract function perform(?x,?y)

#agent Explorer #extends AbstractVacBot
  #implements function perform(?x,?y){
    ~request(Boss, newTask(?t)) & type(?t) &
      agentID(Boss, ?addr) ->
        .send(request, agentID(Boss, ?addr), newTask(?t))
    request(Boss, newTask(?t)) & type(?t) ->
      eis.perform(light(on))
  }
```

Fig. 2 AF-TeleoReactive Program

function `perform` encapsulates the different actions taken by the two subagents once they have reached their destinations.

As with the AF-AgentSpeak solution, the reuse extensions permit significant code - in this case, the *AbstractVacBot* `main` and `progress` functions - to be reused rather than duplicated. The preprocessor also helps to ensure that the program is correct. As noted in the interpreter description above, AF-TR requires that every concrete agent has a `main` function, which acts as an entry point. An *AbstractAFTRAgent* with an abstract `main` allows the preprocessor to enforce this requirement automatically. In our example, *AbstractVacBot* overrides the `main` function inherited from *AbstractEisAgent*, which itself extends *AbstractAFTRAgent*.

5 Conclusion and Future Work

Motivated by the need to support reuse and variation by specialization, this paper has proposed language extensions which add inheritance mechanisms to existing APLs, and demonstrated their applicability by adding them to two different languages and solving an example problem. The use of inheritance significantly reduced the solutions' duplication and complexity. Both solutions strongly resemble the well-known Template Method design pattern; the question of whether any other design or architectural patterns are enabled by these language extensions is left for future work.

The proposed extensions raise the issue of whether the current de-facto APL type system - in which all agents are of a single 'agent' type - is satisfactory, or whether a static type hierarchy for agents should also be introduced. While the benefits of strong static typing in OO programming are well understood, we argue that type errors at runtime in agent and OO programs have fundamentally different consequences: agents ignore unrecognised messages, whereas OO programs raise potentially fatal exceptions. It is therefore our view that the same solutions are unlikely to apply to both paradigms, and much further work in this area is needed.

Acknowledgements. We would like to thank the anonymous reviewers for their helpful comments. This work was supported, in part, by Science Foundation Ireland grants 03/CE2/I303/_1 to Lero - the Irish Software Engineering Research Centre, and 07/CE/I1147 to CLARITY: Centre for Sensor Web Technologies.

References

1. Behrens, T., Hindriks, K., Bordini, R., Braubach, L., Dastani, M., Dix, J., Pokahr, J.: An Interface for Agent-Environment Interaction. In: The Eighth International Workshop on Programming Multi-Agent Systems ProMAS 2010, pp. 37–52 (2010)
2. Bordini, R., Hübner, J., Wooldridge, M.: Programming multi-agent systems in AgentSpeak using Jason. Wiley Interscience (2007)
3. Braubach, L., Pokahr, A., Lamersdorf, W.: Extending the capability concept for flexible BDI agent modularization. Programming Multi-Agent Systems, 139–155 (2006)
4. Busetta, P., Howden, N., Rönnquist, R., Hodgson, A.: Structuring BDI agents in functional clusters. Intelligent Agents VI. Agent Theories Architectures, and Languages, 277–289 (2000)
5. Dahl, O., Myhrhaug, B., Nygaard, K.: Some features of the SIMULA 67 language. In: Proceedings of the Second Conference on Applications of Simulations, Winter Simulation Conference, pp. 29–31 (1968)
6. Daly, J., Brooks, A., Miller, J., Roper, M., Wood, M.: The effect of inheritance on the maintainability of object-oriented software: an empirical study. In: Proceedings of the International Conference on Software Maintenance, pp. 20–29. IEEE, Los Alamitos (1995)
7. Dastani, M., Mol, C., Steunebrink, B.: Modularity in agent programming languages. Intelligent Agents and Multi-Agent Systems, 139–152 (2008)
8. Frakes, W., Kang, K.: Software reuse research: Status and future. IEEE Transactions on Software Engineering 31(7), 529–536 (2005)
9. Madden, N., Logan, B.: Modularity and compositionality in Jason. In: Braubach, L., Briot, J.-P., Thangarajah, J. (eds.) ProMAS 2009. LNCS, vol. 5919, pp. 237–253. Springer, Heidelberg (2010)
10. Mohagheghi, P., Conradi, R., Killi, O., Schwarz, H.: An Empirical Study of Software Reuse vs. Defect-Density and Stability. In: Proceedings of the 26th International Conference on Software Engineering (2004)
11. Nilsson, N.: Teleo-reactive programs for agent control. Journal of Artificial Intelligence Research, 158 (1994)
12. Novák, P., Dix, J.: Modular BDI architecture. In: Proceedings of The Fifth International Joint Conference on Autonomous Agents And Multiagent Systems, pp. 1009–1015. ACM, New York (2006)
13. Rao, A.: AgentSpeak (L): BDI agents speak out in a logical computable language. Agents Breaking Away, 42–55 (1996)

14. Ricci, A., Piunti, M., Viroli, M.: Externalisation and Internalization: A New Perspective on Agent Modularisation in Multi-Agent System Programming. Languages, Methodologies, and Development Tools for Multi-Agent Systems, 35–54 (2010)
15. van Riemsdijk, M., Dastani, M., Meyer, J., de Boer, F.: Goal-oriented modularity in agent programming. In: Proceedings of the Fifth International Joint Conference on Autonomous Agents and Multiagent Systems, pp. 1271–1278 (2006)
16. Russell, S., Jordan, H., O'Hare, G., Collier, R.: Agent Factory: A Framework for Prototyping Logic-Based AOP Languages. In: Proceedings of the Ninth German Conference on Multi-Agent System Technologies, MATES 2011 (2011)
17. Snyder, A.: Encapsulation and inheritance in object-oriented programming languages. In: Conference Proceedings on Object-Oriented Programming Systems, Languages and Applications, p. 45. ACM, New York (1986)
18. Svahnberg, M., Van Gurp, J., Bosch, J.: A taxonomy of variability realization techniques. Software: Practice and Experience 35(8), 705–754 (2005)
19. Tarr, P., Ossher, H., Harrison, W., Sutton Jr., S.: N degrees of separation: multi-dimensional separation of concerns. In: Proceedings of the 21st International Conference on Software Engineering, pp. 107–119. ACM, New York (1999)

A Multi-agent System for Human Activity Recognition in Smart Environments

Irina Mocanu and Adina Magda Florea

Abstract. Activity recognition is an important component for the ambient assisted living systems, which perform home monitoring and assistance of elderly people or patients with risk factors. The paper presents a prototype system for activity recognition based on a multi-agent architecture. In the system, the context of the person is first detected using a domain ontology. Next, the human position is obtained and together with the context forms a sub-activity. The sequence of successive sub-activities is then assembled in a human activity, which is recognized using a stochastic grammar.

1 Introduction

The percentage of elderly people in today's societies is growing at a steady pace [1]. In this context there is a growing need of software systems to support the elderly people or people with risk factors who wish to continue living independently in their homes, as opposed to live in an institutional care. Our goal is to develop an integrated ambient intelligent system for elderly people or people with risk factors, called AmIHomCare, which includes several modalities of surveillance and assistance of such people with special needs. An important component of the system is the human activity recognition module. This papaer presents the AmIHomCare sub-system dedicated to recognize activities of daily living from images captured by a home surveillance camera. In our approach, images are first analyzed and annotated, then context information for the supervised person (location in the room and its position) is obtained. In a second phase this context information is interpreted

Irina Mocanu · Adina Magda Florea
University "Politehnica" of Bucharest, Computer Science Department,
Splaiul Independentei 313, 060042 Bucharest, Romania
e-mail: {irina.mocanu,adina.florea}@cs.pub.ro

F.M.T. Brazier et al. (Eds.): Intelligent Distributed Computing V, SCI 382, pp. 291–301.
springerlink.com

using a domain ontology and a set of rules from a stochastic context free grammar. A multi-agent system is used to support the entire process.

The rest of the paper is organized as follows. Section 2 describes the general structure of the AmIHomCare System in which the multi-agent system for activity recognition will be integrated. Section 3 presents some existing methods for human activity recognition in smart environments. Section 4 describes the proposed multi-agent system used for human activity recognition and Section 5 presents the current evaluation of the proposed system. Conclusions and future works are presented in Section 6.

2 AmIHomCare System

AmIHomCare is an ambient intelligent (an intelligent house) for home medical assistance of elderly or disabled people [9]. The main objective of this system is to develop an intelligent integrated environment for ambient assisted living, which achieves home monitoring and assistance for elderly people or patients with risk factors, controls the environment, and detects medical emergencies.

The system has four main components:

- A component to monitor and control ambient factors such as light, temperature, humidity, as well as home security;
- A component to monitor patient health status by using non-intrusive and intrusive sensors, and send alerts in case of risk values;
- A component to achieve patient gesture recognition and gesture-based interaction with a "robot like" personal assistant;
- A component to achieve activity monitoring, pervasive information access and retrieval based on captured images and patient specific context information.

AmIHomCare also includes a connection to a call center and a home assistance center.

The focus of this paper is to present the activity recognition method used in the development of the forth component of the AmIHomCare. This component is based on a patient supervising system and on a semantic-based image retrieval system.

The supervising system analyzes the images received from a supervision camera, detects an emergency situation and sends a message to the call-center. Alternately, it detects and recognizes a situation requiring assistance and sends a message to the home assistance center. The retrieval system is used for finding medical products or other relevant information related to the patient state of health. The user can make a query based on an image (captured with her mobile phone for example) and the system will retrieve the similar images with the one in the query based on the semantic criteria. The architecture of the supervising system is presented in Fig. 1.

The supervising system captures images from the supervision camera and analyzes these images in order to detect the person in the room. For each image, the position of the person in the room is determined. A sequence of positions together with the supervised person's room context forms a recognized activity. The data

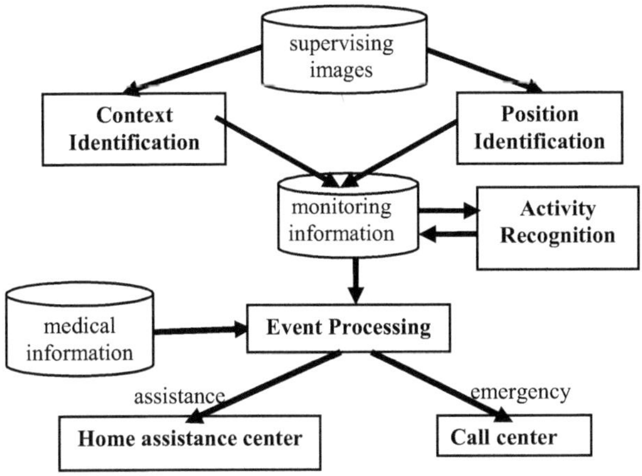

Fig. 1 The architecture of the supervising system component of the AmIHomCare System

resulted from this processing is stored as monitoring information. The monitoring information together with the medical information for the supervised person will be analyzed with the purpose of detecting any emergency situation, or other situation which requires assistance. In either cases above a message is sent to the call center or to the home assistance center.

3 Related Works for Activity Recognition

The goal of activity recognition is to recognize common activities from real life. Standard Hidden Markov Models (HMM) are employed for simple activity recognition as described in [14, 16]. They are not suitable for modeling complex activities that have large state and observation spaces. Parallel HMM [13] are proposed to recognize group activities by factorizing the state space into several temporal processes. Another approach for activity recognition based on HMM is semi-Markov models as described in [4]. Other models are represented by conditional random fields. Conditional random fields are discriminative models for labeling sequences [12]. They condition on the entire observation sequence, which avoids the need for independence assumptions between observations. The Conditional Random fields performs as well as or better than and HMM even when the model features do not violate the independence assumptions of the HMM as described in [12]. Another method for activity recognition uses context free grammar. In [10] is described a context free grammar for description of continued and recursive human activities. Another type of grammars used for activity recognition is stochastic context free grammars. In [3] is described a probabilistic syntactic approach (stochastic context free grammar) for detecting and recognition of temporally extended activities and interactions between multiple agents. In [5] is proposed an attribute grammar which is capable of describing features that are not easily represented by finite symbols.

Using this grammar are represented multiple concurrent events which involves multiple entities by associating unique object identification labels with multiple event threads. A stochastic context free grammar for recognizing activities from kitchen is described in [6]. Other type of methods for human activity recognition is based on hierarchical Bayesian networks [11] or dynamic Bayesian networks [15], which can model the duration of an activity.

4 Activity Recognition in AmIHomCare System

Our goal is to design a prototype system for recognizing human activities in an intelligent house. The purpose of the system is to combine a domain ontology for recognizing the context of the supervised person together with a stochastic context free grammar for modeling the person's activities. The whole process of human activity recognition is presented in Fig. 2.

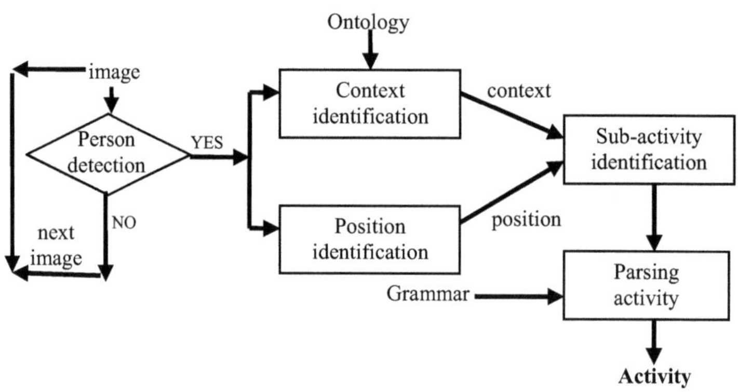

Fig. 2 The process of human activity identification

The main steps of the human activity identification are:

- Person detection in image
- Identification of the detected person's context
- Person pose identification in the recognized context
- Sub activity identification based on the obtained context and on the person pose
- Activity assembling using a sequence of sub activities

Context identification. A miniature supervising digital camera is installed in each room of the house. It continuously takes snapshot images at a predefined interval. Next the supervising image is analyzed for person detection. The image is discarded if no person is detected. If there is a person in the image, two processing steps will follow. First, the identified person's context in the room is obtained. For this purpose the room is divided in interest zones. We will consider the room like in Fig. 3 which shows it divided in three zones: $R1$ the resting zone, $R2$ the reading zone and $R3$ the dining zone. Thus the image is annotated so that image objects will get associated keywords. Image annotation is performed with a genetic algorithm described in [8], which determines the best match between each image region and the corresponding regions of the cluster to which it belongs. An improvement of this algorithm is achieved with a parallel version of the annotation genetic algorithm described in [7]. The spatial position in the room is compared with the spatial relationship of the objects discovered in the image. Based on the comparison result, the zone from the room containing the person will be obtained.

Considering a captured image from the living-room presented in Fig. 3. After the image annotation process, we have a list of objects like: chair, table, armchair. For each recognized object in the room the bounding box rectangle is considered for spatial relationships identification, as well as the bounding box of the person. Comparing the bounding box of the discovered objects (the person and the furniture objects) the objects close to the supervised person will be identified.

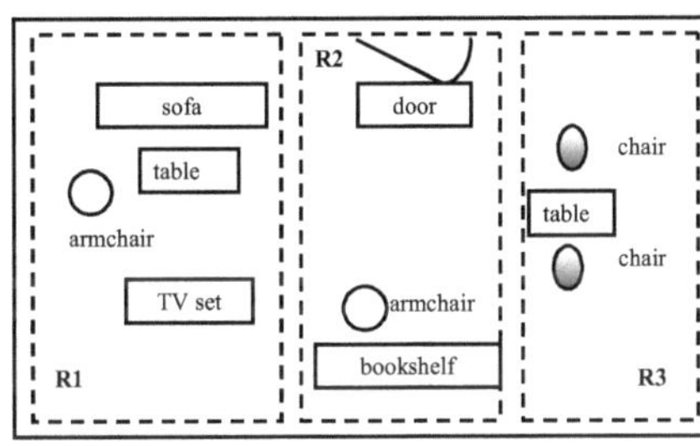

Fig. 3 Areas from the living room

For example we consider the case from Fig. 4: the person is near the armchair, bookshelf and near the TV set. This information will be used for determining the zone in the room where the person is. This has as a prerequisite a model of the house.

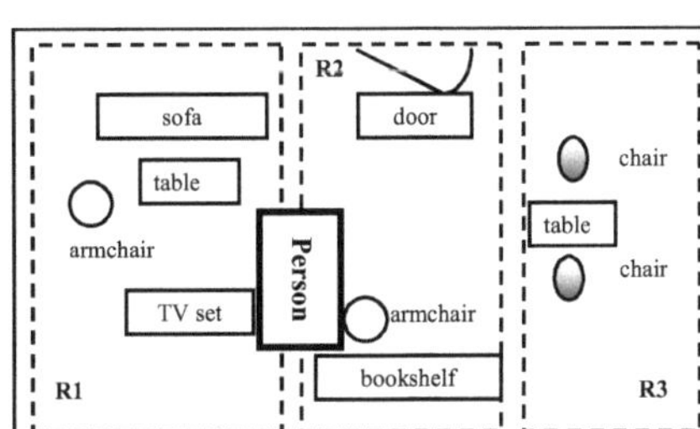

Fig. 4 Person detected near some object furniture

The structure of the house will be modeled in a domain ontology which consists of the rooms in the house, the zones inside each room and the furniture. A part of the used ontology is described in Fig. 5.

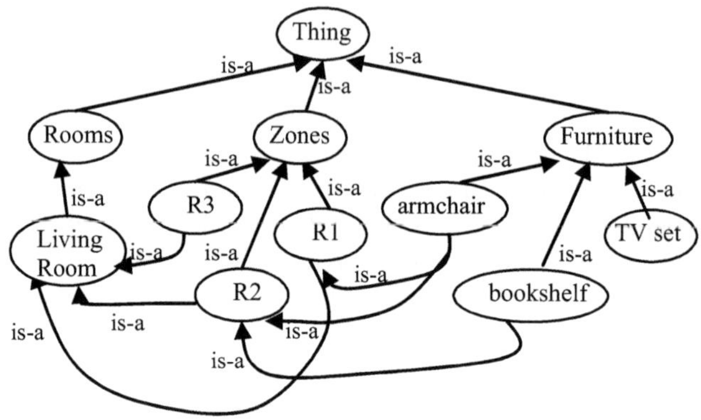

Fig. 5 House ontology

The list of objects obtained by image annotation will be used for ontology interrogation. In this case the list of objects used for ontology interrogation is: armchair, bookshelf and TV-set. From the ontology we get $R2$, (for armchair), $R2$ (for bookshelf) and $R1$ (for TV-set), which are zones from the living room. There are two objects from $R2$ and one object from $R1$. Thus the person will be identified to be in zone $R2$.

But it is hard to discover the zone from the house using only the bounding box of the objects in the image. It will be of great help to have depth information both for each object in the house as well as for the detected person. A distance (ultrasonic) sensor will provide the distance to the person. Also, when the model of the house is created, each object from the house can be represented by a depth using a distance sensor (sonar). Thus each furniture object in the house ontology will have a list of depths associated with it. Each depth value in the list will be measured considering a known viewing angle of the distance sensor. Thus each furniture object O_i from the house ontology will have associated a list of properties $PO_i = \{(dO_i^j, \alpha O_i^j) \mid 1 \leq j \leq n_i\}$, where dO_i^j represents the distance from the sensor taken with the viewing angle αO_i^j and— n_i is the number of distances which characterize the object O_i. Also the detected person P will be described as a distance together with the angle view $(dP, \alpha P)$. The context identification process is described below:

Input: image I annotated with objects $O_1, O_2, \ldots, O_N$, house ontology HO and the detected person P

Output: region from the house R in which is detected the person P

1: interrogate HO with $O_1, O_2, \ldots, O_N$; obtain $PO_i = \{(dO_i^j, \alpha O_i^j) | 1 \leq j \leq n_i\}$ for $1 \leq i \leq N$, and R_i associated with each object O_i.

2: for each i, $1 \leq i \leq N$ $d_i = \min\{distance(dO_i^j, dP) \mid 1 \leq j \leq n_i, \alpha O_i^j = \alpha P\}$.

3: sort ascending all objects O_i based on the obtained distances in step 2, d_i

4: eliminate all objects which have $d_i > \alpha$ (α is a predefined threshold)

5: interrogate HO with the remaining objects from step 4 find the region R which contains the majority of objects obtained in step 4, together with the object in R which is the closest to the person.

Pose identification. The detected person is decomposed in its body components. The human poses are modeled using a stochastic context free grammar based on the body components. The grammar is transformed into an equivalent and-or graph. The human pose identification is performed probabilistically, by constructing a list of parse trees from bottom-up. The parse tree with the biggest probability is chosen from the result list.

Sub-Activity identification. Each activity is decomposed into a sequence of sub-activities. Each sub-activity consists of the human pose and its context. For the human actions we consider three possible activities: (i) watch TV: walk through the living room and sit down on the armchair (from $R1$) or on the sofa; (ii) have a snack: walk through the living room and sit down on a chair near the dining table; (iii) reading a book: walk through the room, go to the bookshelf and sit down on the armchair (from $R2$).

The actions of the person will be modeled with the following stochastic grammar: *G = ({Start, WatchTV, HaveSnack, Reading, Walking, R1Sitting, R1Standing, R2Sitting, R2Standing, R3Standing, R2StandingBookshelf, R2Walking, R1Sitting, R2Sitting, R3Sitting}, {R1, R2, R3, Standing, Sitting, Chair, Armchair, Sofa}, P, Start)* and the set of rules from P are:

$$
\begin{array}{ll}
Start & \rightarrow WatchTV^{0.333} \mid HaveSnack^{0.333} \mid Reading^{0.333} \\
WatchTV & \rightarrow Walking\ R1Sitting^{1.0} \\
Walking & \rightarrow R1Standing\ Walking^{0.25} \mid R2Standing\ Walking^{0.25} \\
 & \quad \mid R3Standing\ Walking^{0.25} \mid e^{0.25} \\
HaveSnack & \rightarrow Walking\ R1SittingChair^{1.0} \\
Reading & \rightarrow Walking\ R2StandingBookshelf^{1.0} \\
R2StandingBookshelf & \rightarrow R2\ Standing\ Bookshelf\ R2Walking^{0.5} \\
 & \quad \mid R2\ Standing\ Bookshelf\ R2Sitting^{0.5} \\
R2Walking & \rightarrow R2Standing\ R2Walking^{0.5} \mid R2Sitting^{0.5} \\
R1Sitting & \rightarrow R1\ Sitting\ Armchair^{0.5} \mid R1\ Sitting\ Sofa^{0.5} \\
R2Sitting & \rightarrow R2\ Sitting\ Armchair^{1.9} \\
R3Sitting & \rightarrow R3\ Sitting\ Chair^{1.0} \\
R1Standing & \rightarrow R1\ Standing^{1.0} \\
R2Standing & \rightarrow R2\ Standing^{1.0} \\
R3Standing & \rightarrow R3\ Standing^{1.0}
\end{array}
$$

Each production rule has associated a probability. The sum of the production probabilities for each non-terminal is one.

Assembling an activity. The ontology and the stochastic context free grammar will be used in a multi-agent system designed for human activity recognition. The multi-agent system will collect all the information which is necessary for both the context identification and the activities components. The structure of the multi-agent system is presented in Fig. 6 together with the use cases for the communication between agents which is presented in Fig. 7.

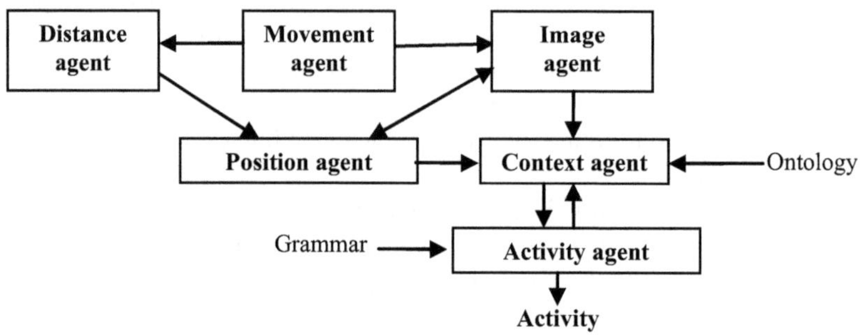

Fig. 6 The Multi-Agent System for daily activity recognition

The Distance agent and the Movement agent collect data from the distance and the movement sensors respectively. The Image agent captures images from the house using the digital camera.

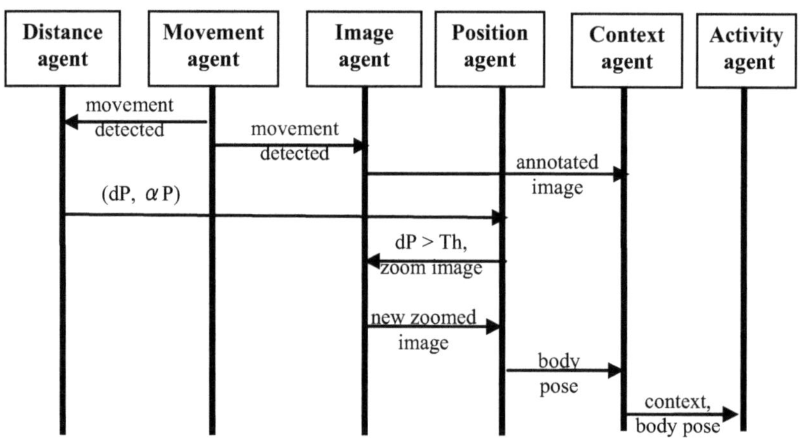

Fig. 7 Use cases for the multi-agent functionality

In case of a movement in the house the Movement agent sends a signal to the Distance agent, which will measure the distance to the moving target, together with the sensor's current viewing angle. Also the Movement agent sends a message to the Image agent, which will eventually align with the detected moving target and capture the image. The Distance agent sends a pair of distance and viewing angle $(dP, \alpha P)$ to the Position agent. The Image agent annotates the image and sends to the Position agent the list of objects in the image. Person detection is also performed on the image. If the distance obtained by the Distance agent is larger than a predefined threshold, the Position agent will send a message to the Image agent which will take another image with an increased zoom value. The Image agent sends to the Activity agent all the objects obtained by the annotation process.mThe Position agent will identify the body components of the detected person together with its body position and will send the recognized position to the Context agent. Next the Context agent finds the context of the detected person and sends a message with the position and the human context to the Activity agent. The Activity agent will assembly a final activity using a set of rules from a stochastic grammar based on the sub activities received from the Zone Agent.

5 System Evaluation

The proposed system was evaluated in a simulated environment. The room (walls, furniture), is modeled in 3D using any 3D modeling software (we used OpenGL software). The supervised human is also modeled using the same tools and technology. The simulated human is walking through the room under the user's control. The virtual sensors' data is automatically generated according to room design and user's movements. The house ontology is developed in Protege using OWL. Three activities have been tested, as described in Section 4: watching TV, reading and having a snack. The results have been compared using the precision and recall metrics (precision indicates accuracy and recall indicates the robustness). The average values for precision and recall for 20 testing sessions are about 91.25%, respectively 94.15%, which indicates a good accuracy for the proposed system.

6 Conclusions and Future Works

The paper presented a prototype system for daily activity recognition in a smart environment. The system is based on a multi-agent architecture, with agents being responsible for different aspects of activity recognition and composition. The agents collect the necessary information for both context identification and identification of activity components. The recognition process of the activity uses an ontology to determine the person's context in the house. A sequence of sub-activities is assembled into an activity using a stochastic context free grammar.

Based on the promising results obtained in case of the simulated environment, the system will be tested using the PlaceLab Data sets [2] and will be integrated in a real world settings in AmIHomCare. Another development will be based on the fact that the Event Processing component of the AmIHomCare system is able to correctly detect an emergency situation or an assisted situation if the recognized activity has an associated duration. For this purpose, the activity will be modeled as a dynamic Bayesian network, which may be obtained from the stochastic context free grammar. We also plan to extend the grammar used for modeling the activities to represent a larger set of activities.

Acknowledgements. The work has been co-founded by the Sectoral Operational Programme Human Resources Development 2007-2013 of the Romanian Ministry of Labour, Family and Social Protection through the Financial Agreement POSDRU/89/1.5/S/62557 and FP7 project ERRIC: Empowering Romanian Research on Intelligent Information Technologies, No. 264207.

References

1. Eurostat (2010), `http://epp.eurostat.ec.europa.eu/`
2. Intille, S.S., Larson, K., Tapia, E.M., Beaudin, J.S., Kaushik, P., Nawyn, J., Rockinson, R.: Using a live-in laboratory for ubiquitous computing research. In: Fishkin, K.P., Schiele, B., Nixon, P., Quigley, A. (eds.) PERVASIVE 2006. LNCS, vol. 3968, pp. 349–365. Springer, Heidelberg (2006)
3. Ivanov, Y.A., Bobick, A.F.: Recognition of visual activities and interactions by stochastic parsing. IEEE Transactions on Pattern Analysis and Machine Intelligence 22(8), 852–872 (2000)
4. Iwaki, H., Srivastava, G., Kosaka, A., Park, J., Kah, A.: A novel evidence accumulation framework for robust multi-camera person detection. In: ACM/IEEE International Conference on Distributed Smart Cameras, pp. 117–124 (2008)
5. Joo, S., Chellappa, R.: Recognition of multi-object events using attribute grammars. In: IEEE International Conference on Image Processing, pp. 2897–2900 (2006)
6. Lymberopoulos, D., Barton-Sweeney, A., Savvides, A.: An easy-to-program sensor system for parsing out human activities. In: IEEE INFOCOM, pp. 900–908 (2009)
7. Mocanu, I.: From content-based image retrieval by shape to image annotation. Advances in Electrical and Computer Engineering (2010)
8. Mocanu, I., Kalisz, E., Negreanu, L.: Genetic algorithms viewed as anticipatory systems. In: AIP Conf. Proc., pp. 207–215 (2010)
9. Mocanu, S., Mocanu, I., Anton, S., Munteanu, C.: AmIHomCare: A complex ambient intelligent system for home medical assistance. In: Proceedings of the 10th International Conference on Applied Computer and Applied Computational Science, pp. 181–186 (2011)
10. Ryoo, M., Agarwal, J.: Recognition of composite human activities through context-free grammar based representation. In: IEEE Conference on Computer Vision and Pattern Recognition, pp. 1709–1718 (2006)
11. Turaga, P., Chellappa, R., Subrahmanian, V.S., Udrea, O.: Machine recognition of human activities: A survey. IEEE Transactions on Circuits and System for Video Technology 18(11), 1473–1488 (2008)
12. Vail, D.L., Veloso, M.M., Lafferty, J.D.: Conditional random fields for activity recognition. In: International Conference on Intelligent Robots and Systems, pp. 3379–3384 (2007)
13. Vogler, C., Metaxas, D.: A framework for recognizing the simultaneous aspects of american sigh language. Computer Vision and Image Understanding 81(3), 358–384 (2001)

14. Yamato, J., Ohaya, J., Ishii, K.: Recognizing human action in time-sequential images using hidden markov model. In: Computer Vision and Pattern Recognition, pp. 379–385 (1992)
15. Zeng, Z., Ji, Q.: Knowledge based activity recognition with dynamic bayesian network. In: Daniilidis, K., Maragos, P., Paragios, N. (eds.) ECCV 2010. LNCS, vol. 6316, pp. 532–546. Springer, Heidelberg (2010)
16. Zhang, D., Perez, D., McCowan, I.: Semi-supervised adapted HMMs for unusual event detection. In: Computer Vision and Pattern Recognition, pp. 611–618 (2005)

DynaMOC: A Multiagent System for Coordination of Mobile Teams Based on Dynamic Overlapping Coalitions

Vitor A. dos Santos, Giovanni C. Barroso, Mario F. Aguilar, Antonio B. Serra, and Jose M. Soares

Abstract. In this work, we focus on problems modeled as a set of activities to be scheduled and accomplished by mobile autonomous devices that communicate via a mobile ad hoc network. In such situations, the communication cost and computational efforts are key challenges. It is well known that keeping information about activities globally known by devices can provide better schedules. However, there are some contexts such as those where activities require startup based on location, whereby information restricted to neighborhood coalitions of devices can still produce satisfactory scheduling. The existing heuristics do not consider this approach. In this paper, we propose a multi-agent system that coordinates the dynamic formation of overlapping coalitions and the scheduling of activities within them. Heuristics for calculating the size of coalitions, as well as for scheduling activities, are proposed. The system is applied to solve the problem of area coverage in a simulated environment, whereby the results show that satisfactory schedules are obtained with lower cost of communication and computation in comparison with the solution based on globally known information.

1 Introduction

We consider the problem of coordinating teams of mobile devices that communicate via a mobile ad hoc network engaged in scheduling activities geographically dispersed in an uncertain environment. The contexts which we are primarily

Vitor A. dos Santos · Giovanni C. Barroso · Mario F. Aguilar · Antonio B. Serra ·
Jose M. Soares
Universidade Federal do Ceara
e-mail: vitor@ufc.br, gcb@fisica.ufc.br,
{mario.fiallos,prof.serra,marques.soares}@gmail.com

F.M.T. Brazier et al. (Eds.): Intelligent Distributed Computing V, SCI 382, pp. 303–308.
springerlink.com

concerned involve problems such as area coverage [1], a problem in robotics with extensive research for the single robot coverage. In its distributed version, the whole area is split in disjointed sub-areas so that covering each sub-area is an activity to be accomplished by a robot within a robot team. Figure 1(a) illustrates this scenario and shows an area with several walls (black points) and a robot team (red points). Depending on the covering approach, many sub-areas may arise over a period of time. Figure 1(b) shows the same area partially covered (light paths) and several sub-areas. Robots need to decide how the sub-areas will be scheduled.

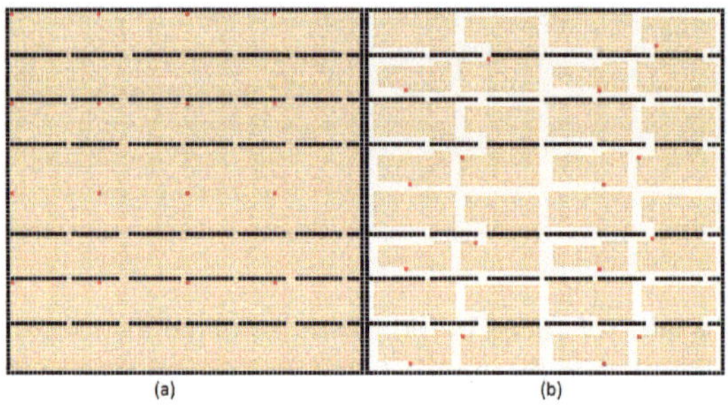

Fig. 1 Environment with horizontal obstacles: (a) with 16 robots in their initial locations; (b) with several subareas.

Distributed control of mobile autonomous devices that cooperate to schedule and accomplish shared dynamic activities is a notably challenging problem. Firstly, scheduling of distributed activities is a NP-complete problem. Secondly, the volume of sent messages may compromise the scalability of solutions.

In order to deal with the problem, we propose *DynaMOC - a Dynamic Multi-agent System based on Overlapping Coalitions*. Each DynaMOC agent is able to organize coalitions and to calculate their own schedules of activities from the knowledge obtained within that coalition. We argue that, for some problems, the organization of agents on coalitions can reduce the computational and communication efforts involved in scheduling.

In the remainder of this paper we show the related works (Section 2), the proposed system (Section 3), the case study (Section 4) and Conclusion (Section 5).

2 Related Work

There are several heuristics to reduce the scheduling search space. Some heuristics reduce the scheduling search space when activities require a separate setup [2]. However, those schedulers do not consider mobile teams and communication costs.

Barbulescu et al. [3] propose a coordination mechanism to schedule activities with setup requirements accomplished by mobile agent teams. However, they do not address communication issues.

Another approach to reduce scheduling search space and communication cost is directed to the formation of coalitions. Early works related to overlapping coalitions in multi-agent systems have been recently developed. Some of these algorithms consider the formation of coalitions for joint accomplishment of activities in the event that such activities require different skills to be used [4, 5]. Our work, however, considers the formation of overlapping coalitions for reducing globally distributed information. Lin and Hu [4] propose an algorithm to build overlapping coalitions to accomplish activities. The algorithm, however, is based upon complete information, unlike the problem domain considered here.

3 Description of DynaMOC

Each DynaMOC agent is composed by three main modules: DynaMOC Data Structure Module, DynaMOC Decision Support Module and DynaMOC Control Module. This architecture is illustrated in Figure 2.

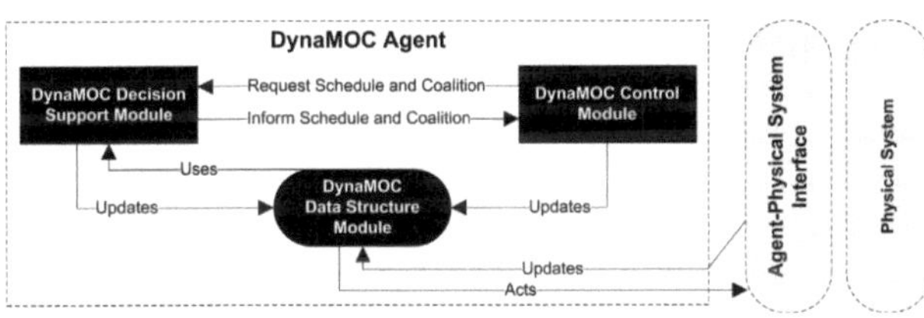

Fig. 2 Architecture of each DynaMOC agent.

DynaMOC Data Structure Module keeps information about states of the agent and activities.

DynaMOC Control Module runs a common *control cycle* in each agent. The first step of the cycle is to update its own coalition through the algorithm provided by Decision Support Module. The second step of an agent is to update remotely its *subjective view* - a term employed to denote the activities known by an agent - with activities not accomplished by other agents in its coalition. The third step aims to locally update the subjective view by removing already accomplished activities and defining new activities, due to divisions of the existing ones. This step is carried out by user-defined functions, as identifying new activities is a domain-dependent task. The fourth step is to schedule activities of the subjective view through the algorithm provided by Decision Support Module. The distribution of the scheduling within the coalition is carried out in the fifth step and the user-defined mechanisms to accomplish the scheduled activities are employed in the sixth step. In the last step, agents of the coalition are notified about the accomplishment of activities and domain data are distributed.

Decision Support Module provides algorithms for scheduling activities and forming coalitions.

In the method for coalition formation (Algorithm 1), the first coalition is the $c-$neighborhood of the agent, where c is an user-defined variable. The size of the next coalition is given by the main idea of the heuristic: each coalition to be formed by an agent i is based on (i) the size of the prior coalition and (ii) the agents who contributed with activities scheduled for i in the prior control cycle. In the algorithm, the value (ii) is specifically calculated as the distance (in hops) from i of the most distant agent which contributed with an activity scheduled for i in the prior cycle. A learning factor l_i in the range [0,1] is applied and provides different weights for these two parameters. Algorithm 1 is shown as follows. Besides c and l_i, it takes the identification i of the agent, the identification k of the control cycle and the activities $A_i(k-1)$ scheduled in the prior cycle.

Algorithm 1. Dynamic definition of coalitions.

procedure CALCULATESCOALITION($i, A_i(k-1), k, c, l_i$)
    if $k = 1$ **then**
        $C_i \leftarrow c-$neighborhood
    else
        Identifies $j \in C_i$ such that $\delta(j,i) = \max_{j \in C_i}(\delta(j,i))$: j was assigned to an activity $A' \in A_i(k-1)$.
        $\theta \leftarrow (l_i)|C_i(k-1)| + (1-l_i)|\delta(j,i)|$
        $C_i \leftarrow$ agents which are up to θ (hops) distance.
    end if
end procedure

For grand (static) coalitions, $c = n-1$ and $li = 1$ and for dynamic coalitions, $c = n-1$ and $0 < l_i < 1$, where n is the number of agents. In our case study, dynamic coalitions were formed with $c = n-1$ and $l_i = 0.5$.

For scheduling, we propose an approach whereby best activities for each agent are alternately chosen. This method, which is presented by algorithm 2. The algorithm takes the agent's identification i, his coalition C, the activities to be scheduled A and the maximum amount of activities k to be assigned to i. The main loop is repeated alternately assigning activities to agents of C until k activities are assigned to i or there is no more activities of A to be assigned. Each activity is chosen for an agent j based on the best outcome considering the activities chosen so far, which is denoted by $R(j,A')$. In each step, the best activity for an agent has to be found in A. Thus, the complexity of the algorithm is, in the worst case, $\Theta(|C||A|^2) = O(nm^2)$.

Algorithm 2. k-Greedy Alternated Scheduling (kGAS).

function KGAS(i, C, A, k)
  $tempC \leftarrow C$
  while $A \neq \emptyset$ and $cont \leq k$ **do**
    $cont \leftarrow cont + 1$
    Find $A' \in A$ and $j \in C$ corresponding to $\min_{A' \in A, j \in C} R(j, A')$.
    $A_j \leftarrow A_j + A'$; $tempC \leftarrow tempC - j$; $A \leftarrow A - A'$
    if $tempC = \emptyset$ **then** $tempC \leftarrow C$
    end if
  end while
  return A_i
end function

4 Case Study

We implemented DynaMOC and applied it to solve the area coverage problem in a simulated environment. The problem consists of finding a path that includes all the coordinates within a given area. One strategy for solving the problem, known as BSA (Backtracking Spiral Algorithm), is based on paths in the form of spirals, whereby the agents use obstacles, walls and areas already covered to build circular paths [6]. We apply BSA to define activities to be scheduled in DynaMOC. Initially each agent has only one activity scheduled, which corresponds to its initial location. However, BSA provides new subareas over time. Figure 1 (b) shows an environment with several areas formed due to spiral paths (light paths).

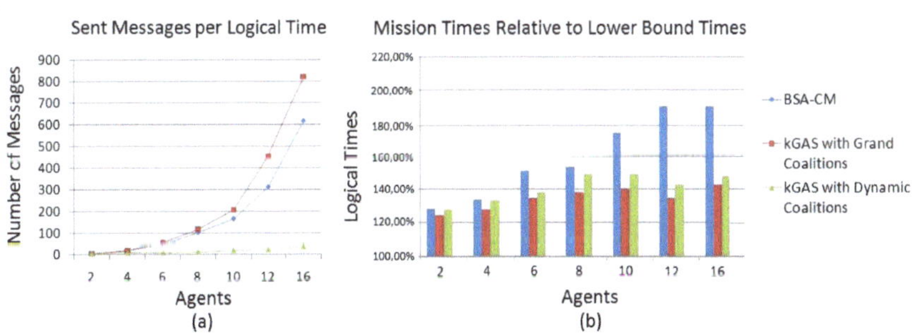

Fig. 3 Results: (a) Amount of sent messages. (b) Mission times relative to lower bound time.

We compared the *kGAS scheduler with dynamic coalitions* to other two versions: kGAS scheduler with static grand coalitions, and BSA-CM version (BSA - Cooperative Multirobot) [6], which is the distributed version of BSA and applies a greedy approach to schedule activities: the choice of the next activity (a new subarea to be covered) involves the participation of all agents and the robot chooses the nearest one, which may be his own or an activity shared by another robot. Coalitions are static and involve all agents (grand coalitions).

Results are shown in Figure 3. The amount of sent messagens is influenced by the necessary number of hops on mobile ad hoc networks communication. Figure 3 (a) shows the sent messages charts.

The time of a mission is the maximum amount of actions (enumerated as logical times) accomplished by a agent. The absolute mission times become meaningful when compared with a lower bound time, which is the shortest time possible for a mission. This value is given by the open area ($6,000$ area units for the environments considered) divided by the number of agents of the mission. Those percentages are shown in the charts of Figure 3 (b).

5 Conclusions

In the computational aspect, DynaMOC could cope with an NP-Complete problem, providing a $O(nm^2)$ solution (the kGAS algorithm). As the number of messages sent was maintained in a scalable level, we conclude that the dynamic coalition algorithm was effective in reducing the coalition sizes without compromising quality of scheduling.

This work can be extended from some other perspectives. The system behavior on the problem of area coverage can be investigated through the application of some conditions not yet considered, such as devices with different speeds and autonomies. It is also possible to evaluate the effect of variating the parameters l_i and β on the coalition definition algorithm.

References

1. Choset, H.: Coverage for robotics a survey of recent results. Annals of Mathematics and Artificial Intelligence 31, 113–126 (2001)
2. Allahverdi, A., Ng, C.T., Cheng, T.C.E., Kovalyov, M.Y.: A survey of scheduling problems with setup times or costs. European Journal of Operational Research 187(3) (2008)
3. Barbulescu, L., Rubinstein, Z.B., Smith, S.F., Zimmerman, T.L.: Distributed coordination of mobile agent teams: the advantage of planning ahead. In: Proceedings of the 9th International Conference on Autonomous Agents and Multiagent Systems, International Foundation for Autonomous Agents and Multiagent Systems, AAMAS 2010, Richland, SC, pp. 1331–1338 (2010)
4. Lin, C.-F., Hu, S.-L.: Multi-task overlapping coalition parallel formation algorithm. In: Proceedings of the 6th International Joint Conference on Autonomous Agents and Multiagent Systems, AAMAS 2007, pp. 211-1–21. ACM, USA (2007)
5. Cheng, K., Dasgupta, P.: Coalition game-based distributed coverage of unknown environments by robot swarms. In: Proceedings of the 7th International Joint Conference on Autonomous Agents and Multiagent Systems, International Foundation for Autonomous Agents and Multiagent Systems, AAMAS 2008, Richland, SC, pp. 1191–1194 (2008)
6. Gonzalez, E., Gerlein, E.: Bsa-cm: A multi-robot coverage algorithm. In: Proceedings of the 2009 IEEE/WIC/ACM International Joint Conference on Web Intelligence and Intelligent Agent Technology, pp. 383–386. IEEE Computer Society Press, USA (2009)

Semantic Management of Intelligent Multi-Agents Systems in a 3D Environment

Florian Béhé, Christophe Nicolle, Stéphane Galland, and Abder Koukam

Abstract. This paper presents a new approach combining the 3D elements composing the environment of mobile agents with semantic descriptors from building information models. Our proposal is based on the IFC standard which is used in the field of Civil Engineering to build digital models of buildings during the design phase. The semantic of IFC objects composing the 3D environment is used to select and set up 3D objects and elements of simulation scenarios. The result of this process dynamically generates the input files to the JaSIM environment that performs the simulation.

1 Introduction

This paper is located in the domain of the simulation of buildings flows with situated multi-agent systems. A multi-agent system (MAS) is a system composed of multiple interacting intelligent software agents [1]. Multi-agent systems can be used to solve problems that are difficult or impossible for an individual agent or a monolithic system to solve. A multi-agent system is situated when the agents are immersed inside an environment. In the domain of buildings simulation, an agent is assumed to be a pedestrian, or any object that owns an autonomous decision-making process. The environment is then everything that is not an agent in the buildings. Three different points of view may be adopted to study the notion of environment in situated MAS [7]: (i) the part of the system which is outside the community of the

Christophe Nicolle · Florian Béhé
Laboratoire Electronique, Informatique et Image - UMR CNRS 5258, IUT Dijon Auxerre, Univsersité de Bourgogne BP 17867 21078 Dijon Cedex, France
e-mail: `cnicolle@u-bourgogne.fr`, `florian.behe@checksem.fr`

Florian Béhé · Stéphane Galland · Abder Koukam
Laboratoire Systèmes et Transports, Université de Technologie de Belfort-Montbéliard, 90010 Belfort Cedex, France
e-mail: `florian.behe@checksem.fr`,
`{stephane.galland,abder.koukam}@utbm.fr`

F.M.T. Brazier et al. (Eds.): Intelligent Distributed Computing V, SCI 382, pp. 309–314.
springerlink.com

agents; (ii) the medium for coordination among these agents; or (iii) the running infrastructure or platform. Weyns et al. distinguish between the physical environment and the communication environment [8]. The physical environment provides the laws, rules, constraints and policies that govern and support the physical existence of agents and the other entities. In the rest of this paper, only this aspect of the environment is taken. Several problems may be solved to properly implement an environment: its topological and geometrical description, its dynamics, and the meanings of each object and zones in the environment. The two first points are addressed by the JaSIM environment model [2]. The last point is addressed by both the integration of semantics in the environment and modification of the agent's algorithms that uses them. Most of the approaches found in the literature are based on the tagging process of the environment. The concept of tagging consists in placing some tags in the environment to inform agents on various subjects. Basically, tags are considered as objects placed in the scene, but they do not have a physical presence. They are invisible for the viewer of the scene and are only seen by the agents. Our proposal includes to describe through tags the usage of some places, i.e. where an agent can sit, pass through, etc. Lugrin and al. [4] place various tags on a single object. These tags are linked together and can also represent information. For example, a glass will be represented by its geometry, a "containing" tag, an "opening" tag, a link between these two tags and a link from the opening tag to the environment. In this proposal, the evolution of an object is done by modifying the tags and links of this object. For example, if the glass is clogged, the link between the opening tag and the environment is deleted and this tag is no more accessible from the environment. Similarly, De Paiva et al. [5] propose to put labels on areas in order to associate a name with a position. However, in the opposite to the previous works, the agents can only go to a location according to its name and not its position. In addition, the authors propose to assign roles to the agents and make them evolving in the environment according to their roles and the current simulation time. For example, a kid is at "school" at 11am and a working adult is at "office" at the same time. This approach is useful to only have agents that are in a given place, i.e. an agent which is supposed to be a student is not in the headquarters of a company, except if his behavior needs it.

2 Environment

The term Building Information Modeling (BIM) has been presented recently as a demarcation of the next generation of Information Technologies (IT) and Computer-Aided Design (CAD) for buildings, which focus on the production of drawings [6]. A BIM system is a central system that allows to manage various types of information, such as the planning of enterprise resources, resource analysis packages, technical reports, meeting reports, etc. A BIM system is a tool that enables users to integrate and reuse building information and domain knowledge throughout the life cycle of a building. The main feature of a BIM is the 3D modeling system with data management, data sharing and data exchange during the life cycle of the building.

Indeed, a building is composed of geometrical elements, which are the basis of a building's design.

The IFC standard was created in the late 90s by the International Alliance for Interoperability (now called buildingSMART). Its goal is to build a common data model for all the actors in the building industry to resolve problems of heterogeneity. The IFC standard is the kernel of the new generation of BIM. Several versions of the IFC have been published, because of the gradual increase of the covered area. The heart of the model was stabilized in 2005 and received the ISO certification as ISO / PAS 16739: 2005. The current release is 2x3TC1. As illustrated by Figure 1 it is an ASCII file containing of the elements of the described building. IFCs enable the exchange of data, either in the form of geometries, but as objects and their structures (walls, doors, windows, stairs ...). IFC files contain a description of all objects in the buildings and their links. The format also describes more abstract concepts such as schedules, activities, places, organizations, construction cost, etc. IFC contains all information required to make simulation of autonomous entities.

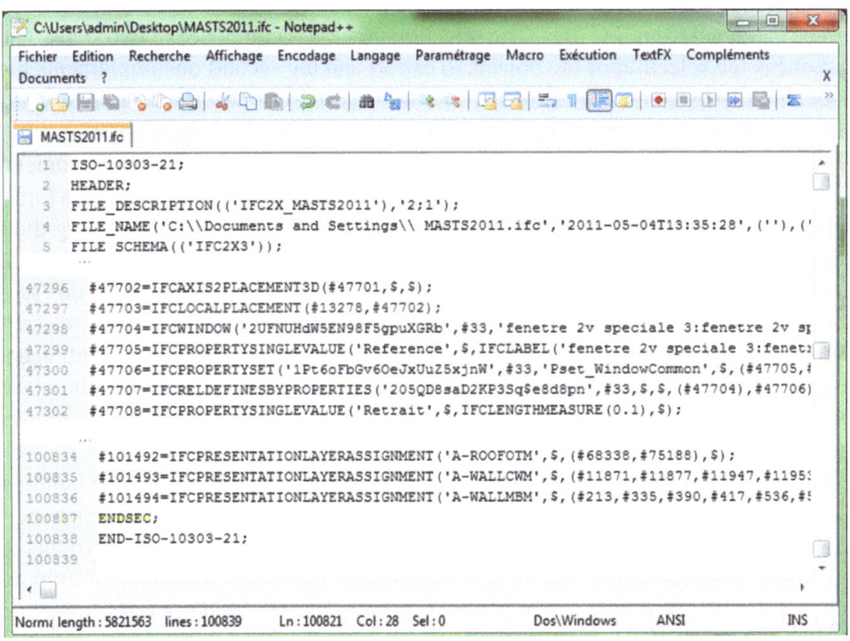

Fig. 1 Screenshot of an IFC file. This ASCII file represents a complete description of a building in 100838 lines.

Nevertheless, this type of simulation in a complex urban system requires dedicated software models. JaSIM platform [2] integrates components which are required to simulation complex environments in 1D, 2D and 3D (in particular, it is used for simulation in virtual reality). This platform integrates several models to reproduce human visual perception in a virtual environment and endogenous behavior of this environment. Thus simulated entities can use the JaSIM platform to perceive and act in a situated system. To run a simulation, JaSIM requires three kinds of input files. (i) A **SFG file** is a XML-based file that can be seen as the scenario description of the simulation. It will contain various informations about the simulation and the environment: definitions of the places, spawning areas for the agents, goals, way points, stochastic generation laws, etc. (ii) **Precomputed structures** are some serialized Java trees that contain various information such as geometry or simple semantics (e.g. "door", "window", etc.) (iii) The last file supported by the JaSIM platform is optional and is only used when the simulation is displayed. The format of this file should correspond to a standard 3D file format such as COLLADA® or 3D Studio®.

3 Proposal

IFC files contain a huge amount of information, more than needed for a MAS simulation. That is why it is required to perform a selection on what piece of information will be extracted from IFC files. The selection process is composed of two steps. The first one is the selection of the objects to export and the second one implements the associations between the IFC objects and the JaSIM concepts. There are two kinds of associations according to their usage. The first one deserves creation of scenario files, and the second deserve creation of precomputed structures of the environment. The first selection step is common to both association types. It consists in performing the selection of objects that will be present in the simulation, including their geometrical structures. Selected objects will be exported in the JaSIM environment and only them will be able to be associated to JaSIM concepts. The rest of this section only concerns the associations made to generate the SFG file. Each selected element can be associated to a JaSIM scenario element such as a way point, agent generator, etc. Associations are made in a semi-automatic way on a selection made according to profiles that will be created following user selections over the software usages. Some of these selections can be generalized to all the instances of a given class. Human selection is done object by object using the tree view of the building (i.e. semantic view) and the same semantic is used to perform the automated association. JaSIM uses the "place-portal" principle to model the floors and the rooms of a building. This kind of association is automatically proceeded, relying on semantics that the IFC provides, e.g. a place by storey. The tree that contains immobile objects is generated with objects selected in the tree view as described above. In this process, the IFC semantics is used to automatically create the JaSIM semantics.

3.1 Architecture

The main part of the proposed software is about the loading the IFC file (Figure 2). Two libraries are used. The first one is dedicated to load the IFC structure in memory according to their specifications. The second one loads the geometry in a directly usable shape structure. A central kernel was developed to manage and launch these

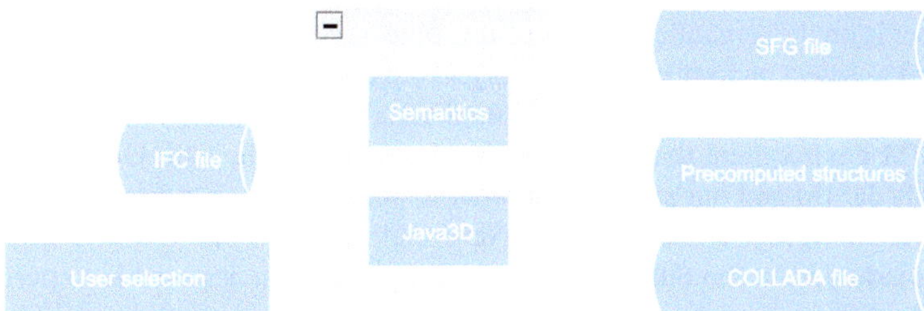

Fig. 2 Global chart of our proposal

two libraries and establish the link between them. This kernel also manages the generation of the JaSIM scenario elements. It also allows communication with the Graphical User Interface (GUI) to update links and other data according to user actions. The application structure is shown on Figure 3(a). To perform the JaSIM file generation, the kernel uses file generating modules shown in Figure 3(b). The COLLADA export module takes as input a Java3D scene graph (built from the IFC files) and generate the corresponding COLLADA® schema. The SFG (JaSIM Scenario Configuration) generator module parses the IFC data structure and converts IFC elements' description into SFG file. The Java tree generator module dynamically generates precomputed structures from both geometry and semantics from IFC files.

(a) (b)

Fig. 3 (a) Kernel centered architecture of ourproposal, (b) File generation process

4 Conclusion and Future Works

This paper presents the use of IFC files to generate a MAS environment. Our main goal was to test the viability of using IFC files in the MAS domain. Using IFC files as a starting point in MAS simulation is beneficial according to the quality of the data, which are always up-to-date. A simulation can be executed as soon as the building is designed to test the quality and the level of compliance of such design. Moreover, using BIM as a MAS input also enables to bring a high level of semantic information to the environment that can be used by the agents in turn. The focus of this paper is on the semantic analysis of IFC objects composing the 3D environment. It is used to select and set up 3D objects and elements of simulation scenarios. The result of this process generates semi-automatically and dynamically input files to the JaSIM environment that performs the simulation at the end. Our next goal is to modify JaSIM and our software in order to split geometry and semantics in two normative file formats, such as RDF. Nevertheless, is a window still a way point if this window is on the 6th floor? An ontology will make it possible to perform a better classification of the elements and to apply several restrictions and filters on the associations.

Acknowledgements. This work is funded by the region of Franche Comté and receive grants from a cooperative project between the Franche Comté and the Bourgogne.

References

1. Ferber, J.: Les Systèmes Multi-Agents: vers une intelligence collective. InterEditions (1995)
2. Galland, S., Gaud, N., Demange, J., Koukam, A.: Environment model for multiagent-based simulation of 3d urban systems. In: The 7th European Workshop on Multi-Agent Systems, Ayia Napa, Cyprus (2009)
3. Gutierrez, M., Thalmann, D., Vexo, F.: Semantic virtual environments with adaptive multimodal interfaces. In: Multimedia Modelling Conference, MMM 2005, pp. 277–283. IEEE, Los Alamitos (2005)
4. Lugrin, J., Cavazza, M.: Making sense of virtual environments: action representation, grounding and common sense. In: Proceedings of the International Conference on Intelligent User Interfaces, Honolulu, Hawaii, USA, pp. 225–234 (2007)
5. de Paiva, D., Vieira, R., Musse, S.: Ontology-based crowd simulation for normal life situations. In: Computer Graphics International 2005, pp. 221–226. IEEE (2005)
6. Vanlande, R., Nicolle, C., Cruz, C.: IFC and building lifecycle management. Automation in Construction 18(1), 70–78 (2008)
7. Weyns, D., Ominici, A., Odell, J.: Environment as a first-class abstraction in multi-agent systems. Journal on Autonomous Agents and Multi-Agent Systems 14(1), 5–30 (2007)
8. Weyns, D., Van Dyke Parunak, H., Michel, F., Holvoet, T., Ferber, J.: Environments for multiagent systems state-of-the-art and research challenges. In: Weyns, D., Van Dyke Parunak, H., Michel, F. (eds.) E4MAS 2004. LNCS (LNAI), vol. 3374, pp. 1–47. Springer, Heidelberg (2005)
9. Yersin, B., Maım, J., de Heras Ciechomski, P., Schertenleib, S., Thalmann, D.: Steering a virtual crowd based on a semantically augmented navigation graph. In: Proc. The First International Workshop on Crowd Simulation (V-CROWDS 2005), pp. 169–178. Citeseer, Lausanne (2005)

Author Index

MIX
Papier aus verantwortungsvollen Quellen
Paper from responsible sources
FSC® C105338

If you have any concerns about our products,
you can contact us on
ProductSafety@springernature.com

In case Publisher is established outside the EU,
the EU authorized representative is:
Springer Nature Customer Service Center GmbH
Europaplatz 3, 69115 Heidelberg, Germany

Printed by Libri Plureos GmbH
in Hamburg, Germany